AF705929

History of
CONCRETE
A Very Old and Modern Material

History of CONCRETE

A Very Old and Modern Material

Per Jahren
P J Consult AS, Norway

Tongbo Sui
Sinoma Research Institute, China

Chemical Industry Press

World Scientific

Published by

World Scientific Publishing Co. Pte. Ltd.
5 Toh Tuck Link, Singapore 596224
USA office: 27 Warren Street, Suite 401-402, Hackensack, NJ 07601
UK office: 57 Shelton Street, Covent Garden, London WC2H 9HE

and

Chemical Industry Press
No. 13, Qingnianhu South Street
Dongcheng District, Beijing 100011
P. R. China

British Library Cataloguing-in-Publication Data
A catalogue record for this book is available from the British Library.

HISTORY OF CONCRETE
A Very Old and Modern Material

B&R Book Program

Copyright © 2017 by World Scientific Publishing Co. Pte. Ltd. and Chemical Industry Press

All rights reserved. This book, or parts thereof, may not be reproduced in any form or by any means, electronic or mechanical, including photocopying, recording or any information storage and retrieval system now known or to be invented, without written permission from the publisher.

For photocopying of material in this volume, please pay a copying fee through the Copyright Clearance Center, Inc., 222 Rosewood Drive, Danvers, MA 01923, USA. In this case permission to photocopy is not required from the publisher.

ISBN 978-981-3145-73-3

Foreword

by Luke Snell

I have been asked to review the proposed addition of China Contribution to the History of Concrete. Mr. Per Jahren previously wrote a book, "Concrete, History and Accounts" that discuss how concrete and cement developed. Since he is Norwegian, the book concentrates on European and North American history and how cement evolved in those areas.

China and India are now the major manufacturers of cements and concrete. China alone produces 60% of the world's cement and concrete. Up to now, its history of development has been ignored by the concrete industry.

To rectify this oversight, Mr. Per Jahren and Dr. Tongbo Sui planed to revise "Concrete, History and Accounts" to include China's rich history in concrete that has led it to become today's major manufacturer of cement.

China's history in cement and cement-like materials goes back 7000 years. Their development of building materials is thought to be similar to that of the Egyptian and Roman materials.

Good engineering is to determine how to use locally available materials to make the structures needed for society. Throughout China's history, they have had engineers that met this challenge and advanced the development of cement-like materials that has led to today's cement and concrete industry. This entire development of China's cement and concrete industry is not completely recorded in the western publications. I have many of the concrete and cement books and China is not recognized for its contribution to our industry.

Most people in the USA and in Europe have seen the popular movie, "Schindler's List."This made people aware of how one person could and can make a difference through his effort: hundreds of people were saved in Europe from the atrocities in World War II.

China has a similar story of "Schindler's List", that is also a part of our cement and concrete history. Mr. Bernhard Arp Sindbery, a Dane, and Mr. Karl

Gunther, a German, were in China working for Jiangnan Cement in Nanjing, at the start of World War II. During the massacre of Nanjing in 1937, they helped get 30,000 people out of Nanjing. This ties the history of cement and concrete to one of history's greatest humanitarian effort. Unfortunately, most of the world knows little of this story. In my opinion, this is a major contribution to our history and widens the audience for this book.

As the western world looks at China's recent engineering development, we see the buildings used in the Olympics and the Three-Gorge Dam. We marvel at the design, engineering and China's ability to accomplish these outstanding projects. China's long history in cement and concrete and these modern day accomplishments are a continuation of their success.

In my opinion, this revised book with China's historical contributions to the cement and concrete industry will be a major contribution to our history. As chair of the History of Concrete Committee for the American Concrete Institute, I welcome this addition to our history and look forward to its publication.

Luke Snell
Chairman
History of Concrete Committee
American Concrete Institute

Preface

by Per Jahren

The original book *Betong- Historie og historier* (*Concrete History and Accounts*), was published in 2011, first in Norwegian and later in English in a somewhat modified version by the Norwegian publisher Tapir Academic Press.

In this new version, the name of the book has been changed to History of Concrete: A Very Old and Modern Material.

More important, Professor Tongbo SUI has added a chapter to the book, about Chinese cement and concrete history.

When I was about 12, and in the 6th grade at school, we were supposed to write a paper. Mine was about concrete.

My father was a site engineer. I grew up near the concrete mixer station at building sites around the country. Some building journals came in the mail every month. As we did not have radio or television, these were often interesting readings for me too. I concentrated on the history and anecdotes of concrete. I pulled out articles and collected them in portfolios, more portfolios, and boxes. I have continued to do this ever since.

In 1967–1969, I had a fixed column in the journal *Ingeniørnytt*, with the title "*Fra Bygningsingeniørens historie*" (from Civil Engineering History). After 27 numbers, the hobby of writing a column on history became too time consuming.

However, the collection of articles continued, with some thoughts that this might be interesting to take up after retirement.

In the early summer of 2009, I started on the more systematic work of writing.

My goal has never been to write a complete history of concrete, but to collect and systemize some central and interesting glimpses from this fascinating history. I have felt a bit like Alex Haley, who went back to his own history and wrote about Kunta Kinte. When he published his book

"Roots" in 1979, there were many of us who found it difficult to lay it down when we had started reading it.

The Concrete History is my roots. After the work started and the interest deepened, the engagement also increased to the same degree.

A retrospect at the history of concrete will have character of the view platform you have chosen. This will also mark the choices and the angles from which I have studied the background material. The goal has been to show the versatility of the concrete industry.

1904 2004

House Resolution 394

Whereas *concrete is the world's most consumed man-made material and second only to water of all materials consumed;*

Whereas *production of concrete exceeded 3,500,000,000 cubic yards worldwide in 2002, more than a half cubic yard for every person on the planet;*

Whereas *production of concrete exceeded 500,000,000 cubic yards domestically in 2002, approximately two cubic yards for every person in the United States;*

Whereas *the ready mixed component alone of total concrete production in the United States in 2002 was enough to build a continuous road ten feet wide and four inches thick encircling the globe at the equator nearly 51 times;*

Whereas *concrete construction provided 2,000,000 jobs in the United States in 2002 during a time of economic recession;*

Whereas *the concrete industry provides employment to numerous skilled employees, including batchers, truck drivers, ironworkers, laborers, carpenters, finishers, equipment operators, and testing technicians, as well as professional engineers, architects, surveyors, and inspectors;*

Whereas *concrete was the predominant material of choice in a construction industry that built $843,000,000,000 of construction in 2001, being used in virtually every construction project;*

Whereas *concrete has an estimated $200,000,000 annual impact on Gross Domestic Product;*

Whereas *the concrete industry is a significant contributor to the economy of every Congressional district in the United States;*

Whereas *the many agencies of the Federal Government rely upon the American Concrete Institute, the technical society for the concrete industry, as a major standards developing organization for concrete design, construction, and repair;*

Whereas *the American Concrete Institute has, through its 18,000-member network of private and public sector volunteer citizens, developed and operated a review system that has provided concrete standards and guides for durable, safe, and uniform construction in the United States; and*

Whereas *the American Concrete Institute celebrates its 100th anniversary of service in advancing the technology of concrete for educational and scientific purposes in order to increase the knowledge and understanding of concrete materials and to support programs that improve concrete design and construction for the common good:*

Now, therefore, be it Resolved, That the House of Representatives—

(1) recognizes the American Concrete Institute—

(A) for 100 years of service to the people of the United States as the technical society for the concrete industry; and

(B) for the economic stability, quality of life, durability of infrastructure, and international competitiveness that the Institute has made possible to the United States; and

(2) encourages and supports the designation of an appropriate day as ACI Centennial Day in recognition of 100 years of service by the American Concrete Institute to the people of the United States.

As mentioned, to show the manifold use of concrete has been an important issue. Let the pictures on the next page be an example on this.

Wine tank no. 14 and 15 of an important wine producer in the Duero valley in Portugal, who produces raw material for Port wine, have been made of concrete. The tanks function very well after being used for many years and no one complains about the wine.

There is also some symbolism in this;- concrete is an aiding material and in this case it aids one of the oldest and globally most important pleasures and essentials for the table.

In October 1975, CEMBUREAU in cooperation with the German '*Bundesverband Der Deutschen Zementindustrie*' arranged a conference in Köln to discuss the role of concrete in the conservation of historic buildings. In the preface to the booklet issued at the meeting, Lord Duncan-Sandys, the President of Europa Nostre says, among others, that

> *Concrete is one of the greatest inventions in the history of construction. Without it the economic progress and the dramatic rise in living standards, which have taken place during this century, would not have been possible.*
>
> *Concrete is one of the most flexible and adaptable of all building materials. It can just as well be used for the construction of modern skyscrapers as that of buildings, which though contemporary in style, harmonize with older neighbors.*
>
> *Concrete also has a major part to play in the task of preserving and restoring existing buildings of historic and architectural interest. In fact, some of Europe's most famous monuments, which were in danger of collapse, could not have been saved but for the skilful use of immense quantities of concrete.*

Per Jahren
Professer of Agder University ,Grimstad, Norway
President of P.J. Consult AS
Former president and honor member
of the Norwegian Concrete Society

Preface

by Tongbo Sui

"Learning from history one will know how things rise and fall".

by Shimin LI (599–649), the 2nd emperor of Tang Dynasty, China

In view of the progress of social and economic development of human beings, cement and concrete as binders have been consistently playing a key role in every era and aspect. As the largest construction materials and the 2nd largest commodity after water in the world, the demand of cement and concrete will be increasing in the future by the year 2050. The higher energy and resources intensity and environment impact for cement making on the other hand result in great challenge in terms of future sustainability. This issue is of significant importance especially for China due to its huge volume of cement output accounting for 60% of the world cement totally.

To respond the challenge, international communities are taking action in different ways. IEA in cooperation with WBCSD CSI in 2009 launched global cement technology roadmap towards 2050 on basis of best available and emerging technologies. An in-depth review of the route that how the binders evolve by generation since dozens of thousand years ago is another wise way in order to enlighten the future roadmap of cement and concrete sustainability.

I am very much honored, upon the kind invitation of Mr. Per Jahren as co-author of this book to deepen the cooperation and friendship with him which started 10 years ago on advances in concrete technology for sustainable development. The more I try to get access to the glorious history of Chinese cement and concrete history, the more I understand that more are needed to learn from our ancestors.

China, as one of the ancient civilized countries in the world together with Egypt, India, Babylon and Greece has also a brilliant history in developing and using cementitious materials. From natural earth materials as Baihuimian, Yellow Mud, to synthetic materials such as lime, Sanhetu, and lime-organic composite binders, all these are precious presents invented and remained for us today and beyond.

Unfortunately this has not so far been well acknowledged and understood by the world. The chapter on Chinese cement and concrete history is an attempt to touch this great, deep and long river and to provide readers with the route how Chinese binders evolved as well as the wisdom and contribution that China has been made to make and use various binders.

The most touching history related to Chinese cement industry is the tragedy took place in Nanjing during the Japanese invasion to Nanjing in 1937-A history of "Jiangnan Cement" version Schindler's List was recorded.

"...Although history cannot be changed, lessons can be learned to face the future..."

by Danish Queen Margret II, in Nanjing, April 27, 2014

As the modernization of Chinese cement and concrete industry today, it is the right time to go back to thousands of years ago to learn the essence and wisdom of harmony on materials and construction in order to create innovative solutions to build a sustainable future of human beings.

Tongbo Sui

Ph. D., Professor

Director General of Sinoma Research Institute

Vice President of Sinoma

International Engineering Co. Ltd.

Acknowledgments

We wish to convey our regards and thanks to the many writers who have been the foundation of this publication. An overview can be found in the reference chapter.

Special thanks are due to our friends around the world, who share our interest in concrete and have sent us materials that we did not have previously.

Our thanks are also due to Rich Andersson, who sent us the data about his father and the ARCO-Barge and the ABAM-book, Gordon D. Brearley for providing information about the American channels, Geir Bjørnbakk who put us on the track to Thomas Edison's concrete houses, Frithjof Garmann for his interesting talks and extra literature, Steinar Helland, who provided us with the publications from Rye and Kolderup, George Hoff for the CD with interesting material from his Sevilla-presentation, Carlos Horstmann for information about Spanish windmills and the trip to the aqueduct in Segovia, Krzysztof Janczura for the information and the book about the fascinating "Jahrhunderthalle" in Wrocklaw, Magne Lysberg for the book about the Namsen bridges, V. Mohan Malhotra for the many emails with personal data and other useful information, P. Kumar Mehta for his mail with CV, Shigeyoshi Nagataki for the book about Japanese concrete history, Adam Neville for interesting talks and mail with CV etc., Per Helge Pedersen for interesting meetings and useful information, Seppo Petrow for the help to find books in the library of the Central

Concrete Organization in Finland, Steven K. Rove for interesting material about lightweight concrete, Kåre Reknes for the material about the lignosulphonate industry, Arne Skjelle for the information about Monierfabrikken etc., Takashi Sumida for assistance in translating Japanese, Hans Henrik Thaulow for interesting talks and additional information about his father, Julie Webb who sent us the copy of the greetings to ACI from the House of Representatives, and Lennart Westerholm for helping us to find information in Finland and for translation from Finnish.

Sincere thanks are also given to:

Dr. Wang Yanmou, Mr. Zhou Zuitian and Mr. Han Changkai for their kind support in providing invaluable pictures about the early period of Portland cement in China;

Prof. Derham Groves, Australia and Ms. Zhu Xiaoqiu, deputy director of Generalissimo Mansion Museum in Guangzhou in providing precious pictures of Guangdong Shimintu (Cement) Factory; Ms. Zhao Xuemei, editor of *China Charm* for providing contact information of Ms. Zhu Xiaoqiu; Ms. Zhou Chunying, CBMA for her kind help in collecting useful information for ancient binders of China.

Special thanks are extended to Mr. Luke Snell, Chair of ACI Concrete History Committee for his kind encouragement and support.

A special and warm thanks to Marit, Airong and Yi for their full support.

Per Jahren
Professer of Agder University ,Grimstad, Norway
President of P.J. Consult AS
Former president and honor member
of the Norwegian Concrete Society

Tongbo Sui
Ph D, Professor
Director General of Sinoma Research Institute
Vice President of Sinoma International Engineering Co. Ltd.

About the Authors

Per Jahren is a semi-retired consulting engineer living in Asker, Norway with half a century of experience from various concrete fields. He is now Professer of Agder University, Grimstad, Norway. Tongbo Sui is living in Beijing, China and working as the director general of Sinoma Research Institute and vice president of Sinoma International Engineering Co. Ltd. As a professor, his focus is in the R&D of low energy and low CO_2 emission cement based materials.

The authors have been friends and work together with cement and concrete sustainability issues for quite a few years and in several countries and continents.

Per and Tongbo met in Hyderabad, India in February, 2005 and Toronto, Canada during the conferences on Advances in Concrete Technology for Sustainable Development organized by Prof. Mohan Malhotra, director of former CANMET for Concrete Sustainability, Canada. The first time when they met in China is due to the International Symposium on Cement and Concrete held in 2006 in Xi'an, Northwest China's capital city of Shanxi Province and ancient capital city in Chinese history famous for terra cotta warriors. One of most important memories for Tongbo as main organizer of the conference is the birthday celebration of Per's wife at the opening dinner of the conference. Per thereafter has become the enthusiastic supporter to Chinese cement and concrete sector by actively participating in series of international symposiums, seminars and workshops held in China on the topics of advances in cement and concrete technology for sustainability development jointly organized by Dr. Malhotra and Tongbo, and giving lectures in many cities such as Beijing, Ji'nan, Lhasa, Mianyang, Nanjing, Shenyang and Xi'an sharing with Chinese colleagues his global view, knowledge and experience.

Per also makes great effort in helping bring the world advanced connection technology from Norway to China in order to promote the development of industrialization of the housing construction.

This book is another enthusiasm of Per Jahren in collaborating with his Chinese colleague and friend Tongbo Sui specifically dedicated to the world largest building materials - concrete sector in pursuing the sustainable development.

Contents

1 Early concrete history

Somewhat mixed opinions can be found about what can be regarded as the first concrete. Various authors have various views about the correct definition of concrete. The composition or mixture design and the production methods of the first concrete were different from what we have today. However, the material has proved its strength and durability over thousands of years.

The earliest binders, or the glue, were often a combination of burnt lime and hydraulic lime. Lime is burned from a reasonably pure limestone, from which the CO_2 is driven out and calcined at 1000–1100°C but does not harden under water. Burnt lime treated with water and mixed with sand gives a lime mortar. The hardened mortar will continuously absorb CO_2 from the air and harden, something that in many ways is the reverse process of burning.

Hydraulic lime, on the other hand, hardens under water like the cements we have today. These are produced from impure limestone containing silicium acids and aluminum and iron oxides and need higher temperatures in the burning process.

It is the product or powder from these sources that is called cement today.

What do the dictionaries say about CONCRETE?

- A simple "English Dictionary" [1] states that Concrete is composed of particles united in one mass.
- A small "Dictionary of Science" [2], states that "Concrete is a building material made of stone, sand, cement and water."
- A more advanced American science and technical encyclopedia [3] states that Concrete is *a "versatile engineering material consisting of a Hydraulic cementing substance (usually Portland cement), aggregate, water, and often controlled amounts of entrained air. Concrete is initially a plastic, workable mixture which can be molded into a variety of shapes. Strength is developed in the hydration reaction between the cement and water. The products, mainly calcium silicates, calcium aluminates, and calcium hydroxide, are relatively insoluble and bind the aggregate in a hardened matrix."*
- An older Norwegian encyclopedia [4] states, as an introduction, that "Concrete is a building material of a cured mixture of sand, stone, cement and water."

Based on the dictionaries and encyclopedias we have a relatively wide range of possibilities to judge if we are dealing with concrete. Other definitions made by modern concrete technologists can be useful, but are not necessarily precise.

One of the great teachers of cement and concrete chemistry, F.M. Lea, in his book "*The Chemistry of Cement and Concrete*" [5] provides detailed definitions. He defines cement as "adhesive substances capable of uniting fragments or masses of solid matter to a compact whole." He says that concrete is "an artificial conglomerate of gravel or broken stone with sand and lime or cement."

Gartmann [6] writes in his book about glue or lime (translated from Norwegian): "We hear about the first lime-burners early in the seventeenth century. The old name for stone that contains more than 50% calcium carbonate is 'limestone' or 'glue stone,' and this we have kept to our time. We find this name again in places where lime has been burnt in old days as for example Limhamn where Swedish cement industry started its activity."

Sven Thaulow writes, by way of introduction, in an article about cement in September 1967: [7] (from Norwegian) "To find material that can bind or glue together stone material to an artificial stone has always been a goal for human beings. The Assyrians and the Babylonians used clay and bitumen, the Phoenicians and the Egyptians used gypsum and lime, and the Greeks knew the art to burn lime. The Romans discovered some natural cements – pozzolans with so-called hydraulic properties, i.e. materials that could harden under water, contradictory to lime that needs air to achieve strength."

The well-known Canadian concrete chemist Pierre-Claude Aïtcin has in the introduction to his book "*Binders for Durable and Sustainable Concrete*" [8] differentiated the glues of ancient times into three categories;

- Gysum (anhydite) based
- Lime based
- Pozzolan based

Even if some of the ancient concretes that are found seem to be combinations of these categories, and some also might contain some natural cements, Aïtcin's systematization seems reasonable. The gypsum-based glue is based on a simple burning process, and Aïtcin claims with reference to Stark and Wicht that the binding properties of gypsum were discovered some 10 to 20 000 years ago. Most of the concretes that were found before the Roman time and are mentioned in the following chapters are lime based. Aïtcin, referring to several authors, claims that lime-based concretes were also found 10 000 to 20 000 years ago.

With reference to Felder-Cassagrande (1997), he mentions mortar found in Nevali Core in Turkey 22 600 years ago. Lime glue has been used by several cultures around the world, but some of them have only been reported on a small scale. Aïtcin for example refers to Deloye, who mentions lime mortar in walls in Bali in Indonesia in 200 BC and to Thornton, who refers to a lime-based floor from 3000 BC in Dadiwan in the Ganow province in China.

The pozzolan-based binders are also in a way lime based, but they are hydraulic, can harden under water, and are more durable. Even if we find some examples of pozzolan-based glue earlier, it is only well into the Roman times that these really got a foothold.

An interesting point in the development of concrete is that it seems that we find long periods when there was very little development in the technology. This might partly be due to the fact that the knowledge about this fantastic material and its use was limited to religious activities. The priestesses in various religions kept the knowledge a secret and used it to strengthen their position. We can also see examples of this in the later millenniums when the development in the concrete environment between the fall of the Roman Empire and until the beginning of the 1700s was nearly limited to churches and monasteries.

How did the use of concrete start?

It is surmised that concrete originated in early times when humans found glue at the edge of the hollow in the ground where they had their cooking fires. When it rained, on the fire edge they found a glue-like substance that could be used to form a stable material. The quality of the glue depended on the material in which the fire hollow was made and the intensity of the fire. This observation had probably taken place in many locations, but the ability to systemize the observation and analyze the material that was found varied and was the foundation of the knowledge that came into existence later. The final product might have been gypsum, lime, lime mixed with a silicium-based pollution or a pozzolan material.

It has been said that the first cultured human being was the one who went out into the field, picked flowers, and put them together in a bouquet. The first concrete technician was he or she who observed that the edge of the fire hollow could glue together sand and gravel, who protected the glue powder from rain, and who understood that the glue formed irrespective of the kind of soil the fires were made in.

1.1 Galilee

In a floor built 9000 years ago in Yiftah El, west of Jerusalem, remnants of concrete have been found. The $180m^2$ floor must have used about $8m^3$ of concrete with about two tons of lime as a binder [9]. The considerable volume of lime must have come from a rather large oven and an organized burning procedure. Fragments of fire-proof material from something that might have been like an oven have also been found in the industrial area of Yiftah El. The concrete that was found shows remarkably high strength, density, and durability properties. Researchers claim that burning of lime was the first use of fire for production purposes.

Jericho in Gallileia is the oldest Neolithic town in history. After the fall of Jericho, according to the Bible in the year 1200 BC, researchers have found several layers of concrete from earlier settlements, among them a polished concrete floor, under the rubble.

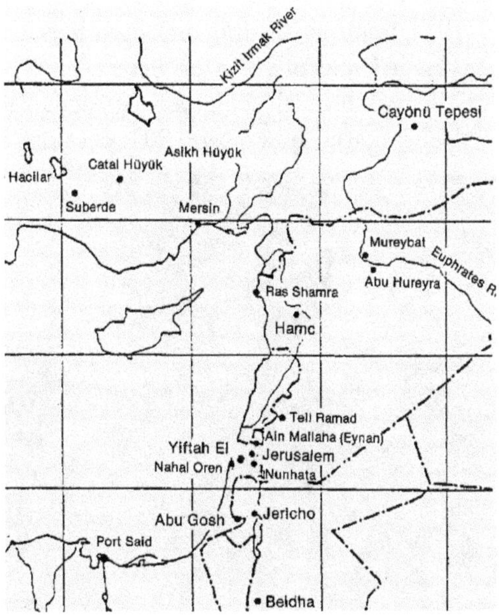

Map showing the Neolithic discoveries of polished concrete floors. [9]

(Neolithic period, when products of polished artificial stone were found – the last of the two Stone Age periods) [10]

The floor from where the researchers later took out samples for testing was discovered while working with a bulldozer during work on a new motorway. In 1886 and 1887, the Israeli archeologist Garfinkel and retired Professor Malinowski from Gothenburg in Sweden, who specialized in concrete, took out samples of the concrete that were analyzed in the laboratory [9,11]. The date of the origin of the samples was found to be about 7000 BC.

Malinowski [11] states that the lower part of the floor consists of lime concrete with a thickness of 6–8 cm. The concrete is beige brown and has a very fine polished surface and in some places is covered with a reddish brown topping. The aggregate consists of crushed rock with sizes up to 10 mm, with very finely distributed mortar around it. The concrete is cast in two layers. The upper layer is a very hard mortar of 3–4 mm in thickness, with good bonding to the bottom layer.

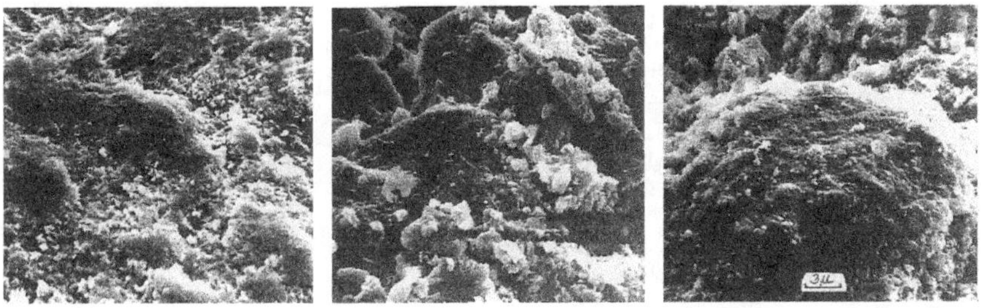

SEM-pictures of the fracture planes of the samples[11]

The laboratory testing, among others, showed:

- High density of 2.2 to 2.4 g/cm^3
- Remarkably high compressive strength of 40 MPa
- Chemical analysis showed 90% of calcite, with small quantities of quartz, etc.

Workers in the Potala temple in Lhasa produce a very durable floor by hand hammering coarse aggregate into a dry and hard mortar of sand, clay and lime.

Picture of the finished floor, probably several hundred years old, in the Dalai Lama's summer palace. These floors are water repellent and nearly crack free, in contrast to "modern" concrete on floors and walkways in the neighborhood, which have a considerable amount of cracking. A contributing reason might be the very low humidity and air pressure (3600–3700 m above sea level), much lower than in cases normally considered.

Malinowski's description of the floors in Galilee reminds me of the visual appearance, method, and use of a much later date. In August 2007, I had the pleasure of visiting Lhasa in Tibet, China. Here are some pictures from the visit.

1.2 Mesopotamia and China

In a paper meant for a conference on concrete in Seville in Spain in October 2009 [12] – the paper can be found in the proceedings though the author never turned up at the conference – Bao Ying tells a fascinating story.

(Bao Ying was born in 1934 in Nanjing in China. When she wrote the paper she was working at the Institute for Building Materials in Wuhan in the Hubai province in China.)

The fascinating story includes several historical speculations and theories. I am referring to it because it, among other things, throws some light on the fact that knowledge about concrete has been shared, but also been kept secret in periods.

Bao Ying, when referring to Bogue in USA, states that the oldest known cement was produced at the Shinar Plains in South-east Mesopotamia about 4500 years ago (2500 BC). The people in Shinar made holes in the ground when they wanted to make their food fires. They observed that due to the high temperature in the hole the surface cracked into small fragments and powder. When this was moistened by the rain, it would harden into lumps of stone. This resulted in the production of natural cement.

With this technology, they produced not only their daily household needs, but also, large building blocks, with their natural cement, sand, and water. The blocks could be up to 50 cm (20 ft) long, 35 cm (14 ft) wide and 10–12 cm (4–5 ft) high. They also used mortar from the natural cement in the joints between the blocks.

Babel's tower was made with this material. The reason for the tower being built only half the planned height was lack of raw material. When the raw material was exhausted, the workers thought that was a message from the Gods, so they decided to stop.

Bao Ying claims that, during King Salomo's time, some 3000 years ago,

Babel's tower in the imagination of the painter Pieter Bruegel the older. The historical art museum, Vienna.

the technology of burning natural cement was not available for everybody. It was a secret that was kept in the temple. She refers to two villages, Succoth and Zarethan, on the east side of the river Jordan. In the temples, there were holes in the ground for the production of cement. With reference to the Old Testament, Chapter 6, Verse 46, she says that between these two villages was the place where King Salomo cast his bronze items in molds of clay. In her opinion, the clay molds would not have been strong enough to hold such a high temperature. She thinks that the molds must have been cast from natural cement.

The Old Testament [13], Second book of Kings, Chapter 6, Verse 46: (translated from Norwegian) states

"It was on the Jordan-plain that the King did the casting, in the solid clay ground between Succoth and Zarethan"

To evaluate Bao Ying's views, it might be interesting to read what the Book of Kings says:

The equipment for the temple

13) King Salomon sent for Hiram from Tyrus.

14) He was the son of a widow from the Han Naftali-tribe, and his father was a

coppersmith from Tyrus. Hiram was a very able man and had great knowledge, so that he could make all kinds of things in bronze. He came to Salomon and did all the work that the King wanted done.

15) Hiram cast the two Bronze columns. One of the columns was eighteen ell high, and a string of twelve ell was needed to measure around it. The other column was of a similar height.

16) Then he made the column tops. They were cast in bronze and were five ells high.

38) Then he made the ten kettles of bronze. Each kettle took forty bat and four ells across. It was one kettle for each of the foundations.

45) The ash bowls, the fire shovels and the sacrifice bowls. All this that Hiram made for King Salomo to the Lord's temple was of shining bronze.

In Bao Ying's opinion, the casting of all the bronze was done in concrete molds. This must at least be regarded as a very advanced use of concrete, even today. Others have reflected that this also meant an enormous amount of bronze metal. Maybe some of the things that were made, at least the internal part of the columns, was cast in concrete? As for the more intricate things that were cast, they could not have been cast in concrete.

We accept these various opinions at least as interesting thoughts.

Then comes the most spectacular and possibly the most speculative part of Bao Ying's story:

In the year 722 BC, a group of the ten northernmost Jewish tribes fled the King's town of Shemer on a secret path, after the attack by the Assyrian Emperor. This group brought with them a number of treasures from the temple in Shemer, among others the secrets about how to produce cement.

When they moved, they were a group of about 20 000. They had sent scouts ahead and after three years of travelling they settled in what is today the Xinjiang province in China. They settled down on the Tarim Plains, on the western part of the Pamir Plateau. Today this is mainly a sand desert, but at that time it was a green area with many rivers. Their main town was called Little Lob, after the old land of Lob. In ancient times, this was a center for the caravan routes and a trading center between East and West.

With their knowledge of concrete, they built a beautiful city for kings and priests. A priest city was built on the north side of the Neah River that runs through the south part of the Tarim Plains. They also built a Jair temple on the north side of the Keneah River. In this temple, they kept God's Arks of the Covenant, which they had brought with them on the long journey. Their culture lived for about 600 years.

About 100 BC the Chinese Han dynasty attacked them with 60 000 soldiers. The descendants of the Jewish tribes got orders from the Chinese cavalry to leave the area within 10 days. With the exemption of the most holy and valuable things from their original land, everything – buildings, churches, etc. – were destroyed. The holy articles such as the Arks of the Covenant were moved to an underground grotto called Shemer Balah. The grotto was sealed with concrete.

The water resources in the Tarim basin were redirected, and what earlier was an oasis was over a period of time transformed into a desert. This is now one of the sources for the sand storms in China.

Archeological discoveries in the area have found ruins that can point back to several of the Jewish tribes in various parts of the area.

The route that the Jews followed to China was probably picked out by forward scouts and was along the old trading routes from East to West. The route was the same that Marco Polo followed many years later.

Marco Polo's route from Venetia to China [12]

- Starst from Venetia 1274
- Persian gulf
- Afghanistan's Bamyan valley of Kabul
- Hindu Kush mountains and Badakhshan mountains
- Wakhan Pamir
- Neza Tash pass and Taghdumbash Pamir
- Tashkurgan's Tajic
- Kash
- Shage
- Horem
- Minfeng
- Jaymuth
- Jorkeam
- Qomkuduk desert
- Hami og Balikum
- Mongolia (Upper capital for the Yuan Dynasty)
- Beijing (the main capital for the Yuan Dynasty) 1279

An important message in Bao Ying's story is that Cush in the Old Testament is not in Ethiopia, but in China. Her reasoning and arguments in this respect have little to do with the history of concrete, so I leave for those who are interested to evaluate this by themselves, for example from the cited reference.

Many authors believe that the land between the rivers Euphrates and Tigris, **Mesopotamia**, is not only the cradle of civilization, but also where the first concrete was made. Another building material, definitely in family with concrete, is bricks or building blocks. Bricks had their origin here [14] – so, why not concrete? With a broad definition of concrete, it is also possible to define the early bricks or building blocks as a kind of concrete.

Mesopotamia, which means the land between the rivers (*Greek; mesos-middle, potamos-river),* includes the modern Iraq, eastern Syria, eastern Turkey, and parts of Iran. These areas have been regarded as the cradle of civilization and have through the years had a number of different cultures and kingdoms, of which the Babylonian and the Assyrian kingdoms are possibly the most important ancient cultures.

The earliest city cultures developed in the 6th millennium BC. Some researchers claim that traces of cultures in this area can be dated back to 10 000 BC; the various cultures developed many new things that are also of importance in the building environment. From this area we find, among other things, Hammurabi's law, probably the first law, that gives rules also for construction activity.

King Hammurabi's (King 1796–1750 BC) law was found by the Egyptologist Gustav Jequier in 1901. The law is written on basalt-slates, that today is found in the Louvre in Paris. The Law also exists in other samples on different clay slates. It consists of 282 law clauses, normally only one or two sentences each.

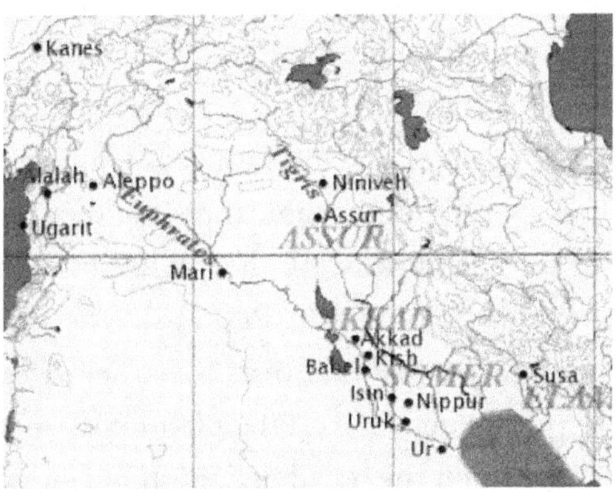

Mesopotamia

We repeat a few of the clauses from Hammurabi's law of 1730 BC (Several versions exist of this text).

> *If a building collapses and no one gets killed, the builder shall build a new building without cost for the owner.*
>
> *If a building collapses and the owner gets killed, the builder shall lose his life.*
>
> *If a building collapses and the wife of the owner is killed, the builder shall supply him with a new wife of similar value (another version says: if the owner's son dies, the builder shall lose his life.)*

Other important developments from Mesopotamia with bearing on the building trade are the use of the 60-number-system. This is the background for present day use of 60 seconds in a minute and 60 minutes per hour, and also the 360 degree angle.

The houses in Mesopotamia were made of mud (might have been clay)–bricks and mud – mortar and plaster. The Mesopotamian pyramids, the ziggurat or step pyramids with a flat top, date from 4000 BC to 600 BC; 28 of them have been found in Iraq and 4 in Iran [15]. The core of these ziggurats consists of sundried bricks, while their exterior has burnt bricks.

Campbell and Pryce [14] claim that the early sundried mud bricks were made in a Mesopotamian Neolithic (Stone Age) culture about 8 000 to 10 000 BC. These early block products were rough in shape and handmade. The use of building stones also created the need for mortar. The first mortar was clay or mud, often mixed with straw and excrements. Bitumen was utilized in the lower part of Mesopotamia where oil was found on the surface. Mortar based on gypsum could easily be burned at the camp fire at only 125℃. Lime mortar needs temperatures in the oven of 1000℃ and was used about the same time that they started to use burnt bricks instead of sundried bricks. All these kinds of mortar were used in ancient Mesopotamia according to Ref. [14]. The earliest building stones used were from 8 300 to 7 600 years BC. Their sizes varied, but they were normally approximately 260×100×100 mm(10×4×4 inches).

Based on the broad definition of concrete, we might say that this is the first use of concrete, even reinforced concrete. Some, however, might say that this is to stretch things a bit too far.

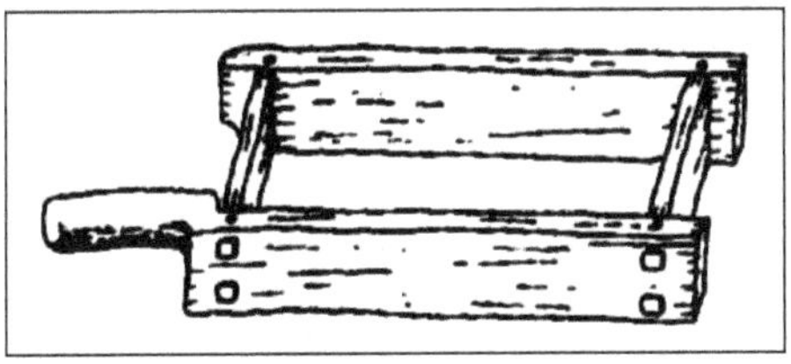

Early building block mold

The early building blocks cast in a mold can be dated to 5 900–5 300 BC [14].

Campbell and Pryce [16] claim that burnt blocks were very rare until the Uruk/Djemdet Nasr period–3100–2900 BC. They were very costly and used only in religious structures. Notes from the third dynasty in Ur (2 111–2 003 BC) state that for a piece of silver, you could buy 14 400 sun dried blocks, but only 504 burnt blocks.

Cement-based building materials also have a very long history in China. In an interesting memo, Professor Sui Tongbo at China Building Material Academy in Beijing has given us some insight into the earliest history [17].

Dr. Sui states that remnants of the so-called **"Baihuimian"** mortar have been found in the floors and walls of caves and crypts used as early as in the Neolithic Yangshao period, 3000–5000 years BC. Baihuimian has got its name from a white powder from a special type of clay called Jiangshi. The mortar used in the Stone Age caves gave a hard and smooth surface to the floors and walls of the caves.

When people started to live in huts and houses of timber, instead of in caves, about 1600 BC, during the time of the Shang dynasty, they also started, in addition to using "Baihuimian-mortar," to use **"yellow-mud"** in the adobe walls. During the Warring States period (403–221 BC), the "yellow-mud" was mixed with cut straw. This technology was used, both directly on the walls and also on wall tiles. Over time, "Baihuimian" mortar went out of use, while "yellow-mud" reinforced with straw has been used up to our times.

Burning of gypsum and clay for use in mortars came into use in China from 3000–1000 BC. In the time of the Zhou Dynasty, about 700 BC, lime was burnt from mussel shells. These shells contained calcium carbon-

A summary of Chinese history [18]

Xia Dynasty		2070BC—1600BC
Shang Dynasty		1600BC—1046BC
Western Zhou Dynasty		1046BC—771BC
Eastern Zhou Dynasty	Spring and Autumn Period	770BC—476BC
	Warring States	475BC—221BC
Qin Dynasty		221BC—206BC
Western Han Dynasty		206BC—25BC
Eastern Han Dynasty		25—220
Three Kingdoms(Wei, Shu, Wu)		220—280
Western Jin Dynasty		265—317
Eastern Jin Dynasty		317—420
Northern and Southern Dynasties		420—589
Sui Dynasty		581—618
Tang Dynasty		618—907
Five Dynasties		907—960
Northern Song Dynasty		960—1127
Southern Song Dynasty		1127—1279
Yuan Dynasty		1206—1368
Ming Dynasty		1368—1644
Qing Dynasty		1616—1911
Republic of China		1912—1949
People's Republic of China Founded on October 1, 1949		

ate, which was transformed during the burning process. Not only was the mortar from this workable and user friendly but the end product was also highly resistant to moisture and water. This technology was utilized in the building of grave chambers from the time of the Zhou Dynasty to that of the Ming Dynasty. In "Tian Gong Kai Wu" (Heavenly creations), a very famous book about technology in ancient China, written in the seventeenth century, very detailed descriptions can be found about the process of burning lime.

Recent Chinese studies indicate that burnt lime has been used in China for 5000 years. Lime mortar walls and floors have been found in various structures typical of the culture of the Longshan period along the Yellow

River, dating back to about 2800–2300 BC.

Using lime was one of the most important measures to strengthen the foundations of ancient building structures. Lime mortar was used as binders in brickwork and as a top mortar in walls and floors. Lime mortar was also used in the Great wall. Work on the geat wall was started in the seventh century BC. The construction of the Wall went on through 20 dynasties, until the sixteenth century. Of the total 6000–9000 kilometers, about 5000 kilometers were built during the Qin, Han, and Ming dynasties. These dynasties developed lime mortar to a higher technological level and used it in the great wall. (see also Chapter 4.5, which describes the early use of admixtures for concrete in China). During the Han Dynasty, block work structures with lime mortar were used in a number of areas, among others a pavilion of several floors.

The methods of lime production are described in the book "*Tian Gong Kai Wu* (*Heavenly Creations*)" from the Ming Dynasty. Burning of lime has also been described in a document from the Qin Dynasty [Ying Zao Fa Yuan (How to build palace-style stone bridge)].

Sanhetu (Tabia) is a kind of mortar or concrete based on three components: lime, clay, and fine sand. This blend has its origin in the fifth century in South China during the North Dynasty period. The other three components that were used were lime, ceramic powder, and crushed rock, or lime, slag, and sand (Qin Dynasty). We also find the book "*How to build palace-style stone bridge,*" belonging to this period, describing Sanhetu, which is made by blending 40% lime and 60% loess with fine sand; this reminds one quite a bit of the "Roman concrete." This concrete was used in the Qin Dynasty for floors, roofs, foundations and, well compacted, also in dams.

1.3 Lepenski Vir

In the middle of the 1960s a Yugoslavian archeologist, Professor Dragoslav Srejovic, from the University of Belgrade, found a hut floor of concrete on the shores of the river Danube in Lepenski Vir in Yugoslavia. The floor can be dated to about 5600 BC. The floor has a 25 cm thick layer of concrete. Some Stone Age fishermen and hunters must have been living in the huts [19].

The lime cement that has been used comes from a deposit 300 km further up along the river. The lime has been mixed with sand, gravel and water to a concrete that resembles the type of concrete of our time [20].

The findings of the hut floor in former Yugoslavia are regarded by many as the first concrete construction.

Lepenski Vir is in East Bosnia, near Danube Milanivac, close to the border of Romania. The first excavation was done in 1965 and in 1966 the site was declared one of Serbia's national monuments [21]. However, it was not until 1967 that the full scope of the findings was understood, when the first Mesolithic sculptures on the site were found.

The excavation was completed in 1971, and the whole site moved up 29.7 meters to avoid flooding from a new artificial lake that was made in the "Iron Gate" ravine through which the Danube runs.

The excavations indicate that the area had been populated from 7000 BC, with a culmination about 5 300 to 4 800 BC, and in several archeological periods. The settlement consisted of a main village and 10 satellite villages. The findings from the excavations include tools of stone and bone, remnants of houses and several religious objects and sculptures.

It is believed that the settlers at Lepenski Vir were one of the first in the European hunting cultures from the end of the last ice age. Archeological findings in nearby caves indicate human activity as far back as 20 000 BC.

The settlement in Lepenski Vir was strongly influenced by the topography of the area. The village is situated in a narrow horseshoe plateau on the bank of Danube, pressed between cliffs and the great river. The village is only a small settlement, possibly of only 4–5 families and of less than 100 people. The most important food supply was probably fish. In later pe-

Hut floor– Lepenski Vir

riods, the place probably became too crowded and people moved away. Some intermediate settlements are also found in the neighborhood, probably used for hunting, or for collection of food and raw materials. There is a distinct plan for the village and the burial place was beyond the buildings. An exemption is an older part of the village, where probably, for some religious reason, the burial place was behind the fire place. All the houses have a distinct geometry, and form exactly equal sided triangles. Every house has a fire place in the shape of an oblong rectangle along the long axis of the house.

1.4 South/middle - America

Some authors claim that concrete was already produced by Indian cultures in South/Middle America by 2500 BC [18].

In an article in *Concrete International* [15] in 1996, Rivera-Villarreal and Krayer discuss the use of pozzolan-based concrete in El Tajin in Mexico, developed between 1100 and 850 BC.They write about unarmed lightweight concrete in top floors and flat roofs. Archeological studies show that the early settlers in the area can be traced back to 3 000–2 200 BC.

Prehistoric Middle America was often divided into five zones. Common for all the zones was the use of lime. It was also known for the distinct development of the use of lime mortars. Several theories exist regarding how the pozzolans came into use. Much of the soil in Middle America had pozzolanic properties. Consequently, 35% of the cement production in Mexico today utilizes natural pozzolans.

Normally, well-rounded aggregate was used in ancient concrete, especially in the case of concrete floors.

In some buildings in El Tajin, the roof had been constructed using pumice with sizes up to 100 mm in the aggregate. The concrete had a specific weight ranging from 1050 to 1100 kg/m^3. The article states that El Tajin at this time had 20 000–25 000 inhabitants and that altogether 250–350 m^3 of this concrete has been found.

El Tajin is in Veracruz in North Mexico, between the cities of Papantla and Poza Rica.

Pozzolan mortar in a decoration in building D in El Tajin [21].

El Castillo in Chichen Itza on the Yucatan peninsula, built in honor of the Mayan God Kukulcan.

Cement mortar was known in South and Middle American cultures independent of the cultures on the other side of the Atlantic Ocean and can also be seen in the joints between the stones in structures in the Mayan culture.

The town was founded about the year 525.

Whitehouse and Wilkins report in the *Roofs of the Civilization* [22] that the living houses in Teotihuacan, over 4000 of them, which were built in "ranch-style," had walls of stone with mortar between the stones or of building blocks made of sundried clay. The floors and roofs were made of a type of concrete, which was made from a volcanic rock from the local district and was crushed and mixed with lime and soil. Both the floors and the roofs were often plastered to produce a shining surface.

Teotihuacan lies in the Mexico valley, about 50 km north-east of Mexico City, and was the largest city in America at that time. In its most glorious days in the middle of the first century, the city had more than 200 000 inhabitants.

1.5 Egypt

The earliest illustrations of what is regarded as some concrete work are found in Egypt and can be dated to about 1950 BC. The illustrations from Thebes' grave chamber show different stages of what have been described as concrete production, from the collection of water in the wells and mixing of mortar, to the use of mortar in the bricklaying. The bottom part of the illustration shows a concrete central wall supported on both sides by brick masonry. We also see the foreman with his whip on his shoulder, keeping a watchful eye on the workers [18].

Illustration from Thebes' grave chamber

Egyptian illustration (Theban picture from 1100 BC) according to [23]

In a book about technical inventions, a similar illustration is shown, where they claim that it is a Theban picture from 1100 BC [23].

There is no mention of concrete production but that the old Egyptians used the water passage and the plumb string is the key point in the illustration. The Italian author of the book does not seem to be much interested in concrete. However, if we look closer at the illustration, we can see that it is an extract from the former illustration.

The Egyptians developed a binder by burning contaminated gypsum. Some authors claim that concrete was used to build pyramids as early as 5000 BC [24]. However, there are somewhat mixed opinions about how and where this concrete was used.

The Egyptian pyramids have been favored objects for researchers as they were built about 4500 years ago. The large Cheops pyramid in Giza, for example, has enormous dimensions with an original volume of 2.5 million m^3, containing 2 300 000 building blocks with an average weight per block of 2.8 tons [25].

In an extensive and very convincing article (12 pages) in *Concrete International* in 1987, Joseph Davidovits claims that the largest pyramids are made of concrete blocks cast at the site[26].

Among others, he also refers to pozzolan concrete from Kamiros in Rhodes, which was built in 500 BC and stands strong even today, 2500 years later. He gives a very detailed chemical explanation of why concrete based on the ancient cements has much better aging properties than

The Giza pyramid in the background, behind the Sphinx, was, with its 146 meters height, the highest structure in the world for 4 300 years.

modern concrete. His main argument is that ancient concrete had much more developed carbonation than the modern one. He claims that the weak link in modern concrete is too much C-S-H-gel.

He also questions the normal classification of ancient cements into two main types:

- Hydrated pozzolan lime cements
- Hydrated hydraulic silicate lime cements

He argues that this classification is relevant only if we judge the cements on the basis of the ones that existed at the time when Portland cement was invented. Studies of 9000-year-old concrete indicate that the weakest concrete type is the one based on recarbonized lime. The better cements or concretes contain variable amounts of silica and silica–aluminum ingredients. He names the final product geopolymerized lime. Davidovits claims that he can prove that the pyramids of Cheops, Khefren, Teti, and Sneferu were produced by geopolymer–lime–cement with aggregate from Turah. The core, however, is block work that uses Giza limestone that contains fossil lime shells. In the original deposits, the shells are found in horizontal layers, while in the pyramids, the shells are spread around in the blocks.

Davidovits article was directly followed by an interview with Davidovits, written by his assistant Margie Morris.

Here is a glimpse of some of the questions:

- *Q: How do you explain the well-known Twelfth Dynasty bas-relief of the ruler Djehutihotep, which shows a giant monument being hauled by numerous workers?* Part of the answer: ··· *is not acceptable evidence with which to explain pyramid construction. In any case, this statue was produced 800 years after the Great Pyramids were built, and it belongs to an era characterized by the abatement of the stone-making technology.*

- *Q: Are you suggesting that the monuments such as the Sphinx and the huge blocks of Egyptian temples were also cast?* Some parts of the answer: *In the Middle Kingdom (2040 to 1640 BC) the princes of Thebes usurped the throne, and the first tomb sculpted directly from cliffs of soft sandstone was constructed. During the New Kingdom (1550 to 1070 BC)almost all monuments were made of carved stone. More were either cut directly into the face of a soft sandstone cliff, so that hauling was required,* ··· *But the Great Sphinx was also sculptured in situ, when limestone bedrock of Giza was saturated in order to disaggregate it for the purpose of making pyramid blocks. I suspect that the head of the Sphinx may be synthetic, but I have not had the opportunity to make an analysis. The bedrock body is very soft and rough and is severely eroded even though it remained burid in sand throughout much of its history. The head, however, remained exposed and has withstood 4700 years of climatic erosion. There is a large crack under the chin which may represent the joint between natural and man -made stone …*

- *Q: Why was this technology lost and unused during the later area?* Part of the answer: *By about 2200 BC, the Sinai mines were severely depleted of the ores required to produce cement. At that point, no structures comparable in mass to the Great Pyramids could be constructed, and the use of the technology gradually abated. As a result Egypt suffered great social instability which was not arrested until bronze was developed and used to cut stone. Over the millennia, the stone-making technology became increasingly more obscure.*

In a book that was published in 1988 [27] the researchers Joseph Davidovits and Majorie Morris write that the blocks were cast on site, where two of the sides where match-cast against the already cast blocks.

The theory about the in situ cast blocks was questioned, among other things, with the argument that

- Laboratory investigations showed a geopolymeric binder.
- The layers in the blocks could come from stops in the work.
- Organic material (hair) found in the block samples.

Dr. Davidovits knew from the demonstrations of another French scientist, Henry le Chatelier, fifty years earlier, that the Egyptians could produce synthetic sandstone [28]. Davidovits, as a specialist in geopolymerization, with more than thirty patent claims on the topic, formulated his first in situ cast stone theories as early as 1974, but it it was the book in 1988 that really started the big discussions.

As a result of the discussions, a group of American experts, with varied views on the subject, went to Egypt in January 1990. In the journal, *Concrete International,* from August 1991, we find articles from participants in the expedition which both confirm and contradict Davidovits' theory [25].

Margie Morris, who coauthored Davidovits' books from 1988, has solid documentation and sticks to her cast-concrete theory and argues her point of view thus:

- A group from the Waseda University in Tokyo constructed a 375 m high model of the Giza pyramid in 1978. As a conclusion to this gigantic experiment, they calculated that it would have taken 1200 years to build the pyramids if the stone theories of the archeologists should be used. Egyptologists, however, claim that the pyramids were built in less than 100 years.
- A geophysical study from Stanford Research International tried to penetrate the two largest pyramids with electromagnetic waves. However, SRI researchers found the masonry to have high moisture content, which was unexpected because the bedrock is dry.
- The large foundation of the Giza pyramid has a variation in elevation of only half an inch (13 mm) from corner to corner. That degree of precision suggests that it may be a cast foundation.

In addition, she has a number of scientific analyses as the basis for her theory.

The archeologists Campbell and Folk [25] base their natural stone view on several arguments:

- One of the two types of stone blocks found in the pyramids has a clear resemblance to the local sandstone.
- The Egyptians did not have the technology to produce such a high quality of artificial stone.
- The stratification of the blocks is natural in this type of sedimentary rock.
- The swamp-like layer in some of the blocks is sometimes horizontal and sometimes vertical. This would not have been the case if the blocks had been cast on site.

In a debate article in *Concrete International* in February 1992, Bob Moores argues strongly against the concrete theory [29].

He bases his view on 18 years of studies and he is in the process of writing a book on the pyramids. He has taken over 2000 color slides of the structures and collected 135 stone samples. Among other arguments, he points out that the Egyptians have produced stone building blocks since the beginning of their culture. The pyramid blocks have no traces of the molds they must have been cast in, like other building blocks in concrete, and all the blocks have different shapes and sizes.

Even if the joints between the blocks are very small (0–2 mm), traces of lime have been found in some of the joints.

Traces of marks from chiseling tools have been found in some of the blocks.

The discussions will obviously continue and for us with too little in-depth knowledge, it is interesting, to see both the way the experts throw light on the problem and the manner in which the experts on both sides try to underline the arguments on the other side with sometimes rather subjective arguments.

Several well-known concrete and cement chemists I have talked to raise doubts about Davidovits' theories and refer to his ideas on cement and the practical tests in the USA, which had not been successful [8].

However, there seems to be some agreement on the fact that the old Egyptians knew how to make concrete.

Aïtcin [8] claims that all the mortars during the Pharaoh period had some kind of gypsum- based background (anhydrite). He avers that only during the time of Ptolemy, 300 BC, lime was used and that it was the Greeks who brought with them the lime technology.

Lea [5] opines that the oldest Egyptian buildings were made with sun-dried bricks or building blocks and that they used wet Nile mud in addition to chopped straw in the layers between the blocks. Burnt bricks was used by the Babylonians and the Assyrians, binding the bricks together with bitumen. This was an efficient method, but limited to the areas with such raw materials.

Lea states that Egyptian mortar, which was used in the pyramids, was referred to as lime mortar by authors. He opines that this can hardly be correct since lime was not used until the Roman times. He claims that the reason why the Egyptians used gypsum and not lime was the lack of suitable burning materials to burn the limestone, which needs considerably higher burning temperature for its calcining.

Though reading about the pyramids might be interesting, not only the stone/concrete aspect of it is argued about. In an article, Per A. Christiansen *"Faraos fjell"*(Pharao's mountains) [30] mentions a number of questions that have been discussed through the ages.

About the question of stone, however, he is clear. About the great pyramid, the Cheops pyramid, he among other things points out that the foundation area is 53 825 m^2, built with 2.3 million stone blocks with an average weight of 2.5 tons. Originally it was clad with shining white polished sandstone. A major earthquake in the year 908 destroyed the cladding. These damaged stones have been used for buildings in Cairo. The stone material for the pyramids was taken from various quarries. From Aswan they got the granite for the grave chambers deep inside the pyramids. From the heights of the Mokattam Mountains on the other side of the Nile, they collected yellow sandstone for the centre of the pyramid. From Turah, near Giza, they got the white granite for the outer cladding. He refers to Herodotus, who states that 100 000 men worked for 20 years on the Cheops pyramid. Later estimates have shown that the workforce probably was closer to 250 000 men. Contradictory to earlier expectations, the workers were not, possibly with a few exceptions, slaves, but free Egyptian men who worked and got paid in periods when there was no other work in the fields.

Possibly, one day we will know the secrets of the old Egyptians. For the ones that favor a well- known comic book figure, maybe it was a bit of each, and both sides may be right. Davidovits has at least indicated the possibility of *"Yes thank you, I take both!"*

***

It is reasonable to believe that knowledge about concrete in ancient times developed in several places in about the same period. The theory that the technology or the use of it developed as a result of the observation of the edge of the fire place hole of the Stone Age people seems reasonable. When the rain moistened the edge of the fire place hole, a material formed, which could glue together sand, earth, and stone to mortars.. The type of glue that formed must have depended on the material in the fire hole, the size of the hole and the burning material, and the intensity of the fire. In some places it resulted in anhydrite/gypsum, in some non-hydraulic lime, and to a lesser extent possibly some hydraulic cement, more like the cement we have today. In some places, the glue might have also been from natural cement–stone and from pozzolanic materials (reactive siliceous dioxide/ glass).

Only the hydraulic lime and the "pozzolan-concrete" had good resistance against rain and water. The other preserved mortars have either been used indoor in the form of plaster or mortar to mend unevenness in stones, blocks or masonry, or possibly have been effected by some carbonation.

With one exemption, the ancient concretes have not been used in large load-bearing structures, and the use is rather limited in its way of utilization. The possible exemption is Davidovits' theory about its use in some of the oldest great pyramids and possibly part of the Sphinx. We will not contradict the theory, but many experts think that the theory is dubious or unlikely.

The use of concrete by the Romans saw a giant step forward in technology and use.

It might also be useful to see the development of the material in the light of other technologies in the same time period [23]:

5 000 BC – the first mural structures – the Dolms.
4 000 BC – the first picture letters in Mesopotamia.
2 800 BC – casting of the first metal items in Al Ufaid in Ur.
2 500 BC – the first rough paddle canoes.
1 500 BC – the stern oars were used as a rudder.
600 BC – the first turning mill stone.

1.6 Roman concrete

The Romans had conquered most of the Italian peninsula by 274 BC and in the next 250 years they increased their influence to rule over Greece, Little Asia, Egypt, France, and the Iberian Peninsula.

The Roman Empire dominated the western world from 27 BC to the

Figure ROMAN EMPERORS YEAR 14—235

14 — 37. Tiberius
37 — 41 Gaius (Caligula)
41 — 54 Claudius
54 — 68 Nero
64 (fire keisere: Galba, Otho, Vitellius, Vespasian)
69 — 79 Vespasian
79 — 81 Titus
81 — 96 Dimitian
96 — 98 Nerva
98—117 Trajan
117—138 Hadrian
138—161 Antonius Pius
161—180 Marcus Aurelius
193—211 Commondus
212—217 Caraculla
218—222 Elagabalus
222—235 Severus Alexculo
235— 284 Military anarcy with at least 20 Emperors
306 — 337 Constantin the great

year 395 and spread their culture and their technology in a manner that had not happened before in history. Their building culture and habits survived long into the Middle Ages.

Roman Emperors

About 150 BC, the Etruscans started to practice a building technique, in which they cast stones together with the aid of a lime mortar. They used molds of brick or wood. Some regard this as the first real use of concrete [31]. This technology has been named *Opus structurae*. Possibly, they may have learned the technology from the water structure art of King Salomo in 1000 BC, where lime mortar made of santorin (pozzolan – stone from the Greek island of Santorini) and other pozzolan or trass were used to make the mortar hydraulic. In approximately 100 BC, the Romans used a similar concrete with hydraulic lime and aggregate *Opus signium.*

The Phoenicians used concrete about 500 BC and the Greeks used a type of concrete to line their cisterns.

Lea [5] claims that the Romans must have learned about lime from the Greeks, who had already used lime mortar on the island of Crete during earlier periods.

There seem to be somewhat contradictory opinions about whether the Romans also learned about the advantage of using volcanic ash (pozzolan) from the Greeks. Volcanic material from Thera, today called Santorin, has been used in Greece since ancient times, but probably in limited amounts.

From an interesting article [32], we come to know about an Arabian nomadic tribe called The Nabataeans, 300 BC, who utilized cement mortar based on pozzolans to line their wells in the desert to make them water tight. Without this technology they would probably not have been able to control their life in the desert. Their secret knowledge made it possible for them to travel much longer into the desert than any other nomads. They could travel where others could not and became an important and rich trading people. Their secret wells in the desert, which were unknown to others, made them superior in desert travelling.

The Romans were empirical in their development of concrete technology and their building methods later spread over the whole world and became one of the most important foundations for the most important building material in the world, but how much they really understood of the theories behind their impressive monuments is a question yet to be discussed.

In a publication celebrating the Finnish Concrete Society's 50th anniversary in 1975 [33], the Danish concrete researcher Gunnar Idorn quotes from the book of Jérôme Carcopino – "Daily Life in Rome in the Epoque of the Emperors";

Blindedly their overdeveloped practical sense and always looking for immediate profit the Romans never saw the advantage of exploratory research.

They could not see what it was about, and it did not attract them.

They compiled results, which others had already achieved, and they took science into their books as a matter of fact, feeling no desire to further develop it or to control it".

Some claim that the first use of pozzolan cement is in the theater in Pompeii, built in 55 BC [3].

Others are of the opinion that it might have been even earlier. Julius Caesar (102–44 BC), might have used pozzolan concrete for the construction of, among others, the Caligula wharf in Pozzuoli harbor [24].

R.J.A. Wilson writes in "*The Roman World*"[34] that the Roman invention of concrete also gave birth to a completely new architecture. The Greeks had started to use arches, but in a very limited scale. With the introduction of concrete, the Romans revolutionized the building culture. Wilson states that concrete was used in the arches in the gigantic warehouse Porticus Aemilia (487m long) in Rome, in 193 BC. He also writes that concrete played a major role in the construction of the Roman bath houses. The first surviving example is from Campania. However, he thinks that the first real impressive development in concrete technology and in Roman architecture, came a little bit later, in the Emperor period.

The Romans developed expertise in concrete, but when it came to the traditional stone material, Rome took help from a great number of Greek stone craftsmen, who migrated to the capital of the world. This blend of Greek architecture and craftsmanship and the new Roman material created completely new trends in architecture.

Cato was the first to write about the burning of lime, about 200 BC. He was very concerned about strict quality control in the process. Pozzolan cement was not in use at that time, but it might be that some of the ancient knowledge about the better lime-based cements also had been spread to people on the Italian peninsula.

Whitehouse and Wilkins write in "*Sivilisasjonens røtter*" *(English: Roofs of the Civilization)* [22] that the Roman invention of concrete is disputable. It seems that they, from the period of the late republic and well into the period of the Emperors, were experimenting with various types of mortars, in particular mixtures that had good curing and waterproofing properties. It seems likely that they did not fully understand the possibilities of the material in the beginning. The reconstruction of Rome after Nero's big fire in the year 64 might have been a breakthrough in the understanding of the use of concrete in new structures and more liberal forms and shapes.

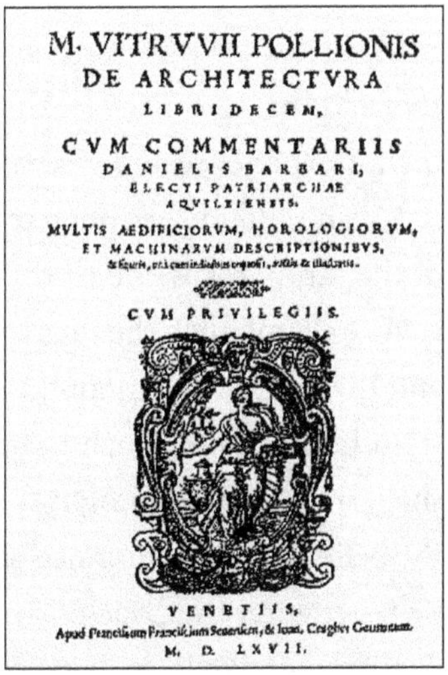

M. VITRVVII POLLIONIS
DE ARCHITECTVRA
LIBRI DECEM,
CVM COMMENTARIIS
DANIELIS BARBARI,
ELECTI PATRIARCHAE
AQVILEIENSIS.
MVLTIS AEDIFICIORVM, HOROLOGIORVM,
ET MACHINARVM DESCRIPTIONIBVS,
[illegible]
CVM PRIVILEGIIS.
VENETIIS,
Apud Franciscum Franciscium Senensem, & Ioan. Crugher Germanum.
M. D. LXVII.

The front page of Daniele Barbaro's famous edition of Vitruvius book of 1567.

Real pozzolan is a red volcanic ash from the volcano Vesuvius, found in the city of Pozzuoli, a few kilometers west of Napoli (Naples) in Italy. (Vesuvius is east of Naples). There was also earth with pozzolanic properties is the Etruscan area, in what is today called Toscana (Tuscany).

The name concrete also comes from the Romans. It comes from the Latin word *concretus*, which means to "grow together" [22].

Hover [33] opines that cement is a misleading translation of the Latin word *Caementum,* which means crushed stone. (He says that the translation is a curios one because this would mean that a cement truck is really a dumper with crushed rock). Hover also avers that concrete comes from the Latin verb *concrescere,* which means to grow or harden.

Many writers about concrete refer to Vitruvius. Let us look a bit more closely on this possibly most cited author on concrete in history.

Many claim that Vitruvius was a very conservative writer, but he seems to have commanded considerable respect during the period. Many Roman authors of his time seem to have omitted subjects that had been covered by Vitruvius in his books, and instead chose to site what Vitruvius had been writing.

The first written description of concrete production was written by the architect and engineer Vitruvius (e.g. Marcus Vitruvius Pollio), about 25 BC. In his publication "*De architectura,*" among others, he tells about the harbor structures near Puteopoli, where concrete was cast in big blocks to be used in the construction. Vitruvius says that *"puzzolan was mixed with lime and clay bricks cannot only be used in buildings, but also for strengthening of structures in the sea, because suddenly it hardens, and it can even hardens submerged in water."*

Vitruvius also gives good advice in his book "*About the origin of all things*". Among others, he thinks that constructors should have more contractor experience and vice versa.

Clean sand in the concrete is important. A simple test is to throw some of the sand on a white cloth. If it does not stain the cloth, it is good. Vitruvius claims that the aggregate must be sheltered from the sun, water, and the light of the moon [35]. That we even today try to shelter the sand from rain and too much heat from the sun is not unusual, but the idea of sheltering from moonlight is today a curiosity. (possibly we have inherited the belief in the importance of the moon as we specify that concrete should be tested after the cycle of a month).

Vitruvius Pollio is often cited by historians of concrete. This quote is taken from Gartmann [6], who refers to the script

> *"De Arhitectura Libri X" "A powder material from the nature can do miraculous things. It is found it area near Baiae and in the living areas around Vesuvius. If it is mixed with lime and crushed rock, it do not only give ordinary building structures great strength, but even masonry made in the sea will be so strong that neither currents nor water can destroy them."*

Dr. Morris Hickey Morgan has translated Vitruvius' book "Ten book on Architecture," which was published by Harvard University Press in 1914. In a preface to the 1960 edition of the book [36], Albert A. Howard discusses Vituvius a bit: "There have been some discussions about when Vitruvius lived, but most researchers think that that he wrote the book towards the end of Nero's time as ruler."

> *(Nero Claudius Drusus Germaricus, born Lucius Domitius Ahenobarbas 17, December [37], inherited the throne in the year 50. He only became Emperor 17 years old in 54, and he took the title Imperator in 66. He died on 9 June 68 [37]).*

Researchers claim that Vitruvius was in no way a great writer. He was probably too much of an "engineer," who has written simple sentences and very formal formulations that most of all seem like some kind of "Contract-language." Toward the end of each chapter, he normally sums up what he thinks is most important. In his "Ten books", he gives advice to the architects over a very wide range of subjects, from education to general town planning, materials, and, not least, the many building types and structures that an architect should be able to build. It is in Book Ⅱ, in particular, that he talks about the building materials.

It might be useful to remember the various topics that the architects of that time should have in command. In Book Ⅰ, Chapter Ⅱ, Vitruvius specifies three types of architectural work;

- The art to build and construct.
- To make time measurement pieces.
- The design of machines.

In Book Ⅱ, we find lectures about the materials. Here are some samples:

About curing technology (Chapter Ⅲ, 2): Bricks for masonry must be made in the spring or in the autumn, so that they get an even drying period. Summer-bricks are weak because they are dry on the surface, but not inside, and they will later crack. Bricks must cure for at least two years before they can be used.

About lightweight concrete brickwork (Chapter Ⅲ,4): In Maxilua and Callet in Further Spain and in Pitane in Little-Asia, they produce brickwork that can float on water. The reason is that they are made of clay from pumice. The masonry is very resistive against moisture.

About Sea sand (Chapter Ⅳ, 2): The sand is no good and can give discoloration that damages the surface.

About River sand (Chapter Ⅳ, 3): Is less suitable for concrete than for plaster.

About burning of lime (Chapter Ⅴ, 2 and 3): When the limestone comes in the oven, it looses its solid properties. It will be burnt out, tired and it obtains open, empty pores. When it comes into water, the water finds its way into the pores. The water feels the action of the fire, and the limestone starts to get warm, before it gets cold again. In the oven, the stone has lost one third of its weight, but mixed with water and aggregate it gives it a solid structure.

About pozzolan (Chapter Ⅵ): The reason for the strength of this material can be found in the fact that the ground in the hills is warm and it has warm springs. This would not have happened if not the mountain underneath had big fires with burning sulfur or asphalt. The fire and the heat from the flames come from the inside through cracks and make the soil light, and the pumice that can be found there is like a swamp free from moisture. In this way, the three substances will be mixed together in a fire.

That it is burning hot in these areas can be proved by the fact that in the mountains near Baiae, that belongs to Cumaeans, there are excavations which have been made into sweat baths. What we call "swamp-stone" or "Pompeii pumice" seems to have been reduced from another type of stone to the type of stone we can find. The type of stone found in this region cannot be produced elsewhere, but can be found in Aetna and in the mountains in Mysia, called "burnt area" by the Greeks. Vitruvius is somewhat surprised that he does not find similar stones with similar qualities in Toscana, even if warms springs can be found there.

In Book Ⅶ, Chapter Ⅰ, Vitruvius writes about floors; he starts with the concrete floor, which is the most important among the polished floors. He emphasizes that the earth floor underneath must be compacted well enough and the wooden formwork must be strong enough. Wooden formwork must be protected, for example, using straw, so that it is not destroyed by the lime in the concrete. The proportion between the aggregate and lime shall be 3 : 1 if we have a new aggregate or 5 : 1 if the aggregate is a reused one.

In the modern sense, Vitruvius was not "only" an architect and engineer, but much, much more, also a geographer and an astrologist. In Book Ⅷ, he writes about water, and shows a very wide knowledge of geography, and propounds theories that probably will have problems in getting accepted today. He claims that winds that come from the cold side, that is the north, are dry. Southern winds are warm and wet. This also has effects on the rivers, the largest and most important of which come from the north. He among others refers to the Ganges and Indus in India, which come from the Caucasus, Eufrat and Tigris in Syria, Ontus in Asia, Dniepr, Bug and Don in Colchis, Rhône in Celtica, Timavo and Po in Italy, Tiber in Mauritania, and others. With respect to the Nile, the story goes thus: Dyris originated in the Atlas mountains and ran west to the Heptabolus lake, where it changed names and ran south into a marshland. This is in the kingdom of Ethiopia, which comprises the rivers of Astansoba, Astaboa, and many more, which run through cataracts in the mountains in the north between Elephantis and Syene and through the plains at Thebes to Egypt, where finally the river is called the Nile. He tries to argue his point about the origin of the Nile by referring to the animal life.

There was a marked development in the knowledge about concrete during the Roman period. The Romans used concrete in many of their structures, but it was far better than all the earlier structures that utilized the knowledge and technology pertaining to the making of concrete.

To envisage this period, lets us look at a story from Rome at the time of Vitruvius

We have taken the story from the Danish building journal *Byggeindustrien* 1924–1967, which refers to the journal *Byggmesteren* 9/67 as its source. The story is called *Boligforhold uændret gennom 2000 år* (Living Conditions Unchanged Though 2000 Years) [38] (translated from Danish).

44 years BC, the inhabitants in Rome were complaining about the many lodging sharks that exploited the lack of living quarters. This lack was also an important reason for increasing crime, divorces and alcoholism.

This short 2000 year old situation report shows how little reality has changed through 20 centuries.

The findings that the archeologists have done in this century in the old ghost city of Ostia and in the heart of the eternal city Rome have thrown new light on the living situation 2000 years ago.

From these findings it is easier to understand cities from the old authors from Rome. Martialis for example tells about people that have to climb 200 steps to get to their apartments (that should be a house of about 10 floors). Juvenalis deplores about some poor school teachers, who have to live under the roof beams, where lovesick pigeons also lay their eggs. Another place he writes about the dangers that might be found everywhere in the narrow streets of Rome, where roof tiles from extreme heights might fall and crush the head of the passing pedestrians. How were the landlords? Cicero owned a few apartments, and for these he received an annual rent of 80 000 sestertier (3 000 USD) in year 44 BC. But we also know today that people were complaining about the sky-high rents they had to pay.

A promise of freedom from paying the rent was an important tool in the fight be-

tween the political parties. César and later Octavianus declared, to the joy of the renters (and the terror of the landlords), that they declared one year of freedom from rent for the poorest people.

The rents for apartments were 5 times as high in Rome as in the nearby villages. Unscrupulous individuals bought complete building complexes and rented them out to usurer rents. Most renters then rented them out again after splitting the apartments up into smaller units. Every interfacing link took its profit. The result was that the original very high rent might have increased by 100% or more from the original for the people who actually were living in the apartments.

The lack of living quarters had many reasons. The population of Rome was under steady increase. The migration from the countryside was significant because it was easier for people to get an income in Rome. Merchants from the Orient opened offices in the Tiber harbor and installed themselves with family, servants and employees. The invasion of foreigners was so strong that the Romans sometimes felt like being foreigners in their own city.

Building of new houses had to take place in the suburbs. Here the satellite towns grew up at a very high sequence, but the need for new living quarters was still high. People did nearly everything to get an apartment. The women (also married) sold their bodies, and men murdered to get a place to live. One man was accused to have tried to poison a complete family to get the apartment they were living in.

The poor living conditions led to crime, divorces and alcoholism, but they did not manage to repair the problem. In the year 187 BC, they thought they had found the solution. The authorities forbid foreigners to start settling in Rome. This declaration was repeated in the years 122, 95, 65 and 40 BC., and it looked like that they might have solved some of the problem.

The Romans could finally breathe, but not the renting-sharks. They saw their position threatened. They tried through legislation to protest against the condemnation of old houses, and the building of new and better ones. If a house declined, the owner had to build a new one within a certain time limit. If they did not, someone else could build on the property. The state also gave favors to those who used inheritance to build new houses.

Worse time come. When Rome in Emperor Diocletian's time lost its role as the centre of the world, the population decreased. At the end of the 500s the population had decreased to about 30000! Then the housing shortage had solved itself.

In his book *Greek and Roman Technology*, K.D. White [39] writes about how technology spreads. He talks about three ways;

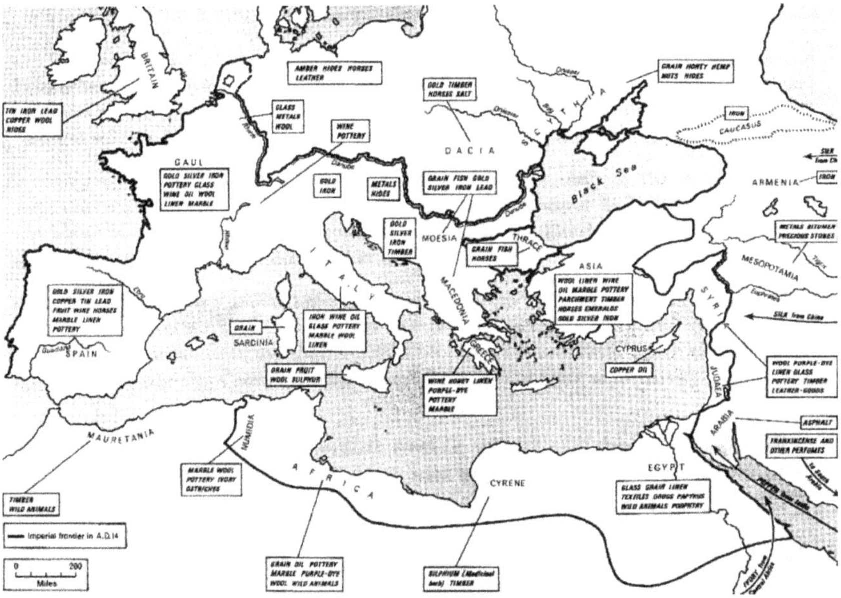

An overview of the raw materials in the Roman Empire [39].

- Transmission through inventors and craftsmen who migrate.
- Transmission through trade.
- Transmission through wars and conquering.

White is rather convincing in his documentation regarding the question about the Greeks not possessing the technology to produce concrete. Even in their early structures, the Romans did not know about the art of building with concrete. The development came gradually, something that is strengthened by the fact that Vitruvius was doubtful about where he could or could not use concrete and the fact that the discovery of the use of pozzolan and its potential for improvements was known only shortly before the turn of the millennium.

Only later, the Greeks started to use lime mortar, but they used too little lime to make their infill walls durable.

In the book *Concrete Progress from Antiquity to the Third Millennium,* Gunnar Idorn, on the other hand, claims [40], with reference to Efstathiadis, that the Greeks used the pozzolanic material (highly siliceous, volcanic) Santorin as far back as 500 to 300 BC.

Hill gives in his book an interesting overview of the raw material sources in

Pont du Gard near Remoulins in Departement Gard.

the Roman Empire. Building materials such as marble and wood are discussed, but nothing is mentioned about lime, cement, pozzolan etc. This does not necessarily mean that these materials were not transported from one part of the Empire to the other, but it strengthens the impression that concrete as a material was not part of an imperial or universal technology. Concrete was used locally if the materials for it were present locally and if they had knowledge about the technology.

We really do not know everything about where the Romans used concrete and where they did not.

The Romans did not use concrete for their water pipes and during the early period not in their water channels either. They used hammered lead in their pipes and masonry in their channels. Because of the hydraulic gradient, concrete was not recommended.

In his book *A History of Engineering in Classical and Medieval Times*, Donald Hill [41] states that the first aqueduct, Appia, was built in 312 BC, but to keep up with the increasing population they had to build more from 100 BC to the year 150. Most of them were two feet wide and five feet high, so that the workers could repair them. The channels got a "lining" with water-tight concrete and a roof to protect the water from the sun. Vitruvius had recommended that the channels should have a vertical fall for every 120 ft (36 meter). Otherwise the fall should be 1 : 200.

Pont du Gard is built on three levels and is 49 meters (163 ft) high. The longest level is 275 meters (916 ft) long,

The top of the aqueduct at Pont du Gard in Provence in France.

Lowest level: 6 arches, 142 meters long, 6 meters wide, 22 meters high.

Middle level: 11 arches, 242 meters long, 4 meters wide, 20 meters high.

Upper level: 35 arches,

275 meters long, 3 meters wide, 7 meters high.

The water channel on the upper level is 1.8 meter deep and 1.2 meter wide, with a fall of 0.2%.

The aqueduct was part of a water system that supplied the Roman city of Nemausus (Nîmes) with 20 000 m^3 of water every day.

It is thought that the construction work was led by Marcus Vipsanius Agrippa and that the channel probably was finished in 18 BC. The bridge itself is built without the use of mortar [42].

Idorn [40] avers that the concrete on the sides of the aqueduct Pont du Gard is covered with a thick durable calcite layer. In his opinion, this is due to the fact that the water that comes from the mountains is rich in lime and through the centuries it has deposited the lime. He has tested similar layers from the aqueduct near Nîmes and he calls this carbonization.

Water was important for the Romans in more than one way. In modern terminology, the Roman "water-building-market" was an extremely interesting market segment. In Sven Thaulow's book about baths and swimming pools, published in 1965 [43], Hjalmar T. Larsen has written a chapter about the history of the baths. We have cited a bit from his history and stories about old Rome (translated from Norwegian).

The Romans scorned in the beginning the Greeks with their overdimensioned thermes. Only in the time of Cato the older and Scipio Africanus the Romans started to take baths the Greek way. Edil Marcus Vipsanius Agrippa constructed in his time of service 170 saunas in Rome, and built the first thermes in Greek style (19 BC).

The baths became popular and the Roman rulers saw the advantage in the possibility of buying the favor of the public by building

Reconstruction of Tepidarium in Caracalla's Therme

baths or thermes, more luxurious than the previous ones. Nero built his therme, "The golden house," in the year 64, about which Matial exclaimed "What is worse than Nero, and what is more beautiful than Nero's baths." Later on, more Emperors built baths or thermes; Vespian (year 68), Titus (year 75), Trajan (year 103), Hadrian (year 120), Caraculla (year 217), Diokletian (year 295), and Constantine (year 324). During Constantine's time, Rome had about 900 official baths and 11 thermes in operation.

The sizes of the projects were impressive. The Titus therme had over 100 bathrooms. Caracalla's therme had a ground area of 124 000 m^2 and a water surface area of 1 300 m^2. For the 2 300 persons who could take a bath at the same time, there were 1 600 polished marble chairs. Diocletian's therme was a bit larger, with 125 000 m^2 ground area and 1 700 m^2 water surface. 3 600 persons could take a bath there at the same time.

Rome in Constantine's time consumed 750 million liters of water per day.

Rome had a very well planned water and sewage system. Already, between 350 and 312 BC, Appius Claudius had built the first major water line.

Aqua Appia brought water from far away and in the end there were 12 such water ways supplying the 1 million inhabitants of Rome.

The Romans brought with them their building technology and art and their knowledge about water supplying structures to many places in their

The Aqueduct in Segovia

big Empire.

Along the Rhine, the Romans worked hard for 200 years, from about the year 70 to 276. They built a 105 km long water line to Cologne which provided the city with 30 million liters of water daily from Eifel. A major part of this structure was of concrete [44].

Not all the aqueducts that the Romans built had concrete in the block work. The best restored aqueduct on the Iberian peninsula, and one of the most remarkable of them all, was built in Segovia in Spain, a mountain town about 100 km north-west of Madrid.

Segovia is situated on a cliff formation high above the local river. The Romans therefore decided to lead the water from the Fuente Fria river in the mountains, 15 kms away to the city. First the water is collected in two water tanks, then it is decanted in sand before it is led 728 meters with a fall of 1% on the aqueduct.

At its highest point, the structure is 28.5 meters. It is built with local granite blocks, without mortar, and has 167 arches. There is no concrete in the arch constructions and there are some uncertainties regarding the use of Roman concrete in the water channel.

In the years 70–80, the 90 km long culvert between Eifel and Cologne was built by the Romans with something that later has been called "Roman cement." As a binder "trass" (German pozzolan) was partly utilized, and cement produced by burning lime sand with some content of quartz and clay was also partly used. The water line was in use for 450 years. Then the material from the structure was used for building blocks in city walls and fortifications. Good examples of the durability of this concrete can be found even today.

Hill also discusses the building of dams with concrete. Two dams in Spain from Roman times are still in use today. Both were built to serve the town of Merida. The dam in Proserpina was probably built in the first third of the second century and was 1400 ft (420 meters) long and 40 ft (12 meters) high. It is built with masonry, but with a concrete inner part. On the top, it is only 7–8 ft wide.

Roman bridges, both for roads to cross rivers and as aqueducts, show a marked development over time. The use of concrete is one of the most important factors in this development. As concrete came into use, the bridges got slimmer and more elegant, and the columns were placed at a longer distance. A typical structure is the stone block work with concrete in the middle (infill).

The Roman concrete technology also spread to the Arabian countries and to Iran.

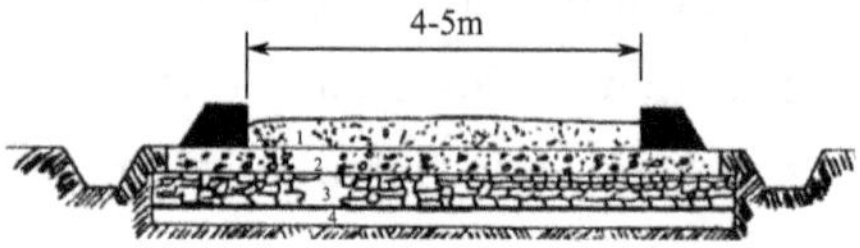

The Roman's building of roads was impressive. Totally they should have built nearly 90 000 km of main roads and 320 000 km of secondary roads.

We know that the Romans used concrete in some of their roads.

Section through a Roman <u>local road</u> **[39]**

1. 30–40 cm rough concrete

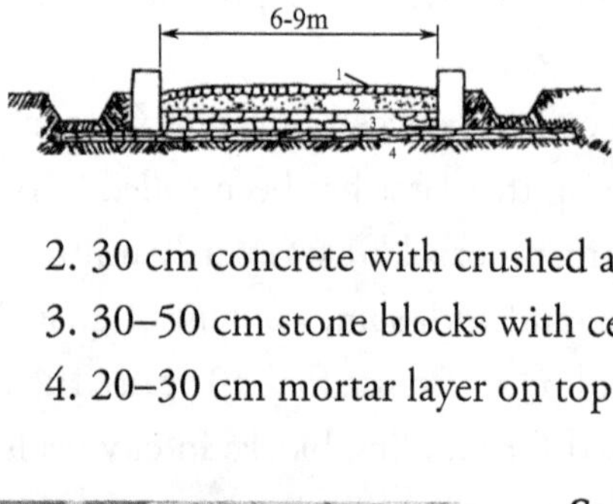

2. 30 cm concrete with crushed aggregate
3. 30–50 cm stone blocks with cement mortar
4. 20–30 cm mortar layer on top of a sand bed

The Pantheon temple in Rome, built in the year 127 [3] [18]

Section through a Roman <u>main road</u>

1. 20–25 cm stone or slates in mortar
2. 30 cm concrete with crushed aggregate
3. 30–50 cm stone blocks in mortar
4. 20–40 cm flat stones

The specifications varied from one region to the other, but where they had hydraulic cement, this was used.

In White's opinion [39], the introduction of pozzolan came about 300 BC, and experience has shown that this gave both water resistant and fireproof concrete. The Romans did not understand why, but their experience told them that a mixture of pozzolan, lime (often burnt), sand, and water gave a good building material.

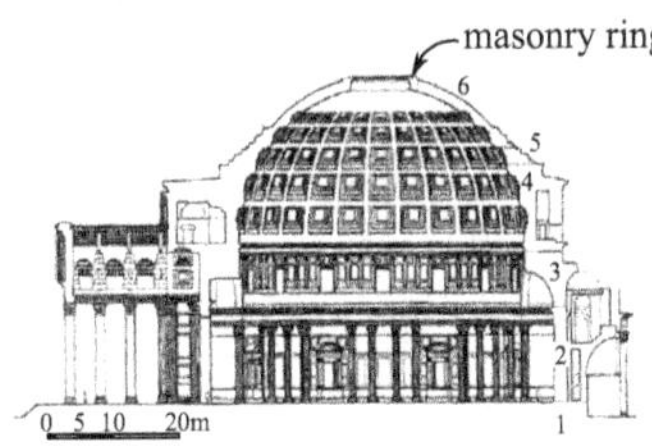

Variation in aggregate in Pantheon [46]

Time has shown that the Romans could make good concrete. Hover Ref. [35] refers to Lamprecht, who studied 80 concrete cores taken from Roman concrete around Europe and registered 2000 years compressive strength of 7 to 14 MPa.

Probably the most wellknown, and most referred to, building structure built from Roman concrete is the Pantheon temple in Rome. The unique dome is made of a kind of lightweight concrete, with crushed pumice as aggregate in the upper parts.

The Pantheon was built during the time of Emperor Hadrian, and the dome has a diameter of 43.3 meters [40].

In Ref. [45], we learn that the dedication inscription over the column front tells that Agrippa, Augustus' son in law, was the builder in 27 BC. Agrippa's Pantheon, however, has disappeared and it is due to the generosity of Hadrian that Agrippa is mentioned as the builder. The clinker from the foundation up to the dome has stamps that can be dated to the 120s.

In addition to developing the lightweight concrete technology, the Romans also found other ways to design light structures. They cast in empty clay pottery and jars in walls and arches.

The structure of the Pantheon is also unique in the respect that they have changed the density of the structures by changing the aggregate density to lighter weights gradually as you get higher up in the structure.

The first century AC was the time for innovation both in architecture and building methods in Rome. This reached its peak early in the second

The Romans preferred to use bricks as a molding material for their concrete[35].

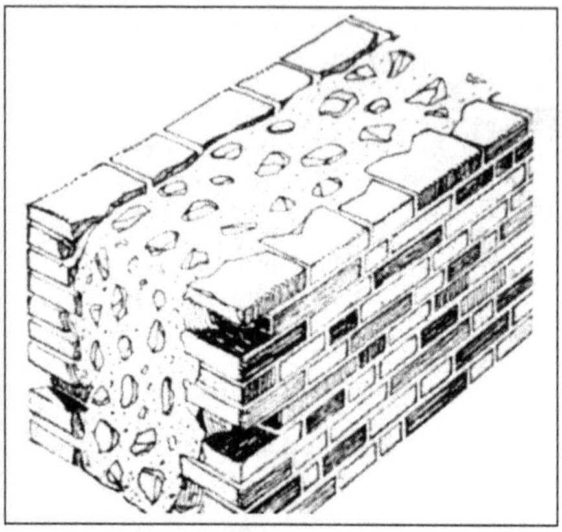

Ruins from the masonry clad concrete walls in the Circus Maximus

century under Trajan. His architect, Apollodorus from Damascus, was a very important person in this respect. Therefore this was also called the Apollodorus period [47]. The new style was based on brick clad concrete. Brick clad concrete became normal both in public and private buildings. A typical thing with this new building culture was the development of new room shapes, both in elevation and in the ground area. The use of concrete and complex arch and dome structures were typical features of this monumental architecture in Rome far into the third century.

We also find lightweight concrete in the arches in the Coliseum [3].

The Coliseum was built by Emperor Vespasian between 72 and 82 BC in an elliptic shape and ground measurement of 188 meters in length and 157s meter in the short axis. The outer wall stands up 50 meters above the street level. The foundation is a 6 meter thick concrete slab over the whole building area. Estimations reveal 22 000 m^3 of concrete in the bottom slab and 2000 m^3 of concrete in the ring wall. The arena could take 50 000 to 60 000 spectators. During the opening in the year 80, between 5 000 and 9 000 wild animals were killed [40].

We know about many structures from the Roman period in which concrete was filled between permanent shutterings of masonry. Not all the masonry has been able to withstand the weathering through several millenniums, so sometimes only the concrete can be found. This is the case in the Coliseum and other buildings in Rome.

Not only the Coliseum, but also the gladiator school in Rome, Ludus Magnus, which is found 60 meters east of the Coliseum, was built of concrete. It was built during the rule of Domitian and completed under

Hadrian and it was constructed in masonry- clad concrete. An underground tunnel led from the Ludus Magnus to the Coliseum. South of the Ludus Magnus was the Ludus Matutinus, which was of approximately the same size. There the venatores were trained. They were the ones who fought the wild animals, so this Ludus was called Ludus Bestiarus. The name Matutinus (the morning) is explained by the fact that these fights normally took place in the morning [47].

Wall painting from the grave of Trebius Justus in Rome showing the building of a house in concrete with masonry lining [47].

Another concrete example from this period is the Circus Maximus [48], the racing stadium. Originally this was built in 500 BC, but Trajan rebuilt the stadium completely. It became an enormous monumental structure, nearly 600 meters long and about 200 meters wide, with a seating capacity for 350 000 people. Along the middle axis was a 344 meter long midfield. This was decorated with statues and trophies and there stood the 17 movable eggs and dolphins, which were round indicators.

We had earlier mentioned the theater in Pompeius, one of the first concrete structures. In the Roman republic, the theaters were regarded as decadent luxury, and were often constructed as temporary wooden structures. The theater in Pompeius was the forerunner to the Marcellus theater in Rome and was opened in 13 or 11 BC. The building was 150 meters in diameter with a seating capacity of 14 000 spectators. The structure was supported by masonry arches partly in stone and partly in concrete. In the arch, in the rotunda behind the grandstand, concrete was utilized [47].

Other examples of interesting concrete structures include apartment buildings up to five floors, grave chambers etc.

Vitruvius mentions in *The Ten Books of Architecture* [36] the several ways of constructing walls. He is concerned about the fact that the durability does not only depend on the materials used, but also on how it is constructed, and, not least, the workers ability to follow the instructions given. He seems to be of the opinion that the most important weakness is that when the infill or the surface stones is porous, and takes away the moisture from the mortar, it will weaken it. A way to avoid this is to provide an air slit or joint in the wall.

We also find some interesting comments on the economics of the masonry in Book Ⅱ, Chapter Ⅷ, 8. *Beautiful wall are depreciated over an 80-year-period in the sense that the cost of the wall is reduced by 1/80 of the cost of the wall each year.*

Roman concrete was not produced in the same way as we do today. A mold was filled with "waste" comprising crushed stone, old masonry, and similar substances. This was then compacted. Then the liquid mortar of sand and the Roman cement glue were poured into the mold.

Vitruvius and others describe this type of wall or masonry as one of five different wall alternatives.

In *Concrete International* in 1997, Kent A. Harries gives an interesting contribution to the understanding of the durability of the Roman wall structures [46]. He gives an analysis of their resistance to earthquakes.

Rome is situated in the middle of an area with strong seismic activity. Since the Roman times this area has had 32 important seismic actions. From the time of the Romans, there have been indications of 14 cases of destructive earthquakes in the first three centuries AD. It is estimated that 15% of these was in the order of 6.0 on the Richter scale or higher. Even if there were many shakes, the historical writings about these are limited. With reference to Newbold in 1982, Harries discusses five historians of ancient times (Thucydides, Diodurus, Livy, Tacius and Dio), spanning a time period of 600 years and history that covers the Greeks and the Romans, but only 43 references to earthquakes are registered. Of these, 42% tells about structural damages, 18% about rehabilitation or repair, and 28% about other matters.

The Romans did not have any scientific or engineering knowledge about seismic activity. The earthquakes were thought of as warnings or revenge from the Gods. Livi, for example, looks upon this as an incident as one that has happened because some-

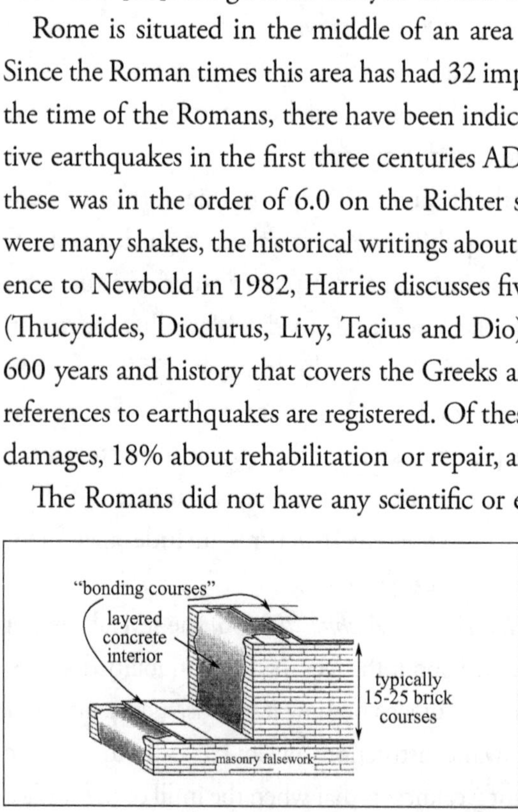

Typical Roman wall with binders

thing had gone wrong in the relationship between humans and the Gods. Dio thinks that this shows that the Gods evince some interest in human life. Diodorus was later clear in his opinion that this was a demonstration of the Gods' anger against human sinfulness. Only Diodorus indicates that there could be some natural reasons behind this. Among Roman Gods, it is Neptune or Poseidon who is the god of earthquakes and water. The explanation is that there is some relationship between the activity of the sea and earthquakes.

The relationship is described by Thucydides who observes that big waves (probably Tsunami!) come with the earthquakes. He does not mention examples, but it might be that he was thinking about the earthquake in 464 BC – the Sparta quake.

As mentioned earlier, the Roman times can roughly be divided into two periods; 200–27 BC, the time of the Republic, and 27–200 BC, the time of the emperors. Between these two periods, we can observe an architectural revolution. While the structures built by the Greeks, Etruscans, and the builders of the first Roman period were weak in earthquake resistance, the later concrete wall structures have a much higher durability and resistance to seismic activity. These structures that, among others, Vitruvius has mentioned, have a concrete nucleus and binders between the outer surface material or mold and the concrete interior. There are various theories behind the reasons for these binders. Some authors think that they are markings for the height during construction, indicating where they could end each day's work. Others claim that there might be other reasons.

Harries attributes the resistance against seismic activity of the Roman wall structures to several reasons;

The foundations of the structures were mostly monolithic, whereas the Greek ones were not so and this was more related to single loadings. This reduced the effect of the movements from the ground.

The walls were monolithic.

Hadrian's wall

John Laing's idea for a modern 'Hadrian's Wall' [49]

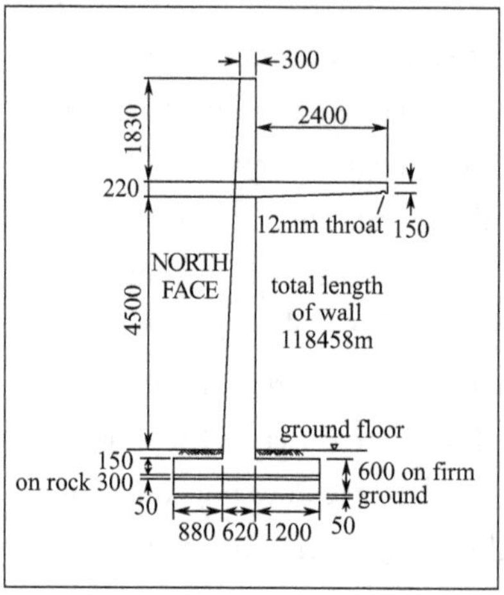

The binders contributed to this.

The Roman arches were very efficient in their load distribution, and the binding system in the wall at the corners.

The binding system in the wall at the corners.

From Roman times we have known about concrete structures in buildings and bridges as far north as Newcastle in England. Hadrian's Wall was started in the year 122 and initiated by Emperor Hadrian. It stretches to 117 kilometers and crosses England at about Newcastle and represents the most northern and most windy frontier of the Romans against the barbarians in the north.

The wall was built by the 3rd, 6th and 20th Roman legions, together with many English workers who were commanded out of. 25 million stones were used to build the wall. Between the half-meter thick outer stones, the wall was filled with stones, mortar, and clay, so that the total wall thickness became 2.4 meters. It is doubtful if we should call the wall concrete, but many authors do so. For every Roman mile (1.48 km), there were living quarters for the guards. Between these they built two towers or signal stations.

In 1974, John Laing Design Associates. Ltd. performed a cost analysis on what it might cost to build such a wall today. The results were reported in an article in *Concrete* in January 1975 [49]. They chose to build the new wall in reinforced concrete.

Brickwork from the Hagia Eirene – a church in Istanbul – with a high quality mortar in the joints.

The proposal for the new wall was modeled on Hadrian's Wall and is 118 500 meters long and 6.5 meters high. The cost estimate for the wall in 1974 was calculated to be 55 million GBP and would require 0.6 million m^3 of concrete and a workforce of 800 men. A similar analysis was made in 1850. Then the cost estimate was 1 million GBP [49].

A note in *Concrete* in 1995 [50], under the headline "*Out of warranty?*", states that a concrete floor had been found in Gloucestershire which could be dated to about the year 200. In the floor, there were marks from boot nails, which probably originated from the compaction of the original concrete.

When the disturbance grew in the Roman Empire in the 200s, the Emperor Marcus Aurelius started to build defense walls around Rome about the year 270. The Aurelian walls were 19 km long and surrounded an area of more than 137 km^2. They were built in concrete with a clay brick surfacing, mainly taken from old buildings. With some modifications, these walls were able to serve as defense for the city of Rome till the end of the 1800s [47].

In 330, Emperor Constantine moved the capital from Rome to Constantinople [now Istanbul – earlier Miklagard (Nordic name) and Bezants]. This East Roman empire lasted until 1453, while the West Roman Empire expired as Italy was conquered by various tribes from the north.

In the Byzantine culture, we find a lot of similarities with the Roman building culture. However, the walls in the monumental structures or buildings are not filled with concrete, but are masonry walls. The joints, however, are very thick, with a mortar that is often mixed with crushed burnt clay, which probably have had a pozzolanic effect on the mortar used for construction.

It seems that a lot of Roman concrete technology was forgotten during the heavy migration period.

Lea [5] is of the opinion that there are several reasons for this;

- The art of burning lime in a proper way was lost.
- They stopped adding reactive materials such as volcanic earth or ground tiles to their mortar.
- The need to hammer or compact the concrete was undervalued.

Lea claims, however, that the technology of burning lime improved again in the twelfth century.

According to [37] part of the cathedral, later the mosque in Sofia, built in 532–567, is built in concrete.

According to [50], cement was used in Salzburg in the 1100s. A major statue of the holy Catarina was cast by the monk Thiemo *aus einer versteinerden Masse* of a "petrified" mass. Thiemo's secret method of producing a product that hardened under water had later been researched by chemically analyzing a small part of the lady's cape. There are indications that this might be pozzolanic cement.

One of the early buildings constructed 1000 years after the Roman period is the Reading Church in England, built in 1127. The walls consist of concrete, originally clad with brickwork. The brickwork has disappeared, but the concrete is still there.

We also find examples of the Norman's concrete building technology in the Salisbury Cathedral, which with its 124 meter high spire is resting on a concrete foundation.

Hill [41], mentions Mariano Taccola (1382–1458), who describes the building of caissons, which are used as underwater foundations. According to a Roman prescription, you take a rectangular box of timber and a wooden wall, two feet on the inside. The space between the walls is filled with concrete. When the caisson is ready, it is towed out by four boats. Every boat has a plug. When the structure is in its right position, the plugs are drawn out.

He does not mention how the structure is completed and how they get the water out.

We also know that the pozzolan "trass" was used for the hydration of lime mortar in Netherland at the end of the 1600s.

An article in *Concrete* in 1975 [51] discusses a concrete recipe found in a monastery in Durham in North England dated 2 July 1413. The recipe was presented in the the journal "Notes and Queries" in November 1855. It might be somewhat complicated to interpret, but several interesting things are found in the text;

> *Take newe brynte lyme and water, and the skarthes of newe tyle, and stampe yam to ye mykilnes of beenes, or sum what gretter, and synders yat comes of ye smtythes' herth and brehe yam o nye same manerlyke ye tyele, and blende yam to gider with water, and take thre partes of lyme; and yen lay two tyle thike in*

lyme and mortar, and yen take all yat es befor sayde and lay itt all above two ynche thikke, an yf itt be thicker itt es ye small poudre oute of ye lyme, and ye tyele, and ye syndres: and ever as ytt es fresh wroght caste ye same poudre above ye werke. And yis most be done in Marche, for ye frost will ells lese itt.

[52] states that Fra. Giocondo used pozzolan cement to construct parts of Notre Dame in Paris in 1499.

Early in the 1500s, the manuscript of Vitruvius' *De Architechtura* was found by Poggio Braccistini in a monastery library in St. Gallen in Switzerland. ([53] claims that the manuscript was found in 1414). One of the enthusiasts, after the finding, was Andrea Palladio. He wrote his book *The Four Books on Architecture* and dated it in Venice 1st November 1570. Our references are from the translation by Tavernor and Schofield [54].

Palladios' book is a study of Roman building art and technology and he has "stolen" relatively essential parts regarding material technology from Vitruvius. Partly, he has also copied the editing of the various chapters. The translator philosophizes about the words concrete and cement. Cement could also mean stone. One reason behind the differences in the views of authors regarding the production of concrete could be the Roman way of producing concrete. They poured their liquid lime glue, containing a hydraulic binder, sand, and water into a mold with stone or rubble. In other words, concrete was both the ingredient and the final product in the mold.

In Book 1, Chapter IX, Palladio mentions the different alternatives that the Romans were using to construct walls. He cites six examples. Several of these have concrete in the middle, and in one of them, some of the concrete is exposed with the river gravel between the masonry layers.

Idorn [40] claims that the use of concrete in "the dark period," from the fall of the Roman Empire until the 1700s, was a secret shared by the church and the monasteries. He mentions that Danish churches from the 1100s until the reformation might have used concrete and an example from 1536 stands out as a monument of important trade knowledge of the highest order. The Danish Norwegian King Christian IV (1577–1648) was a great architect and through Dutch merchants he imported hydraulic mortars to be used in building palaces and fortresses.

In his book *Binders for Durable and Sustainable Concrete* [8] Aïtcin mentions repairs in 1991 of the foundations to the first fortress built in Montreal in 1676.

To the surprise of the researchers, the mortar contained natural cement in the binder and some of the C_2S content had still not fully hydrated. During the repair work, it was necessary to add pigments to the new mortar so that it would look like the old, which has been stained by the soil. Aitcin comments that it was surprising to find that the mortar that glued the big limestone blocs together either was a hydraulic lime of high quality or a natural cement. The raw materials on Montreal island have a chemical combination that looks very much like the materials from which we produce cement today.

In 1678, Joseph Moxon wrote about latent heat in burnt lime when water was added.

1.7 Norway's first concrete—The Farris dam

In the spring of 1764, the work on the Farris Dam in Larvik was started. A special print of the publication *Vestfoldminner* (*Vestfold- memories)* 1959 – Asbjørn Bakken: *Farrisdammen – Et bidrag til sementens historie (A Contribution to the History of Cement)*[55], among others states the following:

The dam at the end of the Farris Lake had existed for many years, but it had collapsed several times and created serious problems and small catastrophes. On the 10th of September 1760, an important meeting was held to discuss how a new dam could be built. The discussion was serious and about whether the dam should be built in timber, as the old one, or of stones – large rectangular stone blocks. The cost was calculated to be 2135 riksdaler for a timber dam and 2741 riksdaler for a stone dam.

Oberinspektør (Over-inspector) Lars Fabricius claimed that he wanted a stone dam with a front of pine, and both sides covered by turf meadows, "and for the joints to use cement, which will make this wall more exclusive."

We do not know where this idea came from because cement was unknown in Norway and the building masters in Larvik had never heard about it. There is reason to believe that it came from the son of the Oberinspektør, Fredrik Ludvig, who at

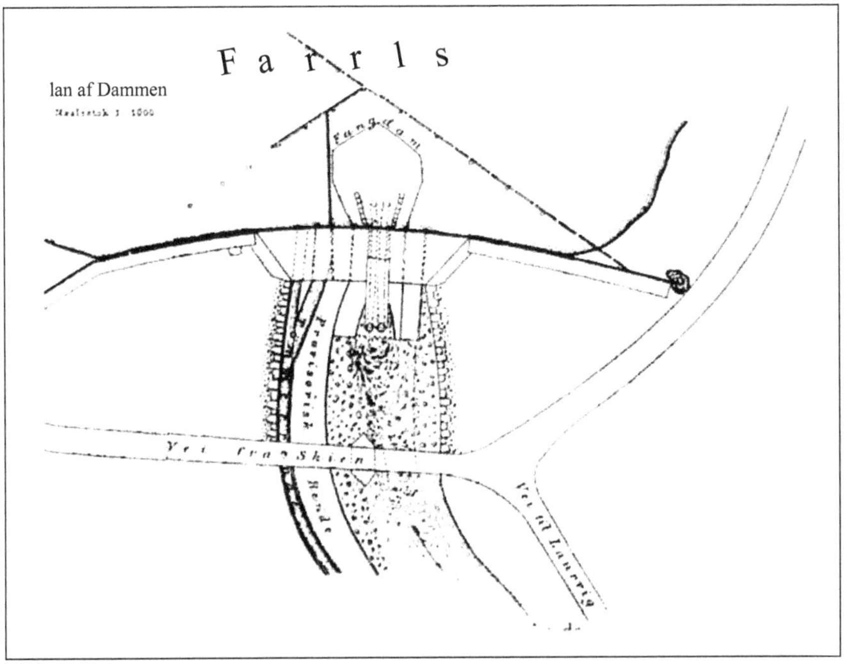

Drawing of the Farris Dam in 1886

that time had his education in Copenhagen in Denmark. In a letter about travel around Europe, he writes from London; "I have through Captain Balzer Hansen received the letter regarding the type of cement used for masonry that can stand in water, that is called tarraz(the pozzolan – trass). This cost here at this place, of the best type is 3s6 pr. Busheller, or 2 Norwegian barrels, but from Holland from where it comes, it can without doubt be acquired cheaper"···

The Earl, Fredrik Ludvik Danneskiold-Laurvig, decided later that a stone dam with cement should be built, with the recommendation of the Oberinspektør because of··· "it is more durability and will be more useful for your Highness"

The dam was decided to be 12 alen wide at the bottom and 5 wide at the top. The new cost was calculated to be 2900 riksdaler, plus the cement, the price of which they did not know.

The signal for the start was given by the new Earl, Christian Conrad in Copenhagen, on 21 August 1762. The work started and the stone blocks were cut out 1 alen 18 inches long,16–18 inches high 12–18 inches in width. In April 1763, 331 big blocks had been brought to the dam site.

In June 1763, the drawings for the dam were again modified and the new price was calculated to be 6866 daler. This included 60 barrels of cement for 200 daler and 56 loads of lime.

In the summer of 1764, the eastern part of the dam was finished, but there were major problems due to lack of workforce. Already in 1763 the local sheriffs proclaimed in all the churches in the area that they wanted volunteers for the work, for good payment.

In the spring of 1765, the lack of labour force got critical, and on the 24th of May an order was issued by the Oberinspektør to all the sheriffs in the County: "labour must be found to work on the dam, where work now has to go on night and day. Every week there shall be recruited 40 men, and the sheriffs shall by force write out 10 men from each church district. They shall start immediately. The district of Tjøme shall not be included and preferably not the districts of Kjose and Kvelle either. There the farmers shall produce wooden coal for the iron work" ···

THE CONCRETE:

The Cement was bought from Amsterdam, from where they also got the lime. Three parts of cement were mixed with one part of "Muschal lime." This mixture was blended with good beach sand.

About the work and testing methodology: "···*have to be worked together in the lime bench, with the lime whip, that was of iron, without adding too much water, as it should not have more liquid than it has by itself? Two men must beat the mixture a whole day in each lime bench, but the mixture prepared must not be more than what is used regularly. If there is a surplus, it has to be covered, otherwise it will lose its power"* ···

Pictures of the Farris Dam 1978

"It can be tested, when it is sufficiently worked in the lime bench and is usable, ,this way; after one day of whipping/beating, a piece about the size of an egg is taken in the evening. This is rolled into a lump and placed in a bucket of water over night. If you find it hard in the morning, it is done and can be used, but if it is loose or disintegrated, then you need more time with the whips, until it passes the prescribed test.

The Farris Dam was both a success and a sensation. Already in 1773 there was a major flood, and the dam stood the test.

Testimonies from important visitors to the dam seem to prove that it was the only one of its kind. In 1773, the Prince of Hessen visited Larvik and the main attraction was the dam. In 1778, the Crown Prince, later King Fredrik VI came to Larvik. During his visit on 2 August to the Jarlsberg Estate, the mood of the Crown Prince was not good. Of the many excursions he made, only the visit to the dam was to his satisfaction.

The Farris Dam was repaired and modernized in 1947.

THE CONCRETE IS STILL GOOD:

I wrote an article about the concrete in the dam in the middle of the 1970s, and afterwards I was provided with a piece of concrete from the dam. In the summer of 1990, this was analyzed using the thin slice technology by Inger Gaupset at Norges Byggforsknings Institutt (Norwegian Building Research Institute in Oslo). The analysis of Inger Gaupset enlists among others [56]:

- Good homogeneity in the binder.
- *w*/*c*-relationship = 0.5–0.6 (this is based on a qualitative comparison to modern concretes)
- High degree of hydration
- Air content = 0.33%, with all pores under 0.35 mm
- Spacing factor = 0.68
- There were very few air pores and they were all mainly circular. The pores were all evenly distributed in the binder.
- The thin sections show a high portion of calcium-hydroxide $Ca(OH)_2$ in the binder.

Thin polish analysis of Farris concrete, Norwegian Building Research Institute 1990

- Alkali-silica-reactions are not present.
- There are signs of some carbonization of the binder. (This reaction, however, might have taken place in the many years during which the sample was taken out before it was tested).
- Some micro cracks can be seen in the aggregate, but cracks are not found in the binder.
- The aggregate was mainly rounded grains with an even distribution.
- Binder volume = 40.8%

In total, we can categorize this concrete, which is well over 200 years old, as *very good.* Even with a valued water/cement ratio of 0.5 to 0.6, it should be usable for most of the areas in which we use concrete today. Pos-

The Farris Dam 1978

sibly we have something to learn from old formulas – their carefulness in proportioning, workmanship, and curing.

***

The Roman concrete technology before about the year 0 was not so different from the ones in the previous years. Four main factors brought the Roman concrete technology a big step forward.

The systematic documentation and the description of the methods, among others through Vitruvius books, made it possible to compare alternatives and develop solutions that are adapted to local needs, materials, and traditions.

The discovery of the durability advantages of the pozzolan concrete.

The new widened functions of the material; from harbor structures to buildings and temples, advanced infrastructure projects such as bridges and aqueducts, and the advanced Roman road system.

The size of the Roman Empire led to variations in the material challenges and the technology spread to many people and countries.

It is important to have in mind that the technology was gradually developed over a few hundred years, and there are differences in the use of the technology between early and late Roman time and from one region to the other. The last point is probably both due to the challenges from the spreading of technology and learning and the differences in the materials used.

After the fall of the Roman Empire, we get a period of well over 1300 years, a periode during which there is very little development The use of concrete was seldom and restricted to cases where the church was involved. The Roman concrete secrets seem to have been well-protected in the archives of the church.

Then in the middle and late 1700s something happened. We observe that the knowledge about concrete had become the property of a wide spectrum of people and the increased use led to new developments in cement and concrete technology. Some of the new developments were just freshening up of old Roman wisdom, but we also find new developments. At the end of the 1700s and the beginning of the 1800s we see the development and patenting of quite a few new types of "glues" and new cements.

The development of Portland Cement (Patent in 1824) is said to be a giant new step forward, but it might possibly only be a border that is crossed in terms of burning temperatures.

2 Portland concrete history

2.1 Pre-portland

The beginning of the history of "Portland concrete" is really the history of the advancements in the Roman technology.

Some authors name the English Engineer James Parker as the father of modern concrete [56]. Toward the end of the 1700s, Parker developed a rapidly hardening lime concrete with the addition of pozzolan. It hardened in 10 minutes with a heat development of up to 60 degrees Celsius. The Parker cement was also used outside England and was probably used in the foundation of Norway's first concrete structure – the Farris dam. The Parker cement was a market leader until the 1830s. Parker patented a natural hydraulic cement in 1796. Among others, this was used by William Jessop for the West India Docks in London [57].

Many authors claim that the new developments in concrete really took off in 1756 when John Smeaton got the assignment to build the third lighthouse at the Eddystone cliffs – an extremely weather-beaten cliff, 14 km outside Plymouth in England. He experimented with different types of cement and ended up with a blend of lime from Aberthaw and pozzolan from Italy [57]. The pozzolan came from Cicitavecchia outside Rome. The lighthouse was in use until 1882. Then it was taken down and moved to Plymouth. There it still stands, watching over the harbor from Plymouth Hoe. In his experiments, Smeaton found that he got the best cement of his time by mixing clay and lime in proportions varying between 3 : 14 and 1 : 17 [18].

Smeaton's lighthouse has an extremely interesting prehistory [35,58]. The lighthouse not only marked a breakthrough in the use of concrete, but was of wide social importance. Building of lighthouses, not only in England, but also in countries like Norway meant that concrete as a material was tested in practically most extreme conditions of weather and water. Lighthouse development during this period of our history was of significant importance in the development of shipping and ship transport. At the beginning of the 1700s, an average of one ship per day wrecked in the English merchant marine, most often because they could not find their way into the harbors. During the last 25 years of the sail-ship-period, up until 1880, nearly 50 000 British ships went down, and as late as in the year 1881 nearly thousand ships were wrecked. After that it got better. The reasons behind this were partly steam ships and partly improvements in the lighthouses.

Eddy means backwater and the cliff close to it bears its name with honor. A lighthouse was first constructed on the cliff in 1698. Henry Winstanley and his crew built a 36 meters high tower anchored by iron rods drilled into the cliff. This tower was destroyed by a winter storm five years later, in which among others Winstanley drowned. A fascinating story from the time of the building of the lighthouse is that a French war ship passed through the cliff in 1696. They thought the light from the cliff came from an enemy warship, so they landed on the cliff and took Winstanley and his crew as prisoners and brought them to France. Winstanley was brought to the

Winstanley's lighthouse

Sun King, Ludvig XIV. There Winstanley had to explain to the king what he had been doing. As a result, the King tried to persuade him to stay in France and work for him. As this did not work out, they were allowed to go home for as the King said he was at war with England, but not with humanity.

The hurricane that destroyed the lighthouse in 1703 was violent. It was reported that the roofs of the churches flew like carpets through the air. Four hundred windmills were taken away by the wind, the damages on Westminster Abbey were severe, and the frightened Queen Anne sought shelter in the basement of St. James's Palace. A new lighthouse was ready in 1708 and was known as Rudyard's tower. This oak tower was well secured with anchor bolts and blocks of stone, but it was burnt down during a winter storm in 1755.

Smeaton's tower was completed in 1759 and was in service until 1882, when the foundation weakened. John Smeaton was only 35 years old when his 34 meter high tower was ready. He was also the first to called himself a civil engineer.

John Smeaton was born in 1724 and the great engineers at the end of the nineteenth century considered him the great master. His father was a wealthy lawyer from Leeds. John was not particularly knowledgeable in the subjects at school,

Rudyard's tower

John Smeaton

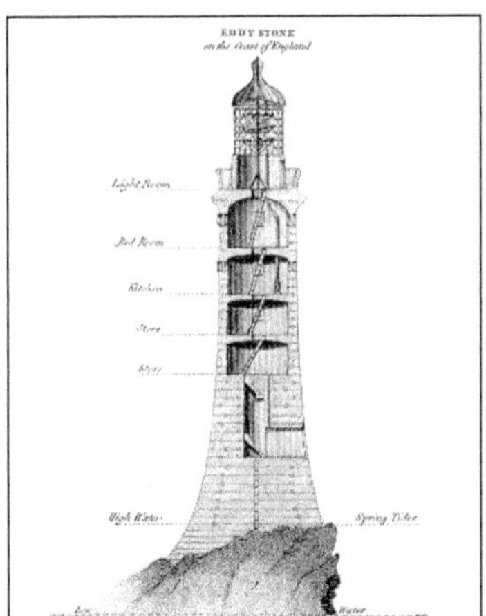

Smeaton's tower

with the exemption of mathematics. He was thought of to be given a career as a solicitor, but after a few months of studies in London, he wrote home to his father and asked for permission to study engineering. Engineering was counted nearly as some kind of study of arts, so it must have caused his father a lot of concern before he gave his permission. Thanks to his wealthy father, Smeaton got admission to the Royal Society very early. He also early gave lectures at the society. The first paper was about an improved compass. Later, he showed them more developments and inventions. He got a scholarship from the Royal Society and studied channel building and the use of cement in Holland during a year in Belgium and Netherland in 1754. Before this trip, he had an instrument shop in London, but when he came home, he closed the shop and concentrated on various water construction projects based on what he had learned in Netherland.

The Eddystone lighthouse is famous in more than one way. W.G. Corley [59] refers to an old chantey that goes like this:

My father was the keeper of the Eddystone light,
And married a mermaid one fine night,
Out of the match came children three,
Two was fish and the other was me.

In the jubilee book of the Swedish cement producer Cementa in 1972, Alf Åberg [60] mentions about interesting activities using cement in Sweden in the 1700s. Christoffer Polhem worked on the production of cement from Swedish raw materials, without the use of volcanic soil. When they built the locks at Trollhättan in 1718, he used alum schist that was burnt and ground and added to the mortar. The experiments were continued by Sven Rinman, at the Garphyttans Alum works, who made cement by blending rock lime and the powder from burnt alum schist. This activity using cement started in 1779 and went on well into the 1800s. The

The restored lighthouse outside Plymouth

cement was supplied to mines, sluice works, and some buildings. On an average, they delivered 300 barrels (50 tons) a year. When the Göta Canal was about to be built in the 1820s, Professor Gustav Erik Pasch got an assignment to develop a good mortar that could be used under water. He also concluded that the combination of burnt lime and alum schist gave the best result. An old report from Garphyttans Alumworks in the 1870s states that the cement that was produced thus was regarded as good as pozzolan cement.

Probably, independent of each other, the German Professor John in 1817 and the French Engineer L.J. Vicat in 1818 found cement systems based on burnt clay. The burning, however, took place at relatively low temperatures. (Ref. [61] claims that this was in 1812–1813). Vicat's cement was used for the first "modern" concrete bridge in Souillac in France in 1816 [53].

Vicat's experiments with hydraulic lime led to the making of a mixture in which he calcined a blend of limestone and clay, which was ground together in a wet mill. Some claim that this really was the forerunner to Portland cement [5]. The interest in concrete had grown in France, with experimental works by Cointeraux at the end of the 1700s. Later, at the beginning of the 1800s, these were followed by the work of Lebrun and Francois Coignet. Coignet was the first who managed to produce concrete in a reasonable volume with good workability [62].

Aïtcin [8] mentions about the first important factory for hydraulic lime, which was built in 1830 in Le Teil in South France by Pavin De Lafarge. In 1839, this factory had two five meter high vertical kilns that produced 3000 tons of lime. In 1865, the Lafarge factory in Le Teil got an order of 120 000 tons of cement for the construction of the Suez channel. (The 161 km long Suez channel was opened in 1869).

Around 1860, the Lafarge factory also developed slow hardening hydraulic cement.

In 1779, Bry Higgins took a patent for the hydraulic cement used for external plastering and in 1780 he wrote, "*Experiments and Observation Made With the View of Improving the Art of Composing and Applying Cancerous cements and Preparing Quicklime*" [61].

In 1810, Edgar Dobbs got the patent for hydraulic mortar and stucco [61].

In 1822, James Frost produced hydraulic lime from Vicat's method and called it British Cement (Lea [5] opines that this was in 1811).

***

This was the "Century of Inventions." The world was changing at a rate much faster than before. In a perspective with other inventions, we remind about that [23];

1629: Italian Giovanni Branca made the drawings of "Le Machin" – a steam engine.

1728: A mechanic developed the data card to improve the weaving chair.

1763: French Cugnot constructed a steam driven car in Brussels.

1768: Cartwright and Kay constructed the spinning machine that could be driven by a steam engine or an animal.

1770: Successful experiments were conducted with a steam engine on a ship.

1788: Castlewright produced the first mechanical weaving chair.

1828: The railway between Liverpool and Manchester was laid and there were some advancement in the materials used.

1825. H.C.Ørstedt showed how to produce aluminum.

2.2 Portland cement

In 1924, the English bricklayer Joseph Aspdin got his patent on Portland cement. He ground together burnt limestone and clay to a fine powder. He called this product Portland cement because of its resemblance in color, hardness, and strength with the natural stone from the peninsula of Portland in Dorsetshire.

Joseph Aspdin's cement kiln at Wakefield in 1824 [63]

The public was made aware of Aspdin's invention through a small notice in the local newspaper, Leeds Mercury dated 6 November 1824, which stated "we hear that Joseph Aspdin, a bricklayer in this city, has got a patent on a superior Cement that looks like Portland stone" [18]. The name of the patent was "An Improvement in the Modes of Producing an Artificial Stone" [64], and it had the British Patent No. 5022.

It is not unlikely that Aspdin had got the idea for the name 'Portland' from John Smeaton (Eddystone lighthouse). Smeaton wrote a book about the lighthouse in 1813, in which he categorized his concrete as being as strong and durable as Portland stone. Another theory is that Aspdin, through his work as a bricklayer, had been working with Portland stone and got the idea for the name from this.

As mentioned earlier, there were also other cement types in use at this time. James Parker produced his cement based on nodules from London and clay from the island Sheppey, and called it "Roman cement." This product was for many years Portland cement's most important competitor.

Aspdin built his cement kiln in Wakesfield, and it is possible that it was from there that Brunel got his cement for the construction of the Thames tunnel in 1828. Aspdin's cement cost 110 pence per barrel plus transport, while the competitive "Roman cement" was priced 60 pence including delivery. Brunel might have known that the Portland cement was superior in quality.

However, there have been differing views on what type of cement Brunel used. Ref. Chapter 3.1.2. "The Thames tunnel" [65].

Joseph Aspdin died on 18 March 1855 and has a modest place in the churchyard in Wakesfield.

In a long article in *Concrete* in August 1974 [66], R.J.Barfoot writes about what he called the *Jigsaw* about the Aspdin family. Barfoot has made thorough studies and finds quite a few contradictory details in later descriptions, among others, regarding confusions about the names.

Joseph Aspdin's death certificate dated 20 March 1855 states that he was 76 years old when he died. This is probably wrong. The correct birth date is probably 25 December 1778. We also know that Joseph's father Thomas was a bricklayer and that he died in 1800. Thomas had brothers, who had sons and many of them were bricklayers. It seems to have been a large Aspdin-bricklayer-clan in the area at that time. Joseph's brother John was also a bricklayer. They shared a close relationship and were the best man in each other's wedding. Moreover, both of them lived at the time of the

Portland cement – patent in neighborhood houses is something that might also have caused some confusion with respect to names and addresses.

Joseph Aspdin married Mary Fotherby on 21 May 1811 and they got two sons and four daughters. The sons James and William were baptized at the same time in 1816. James was the oldest son, born on 23 August 1813 and not William as mentioned in one of the articles on concrete .

When Joseph started his work with cement, he went into partnership with a local precious metal smith, John Beverly, which seems to have provided the needed capital. The name of the company at that time was Aspdin and Beverly, Patent Portland Cement Manufacturers, and the addresses were 68, Briggate, Leeds, and Kirkgate, Wakefield.

The cement kiln in Wakefield was removed in 1837 to make room for Lancashire and Yorkshire Railroad Company. In 1838, production resumed at Ings Road in Wakefield. The name of the company was now Joseph Aspdin and Co., and Joseph had taken his sons into the business.

Joseph's oldest son James was the book keeper in the company and in 1841 it seems like he had taken over half of the ownership of the company. Joseph continued to work in the company in Wakefield until 1853, when he was 75 years old.

James, who died on 21 December 1873, had five sons and one daughter. Two of the sons, William and Thomas, went to Canada around 1868 and founded the town of Aspdin in Ontario.

Joseph's youngest son, William Aspdin later improved the quality of the cement by increasing the burning temperatures. He took out a patent on this in 1852. Many regard him as a great cement producer of his time. William has been described as both strong on initiatives and easy to get into conflicts. Some also think that later alcohol got a bit too much importance in his diet.

From 1841 until his death in 1864, when he was 48 years old, he either started or contributed to the establishment of five cement companies and six cement factories.

1841 seems to be the year for changes in the Aspdin family. William, 25 years old, married Jane Leadman. The same year he left his home in Wakefield and went 300 km south to London. Most probably, there had

Drawing of the factory in Gateshead in approximately 1852

been arguments in the family. We know that his brother James took over half of the company from their father the same year. However, we do not know if it was the hen or the egg that came first. Did William leave home because James received special privileges, or did James get the privileges because William left home?

The discussions on the enormity of the misunderstanding in the family have also been questioned. Some claim that the discord was total, while others state that William's move to London was planned by the family.

When he was 26 years old, William managed to convince the 65-year-old shipping and insurance broker, John Milthorp Maude, to support him financially and the company J.M. Maude, Son & Co. was established. The cement factory was built in Rotherhithe. The factory in Rotherhithe must have done well and in 1846 the name was changed to Maude, Jones and Aspdin. The factory was then moved to Northfleet in Kent. In 1847, the partnership was broken and from 1 January 1848 the name was changed to Robins, Aspdin & Co. The company went on very well and in 1851 they produced 100 tons of cement per day in barrels. In November the same year, the partnership was broken up because William was accused of misusing the company's fund.

In September the next year, William started a new partnership, Aspdin, Ord & Co., with an office in New Wharf, Abingdon Street, Westminster and a factory in Gateshead-on-Tyne. There they had a production capacity of 3000 barrels per week.

A piece of precast concrete from W. Aspdin's house, height 11 cm [67]

A brochure from the company in 1853–56 states that at that time William Aspdin was the only producer of Portland cement, but there are reasons to believe that his brother James still produced cement in the old factory in Wakefield.

But things went wrong for William in Gateshead as well and German registers mention that William arrived in Hamburg with his family on 9 November 1857.

William had been for a long time considering the possibilities in Germany, while he worked in Northfleet, and had the company establish distributors in Berlin to cover Preussen and Saxon. Another agency had been established in Cologne for the west and south of Germany.

In 1860, he got an assignment from the Heyn brothers to build a Portland cement factory in Luneburg. This factory survived until 1926 under the name Portlandzementfabrik Heyn Gebruder. William lived in Luneburg until 1863. At this time, he met another English emigrant, Edward Fewer. They established the company Ed. Fewer & Co., with the intention of building a factory in Lagerdorf in Holstein. In July 1863, however, William had stated in a newspaper article that the partnership had been broken and that he would no longer take responsibility for the production of cement in Lägerdorf.

On 11 April 1864, William died in Itzehoe and was buried there. It is assumed that he at this point in time was involved in the establishment of Zementfabrik O.F. Alsen und Sohn in Itzehoe.

In 1983, Idorn and Thaulow wrote an interesting article in *Cement and Concrete Research* [67]. They had received a piece of concrete from Portland Hall in Gravesend in Kent (William Aspdin's house), which was built in 1847.

The piece of concrete has been taken care of from the time of the demolition of the house, and this is possibly one of the oldest productions of precast concrete that we know of in modern times. From the analysis that took place and the reports, we among others learn about

- Lime-based aggregate that could have also been acceptable today.
- Signs that might have been caused by addition of air.

- The original concrete that had a considerable amount of unhydrated cement.
- Low water/cement ratio.
- How modern concrete can be produced with many hundred years of durability.

The quality of Portland cement was considerably improved in 1845 by Isaac Charles Johnson who increased the temperature at which limestone and clay are burnt. Johnson, who was born in 1811 and died in 1911, 100 years old, had observed that the "over-burnt" lumps he found in the kiln, even if they had a longer setting time, had better quality after grinding. He also showed that the raw materials should be mixed together in a specific ratio and that the materials should be heated to above 1400℃, which is the temperature at which the materials are sintered to clinker. (Clinker has a specific sound when dropped on the floor, hence the name). When the clinker has cooled down, it is ground together with the other additions, for example gypsum, before you get the final cement product. Johnson started his work in Rochester, but later he took over the old factory of William Aspdin in Gateshead [5]. Johnson mentioned that William Aspdin had tried to keep his process a secret by carrying barrels of copper sulfate into the kiln, thereby hinting that the process were based on salts.

Many, in particular, cement technologists claim that the modern developments in concrete did not really start until Johnson brought about his developments in 1844–1845.

Who is really the father of modern concrete?

- Smeaton, who started the development?
- Inventors between Smeaton and Aspdin who further developed hydraulic lime-solutions?
- Joseph Aspdin?
- William Aspdin who improvised on his father's solutions?
- Researchers who gave us the theoretical answers?
- Johnson who burnt the ingredients to a sufficiently high temperature to get clinker?

- Blake Stonebraker who in 1862 introduced shaft crushers that could crush the clinker in England?
- Plus possibly about half a dozen other fathers.

Probably we should talk about several fathers, - a child might have many fathers, but only one mother, - in this case limestone.

In England, quite a few cement factories were built, but the really big improvement in quality came in the 1880s with the introduction of the rotary kiln.

Clearly, England was leading the early development of modern concrete.

The first concrete house that was built was a two story construction in Swanscombe in Kent, built for Mr. J. Bazley White in 1835.

The reinforced concrete structures did not come up until 1854, but the possibility was mentioned already in 1830 by a building master from Newcastle, William Boutland Wilkinson. He took out a patent on reinforced concrete floors with flat iron rods and with the vertical wide side as the main reinforcement and wire as secondary reinforcement in 1854: *Improvement in the construction of fireproof dwellings, warehouses and other buildings.*

Though Francois Coignet had included iron in the concrete in 1852 and Joseph Lambot had used iron in his boat in 1855 – both in France – it is the general opinion that Wilkinson was the first to use reinforced concrete in building structures [68].

Roger H. Harper mentions in an interesting article in *Concrete* in 1976 about what he calls *Concrete's battle for London, 1867–1886* [62]. The article narrates the many opposition and barriers that the early concrete enthusiasts had to cope with.

Joseph Tall had learned from François Coignet and further developed the concrete technology. In 1864, he received a patent on how to cast a concrete wall by using a movable shuttering system. With this system, he built a few smaller experimental structures in Bexleyheath in 1865. Then in 1867 he tried to enter into the London market.

The building activity in London was regulated by the Metropolitan Building Act of 1855. Reducing the spread of fires was an important factor and in many ways they were not allowed to use wood or timber in walls. Masonry was the dominant solution. The first two rules in the first chapter approximately stated this; "*Any building shall be surrounded by walls of masonry, stone or similar materials, which shall be bound acceptably together by mortar or cement, and all adjoining walls shall be sufficiently bound together.*"

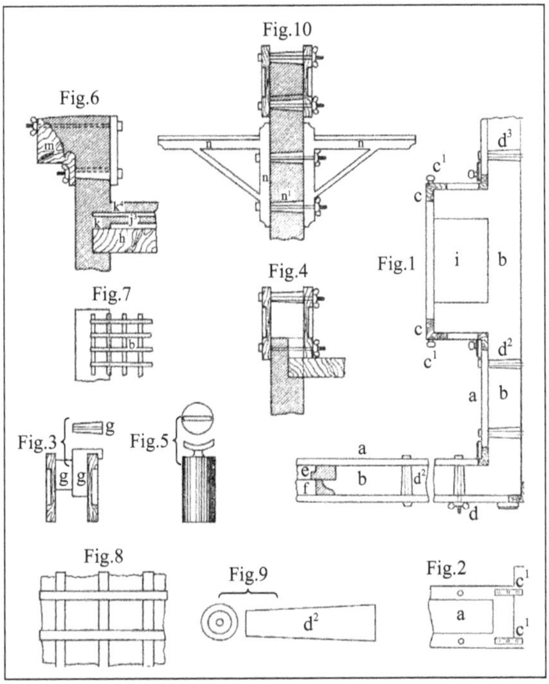

Tall's patent figure [69a] from "The Civil Engineer and Architect's Journal" (concrete, sept. 1974)

At the same time it should be mentioned that the professor at King's College in London, and a very important person in building legislation, William Hosking, had said in 1847 "*concrete cannot be regarded as a completely acceptable substance in walls in cities.*"

In September, Joseph Tall wrote to the weekly journal *The Builder*. He stated that he wanted to build a smaller concrete house and he asked the editor if this was allowed according to the Metropolitan Act because the local building official seems to have denied this. The editor answered that he could not see any obstacles in the law. The Editor, George Godwin, was not only an architect but also an authority in the matter. (Godwin had, 30 years earlier, received a silver medal from RIBA (Royal Institute of British Architects) for his article "The Nature and Properties of Concrete").

But both Tall and Godwin were wrong, and it took over 20 years and a lot of trials and discussions with authorities before the rules allowed the use of concrete on a general basis. The "District responsible," who refused to permit Tall to build a concrete house in the beginning, reasoned his refusals with the argument that concrete was not bonded together as it was required in the regulations.

After some time Tall got a special permit, and in October 1972 license was given to the contractor Phillip Brannon for his patent *Monolithic Fire Proof System*. Brannon was probably a bit too optimistic and creative in his development. His company, "The Monolithic Fire Proof and Sanitary Construction Works Ltd., had a project near New Bunhill Fields, where they cast a flat roof with the span 57×37 ft. (17,1×11.1 meter). The roof was reinforced with wire, but it fell down and killed one of the controllers of the building authorities.

After this incident, it is no wonder that the control authorities in London were skeptical. In October 1879, they issued a note that stated that concrete needed a special permit to be used.

However, the use of concrete increased, and RIBA published a number of articles on the subject;

- Thomas Wonnacott: *The use of Portland Cement* in 1871.
- Alexander Payne: *Concrete as a building material* in 1876.
- Arthur Cates: *Concrete and fire – resistance structures* in 1878

In January 1886, the Metropolitan Board revised their view on concrete and we got what probably was the first standard or building regulation for concrete. In this standard, we among others find rules with requirements about

- Use of Portland cement
- Use of clean aggregate – *Thames or pit* – aggregate or crushed stone or masonry that passes a 2 inch mesh in the sieve.
- Mix proportions of 1 : 2 : 3 for cement : sand : coarse aggregate.
- Thorough mixing by machine or hand.
- Formwork by parallel frames at equal height.
- Minimum wall thickness, the same as for masonry.
- Walls and chimneys over the roof to be plastered with Portland cement.

As part of this development story with respect to building rules, it must be mentioned that accidents were not unusual at the time;

In 1868, four enthusiastic workers started building a house with four stories in Twickenham using Tall's patent. After three days, they had reached a height of 190 cm. They continued up to a height of 12 meters, but without any shoring of the mold. The journalist from *The Building News* gave among others the following comment on the result of the stunt *"the garden wall* ..." ... *Looked very well, but the house itself had fallen down."*

In Chapter 3.1.2., "Historical structures" we have mentioned an example of one of Tall's houses from 1866.

It has been told that Tall got a contract from the Emperor of France to build several five story houses for the workers at the Paris exhibition in 1867 [69].

It was not in buildings that cement had its most important market in the beginning. According to Fidler [70], a considerable amount of cement was used in the London sewage system in the years 1859–1867. Up to the beginning of the 1860s, they might have used nearly 70 000 tons of cement for this purpose.

Export of cement to USA was also an important growth factor. In 1871, 500 barrels were exported which had increased to three million barrels in 1896. (USA started producing Portland cement, in the 1870s, the first time in Coplay in Pennsylvania by D.O. Saylor).

However, it was not until sometime between 1890 and 1896 that the production of Portland cement in USA exceeded the production of natural cement.

The contractor François Hennebique built the first building with concrete frames in 1897, Weavers Mill in Swansea. That was probably the real starting point of the development that happened later with reinforced concrete.

Norway was one of the first countries that began to use Joseph Aspdin's patent for Portland cement. With clay infected lime nodules as raw material, partly from the islands in the Oslo fjord, cement factories were started at Nordre Langøy (North Long Island) in the Bunde fjord and at Bakaas Brug at Akerselven (Aker River). Both factories also got diplomas from the world exhibition in Paris in 1855, where cement as a building material was declared a revolution in the building industry [6].

Producer Hole at Langøen in the Bunde fjord did not have any big factory. He had eight men at work and a steam engine of 6 hp. that serviced the ball mill. The production was at its highest, 10 000–12 000 barrels

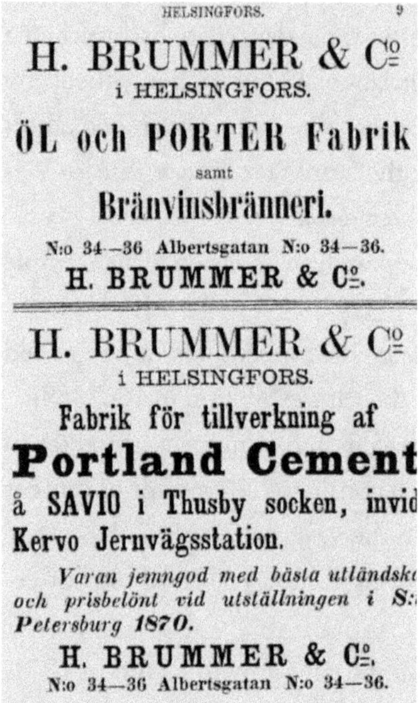

Advertisement from Brummer & Co in 1871 [69b] (It is interesting to note the cultural change from Norway. While the Norwegian producers were proud of medals from Paris, it is St. Petersburg that Finland looked up to)

(approximately 1500 tons) per year. Producer Tobiesen at Bakaas produced a similar quantity of barrels with 10 men. His machine was water driven with 6 hp. and water from the Aker River [6].

Bad economic times, however, led to the closure of both factories in the 1860s.

The beginning of modern cement production in Norway happened through the establishment of Christiania Cement Aktiebolag in Slemmestad in 1888 (see Chapter 5).

Idorn [40] describes about six factories producing "Roman cement" at the island of Bornholm in Denmark in 1855 and opines that the first Danish concrete structure with "Portland Cement" was Fort Prøvestenen outside Copenhagen, which was built in 1859–1863. Before the building, they had tested the advantages of concrete against artillery grenades and masonry structures.

Drawing of Brummers cement factory

Portland cement was for the first time mentioned in the import statistics of Finland in 1856, but concrete had been used a few years earlier in the Saima Canal. Lime from Narva in Estonia had been burnt at the building site to produce cement of the specified quality. Swedish *alun* was burnt together with lime for the final "glue powder" (Ref. also Chapter 5.1.). In 1869, Herman Brummer founded the first Portland cement factory in Finland. The factory was run at a rather modest capacity in the 1870s, with a foreman and ten workers. In 1876, they presented an impressive fountain at the industrial exhibition in Helsinki. The same year, they had plans to expand the factory, but this was not done. The production had to close down from 1887 to 1891 and in 1894 the whole factory had to be closed down.

It was not until 1914 that we got the first real establishment of Finnish cement industry, with the establishment of Paraisten Kalkkivuori Oy's factory, known by many as the company PARGAS [71].

The first Portland cement factory in France came into existence in 1840 in Boulogne-sur-Mer. In 1898, the French production had reached 400 000 tons from 16 factories. In 1905, this volume had grown to 550 000 tons. In addition, we also have to count the 400 000 tons with natural cement and 60 000 tons with slag [8] (from Papadakis and Venuat according to Aïtcin).

The Frenchman Monier took a patent on a reinforcement system for concrete in 1867, when he built large flower pots with rods and wire as reinforcement, and another Frenchman, Lambot, built his boat with mortar pressed in between a skeleton of iron rods. The boat was displayed at the world exhibition in Paris in 1855 (see also Chapter 4.1).

In 1855, a cement factory was built in Stettin in Germany. There the German pioneer in cement and concrete, Dr. Michaelis, had his base. He also played an important role in the start of the cement production in Sweden.

Cement was first produced in USA in 1818 (see Chapter 3.1.3.), and Portland cement was first produced in the country, according to Heun, in 1875 [71]. Dabney [72] is of the opinion that the first Portland cement in USA was produced in Coplay, Pennsylvania in 1871. Both before and after this, considerable amount of Portland cement was imported from Europe. To bring the freight cost down, the barrels of cement partly came as bal-

The Asano factory about 1890 [73]

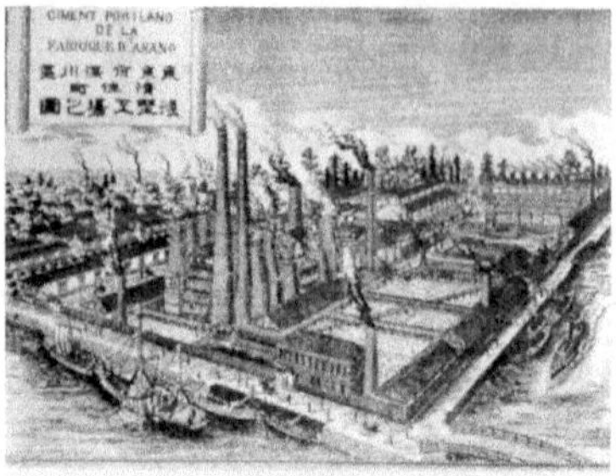

明治23年ころの深川工場

last on the return transport, as the main transport was from west to east.

The first production of hydraulic cement in Canada was in 1830.

The first patent on Portland cement in USA was issued to David Saylor when he showed the importance of the clinker [61].

It took over 50 years for the shaft kilns to be changed to kiln types with higher capacities. The first rotary, horizontal cement kiln was constructed in the 1880s by the English inventor Ransome. The kiln was 8 meters long and could produce continuously [6].

In 1901, the Atlas Cement Company in Pennsylvania, USA, had 51 rotating kilns with a diameter of 2 meters and a length of 20 meters. Each kiln produced 25 tons of clinker per day.

When we move into the twentieth century, the production of Portland cement across the world had reached 10 million tons per year. The production of natural cements is starting to be marginal and represent only 2% of the total cement production [8]. The early rotating kilns are about to take over the old vertical kilns and the early ball mills have come into use.

The first blended cements were slags cements and they were first produced in Germany in 1901. The Germans had reached a high level of their cement technology and organization, and even in 1877 they established the first association for cement producers – Verein Deutscher Zementwerke (VDZ). This move was one of the reasons why the German industry, through joint efforts, managed to get a strong reputation for their cement and slowly outmaneuvered the English cement industry in the export market. In 1884, Dyckerhoff exported 8 000 barrels (1 360 tons) of cement from Germany to build the foundations for the Statue of Liberty in New York. Other famous buildings in USA built with German cement were the Metropolitan Opera House, Waldorf-Astoria Hotel, and the Stock Exchange on Wall Street.

Cement Associations also appeared in more countries, for example,

- France: Syndicat des Producteurs de Ciment in 1900;
- USA: American Cement Association in 1902;

- England: British Cement Association in 1935.

An example from the other side of the world; the first Japanese industrial production of modern cement started on 19 May, 1875, in Kiyozumi-cho in Fukagawa in Tokyo (about where the Tokyo City Air Terminal is today). The manager who established this government owned factory was Saburo Utsonomiya (1834–1902). The name of the factory at the start was only "Cement Factory." The factory was later sold to a private investor and was known as Asano Cement Company [73].

2.3 After Hennebique

The development of concrete that started in the last half of the 1700s picked up speed continuously and toward the end of the1800s it got a new dimension, the reinforcement. But it was not just the reinforcement that speeded up the development; the knowledge about cement and concrete had spread to a number of countries and it is not meaningful any more to look at the development in a chronological order alone. Knowledge about cement and concrete began to be considered "general knowledge" and some of the fields of skills were permutations of new areas of skills. The versatility of the concrete industry has been increasing since then.

The development in some countries and in various sectors in various places in the world partly happens independent of each other and partly influenced by each other. We also see that a person in an important position can either accelerate the development in their sector or be a major hindrance.

"After Hennebique" is neither any exact notation in time nor does it refer to time with respect to the use of reinforcement as such. However, Hennebique developed a type of construction that has a strong resemblance with the way we use concrete today, and possibly more than anyone else of his time did.

During Hennebique's time, concrete changed from being a "simple" building material to something more complex and versatile. The technologist became more conscious of its varied uses. The theoretical background of cement chemistry and concrete technology is the principle of construction and design models, and they utilize structural principles that take advantage of the special characteristics of the material.

Lambot's concrete row boat from the museum in Brignol

The development of concrete became part of the industrial revolution.

In the following pages of this chapter and the chapters that follow, we have tried to narrate segments from the concrete history, combined with interesting stories from the childhood of modern concrete and partly from our time period. This might help in trying to throw some light on the versatile and great technological and industrial area that concrete has developed into.

It is no longer meaningful to claim that the development takes place only in a single country or a particular environment. The technological area is getting too big and versatile, and the development is too very inhomogeneous. The development in different sectors, and in single countries, might be accelerated by political development and wars – from the big "mega" building-projects, from the ideas and initiatives and positions of single persons, by increased communication and education, by accidents and natural catastrophes, by increased social development, by the increased populations, and by casual incidents.

We will find examples of all these developments – more – in the following chapters. Sometimes, it has been necessary to go back a bit in history to see the connection.

Many recognize Hennebique as the father of reinforced concrete and with him started a new era. But it was more than his work that was responsible for the development. My concrete professor at NTH (Norwegian Technical University in Trondheim), Inge Lyse, states in his book about *BETONG (Concrete),* written together with N.J. Wiig [74], among others.

The use of reinforced concrete can be dated back to about 1850, when the Frenchman Lambot built a small boat in this material.

In England W.B. Wilkinson patented a real reinforced concrete floor in 1854. Seven years later François Coignet showed his principles for the new construction. In the same year (1861) the French gardener Joseph Monier used metal frames as reinforcement for garden pots and containers of concrete. He had taken out quite a few patents on similar things before 1870. But, the use was limited until the German engineers Wayss and Bauschinger investigated and reported about the Monier system in 1887. After this the use of reinforced concrete spread quickly. Major progress in theory and practical use were done by Austrian engineers. The Melan-system with stiff reinforcement, profiled iron used as reinforcement, was developed in the beginning of the 90s. At the same time, the Frenchman Hennebique developed reinforcing systems that in many ways are close to the design practice we have today. A report that also gave valuable information about test results about this new building material was published in 1877 by the American Thaddeus Hyatt.

Albrecht Eika wrote a small article to pay homage to **Monier** under the title "Et hundreårsjubleum (A hundred years jubilee)" in Teknisk Ukeblad (Technical weekly journal) dated 28 September 1967 [75], which celebrated the 100 years since Monier's first patent on reinforcement. Eika had mentioned that already 10–15 years earlier, several persons independently of each other had mentioned reinforced concrete constructions.

Monier made license agreements for his inventions, among others, with Wayss, Freytag, Dyckerhoff, and Koenen in Germany. The first concrete structure in Germany was a dog house built in 1884, which today can be seen in a museum in Munich. In the course of time, Monier got an honors diploma at the world exhibition in Paris in 1869.

Monier had ideas, but also financial problems, and a very bad personal economy. In 1902, his old competitor Hennebique sent around an appeal to his colleagues to collect money for the man whose ideas had meant so much for them. The interest that Monier received from this initiative disappeared very quickly. He died four years later and there were only a few that followed him to his burial place. François Coignet got the approval for his patent on reinforced floors in 1855. He designed and made a two way floor with spans of 7 and 9 meters with a plate thickness of 27 cm. The reinforcement consisted of 12 cm high I-formed iron profiles at 1 meter intervals. The load bearing capacity was claimed to be 1500–1800 kg/m^2. Coignet also introduced a dry compacted concrete with a very low water/cement ratio that gave higher strengths than normal. Coignet also developed concrete mixers and formwork systems.

Hennebique was not the first to design a reinforcement, but his way of

From one of Hennebique' reinforcement drawings from early 1900s.

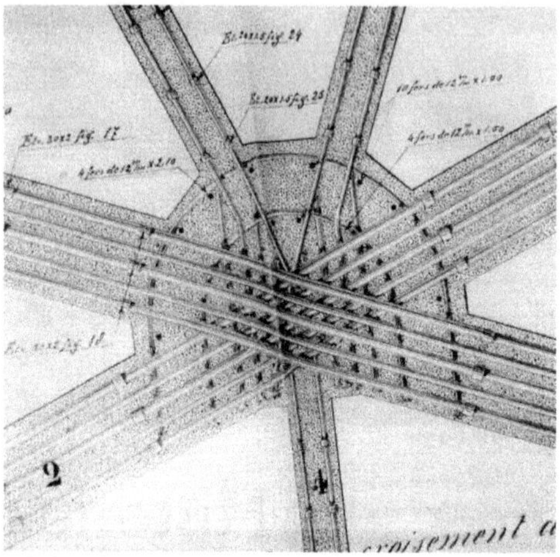

designing it looked very much like what we use now, 100 years later. In an article in *Concrete* in March/April 1992 [76], James Sutherland wrote an invitation to the English concrete enthusiasts to send old historical concrete drawings to a work group in The Concrete Society. It has become more common to use microfilm and electronic copying of older drawings, otherwise older drawings would get lost in the future. The early drawings were very accurate, and some even at a scale of 1 : 1. (This is probably something iron binders would have liked today in the more congested cases). Some drawings even had shadowing. He illustrated this with one of Hennebique's drawings from the early 1900s.

Hennebique's English agent was Lois Mouchel and his "monopoly" created a kind of damper on the work using concrete in England. Mouchel had license offices in six cities and contractors in twelve [68]. From 1897 and until he started his contracting company in 1911, there were all in all 1073 projects completed with Hennebique's "ferro-concrete" system. This

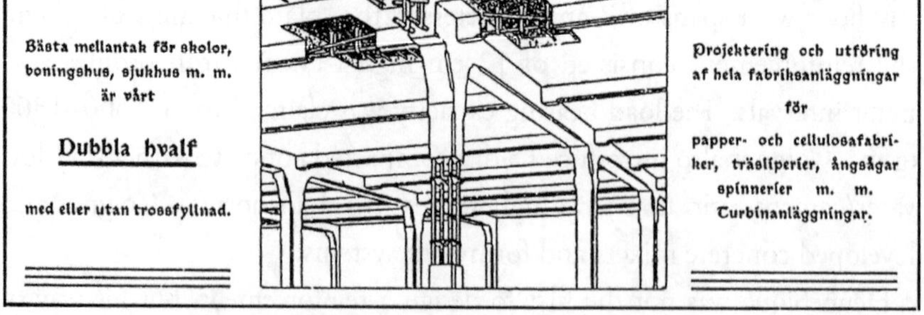

Hennebique's advertisement from Helsinki, Finland in 1907.

might sound a big number, but it is small compared with the 20 000 jobs that was started up in Europe in total.

CWS soap factory in Dunston near Tyne (1907–1915)

In England, Hennebique had a particularly good market share in the northeast i.e. the Newcastle area. He got good support from two important builders: the postal service and Cooperative Wholesale Society (CWS).

CWS building in Newcastle (1900)

Moritz Kahn established his "Trussed Concrete Steel Company" in 1907 and brought in expertise from USA to compete with the Hennebique–Mouchel system in France. Khan's first job was the YMCA building in Manchester in 1908 [68].

Fidler [70] writes that the pioneers in USA reinforced concrete field, during the industrial revolution, had their main activity in the Northeast and around the great Lakes. E.L. Ransome (1844–1917) introduced anchor bolts of iron instead of the tension flange in standard iron profiles and used concrete in the compressive flange. To secure the anchorage, he introduced twisted bars. In 1888, he built a reinforced concrete building with concrete beams and arches, but with cast iron columns. The next year, he introduced beam/ribbed floors. Ransome built his first building with a complete reinforced concrete load bearing system in Greensburg, Pennsylvania, in 1903.

Another pioneer from USA was Thaddeus Hyatt (1816–1901). He also received a patent on reinforced concrete in 1878. He was the first to find out that the temperature-expansion coefficient for iron and concrete was approximately the same. He therefore argued that reinforced concrete was particularly resistant to fire. He also showed that bars was the most economical way of

Westminster Cathedral

reinforcement, but never got around to try out his ideas in real structures. But there were still those who did not believe in reinforcement.

Westminster Cathedral

The first stone for the Westminster Cathedral was laid down in 1895 and it was opened in 1903. The architect for the structure was John Francis Bentley. He was born in Doncaster in 1839 and died on 2 March 1902. The building of the cathedral was a great success at the time.

In June 1900, RIBA (Royal Institute of British Architects) held a congress, and during the dinner, two days after an excursion to the building site, the main speaker said that he seldom had been so impressed as when he entered into this special building [77]. Shortly after the visit, representatives from RIBA visited Bentley to ask him if he would accept a nomination for the RIBA Gold Medal. In February 1902, it was announced that Bentley would receive the Gold Medal. He had never been a member of RIBA, but he appreciated the honor very much. The nomination was supposed to have been confirmed on the 3 of March, but *Bentley* died the day before. In 1952, fifty years after his death, Bentley was honored with a name and relief in a plate of limestone on a wall in RIBA's head quarter, together with all the other Gold Medal winners [77].

(We checked the list of RIBA's Gold Medal winners on Internet. Bentley is not on the list. The Gold Medal was not given in 1901.··· owing to the death of Queen Victoria). Among the other winners we find Frank Lloyd Wright–1941, Eliel Saarinen – 1950, Le Corbusier – 1953, Hugo Alvar Henrik Alto – 1957, Pier Luigi Nervi–1960, Kenzo Tange–1965, Ove Arup–1966, Richard Buckminster Fuller–1968, Jørn Utzon–1978, and Oscar Niemeyer–1998).

In 1979, Martin Hunt wrote an article in *Concrete* about the concrete in the cathedral [78];

During the foundation work they discovered an old concrete slab with an average thickness of 2.7 meters. The slab had been the foundation of a prison built in 1834. On this, they cast an additional 3 000 m^3 of unreinforced concrete. The concrete was made of Thames aggregate and Portland cement, thoroughly hand blended three times. In the third round, the water was added. Then the concrete was wheel-barrowed to the foundation ditches and placed in a 2 ft. (61 cm) layer and leveled as a foundation for the masonry above. The cement had been chosen after thorough testing and the brand "Goliath" was chosen for its extreme fineness. For the foundation, they used 600 tons of cement, 454 m^3 of water, and 3 815 m^3 of aggregate. This would mean a mixture proportion of 1:12 and a water/cement ratio of 0,74.

Architect Bentley prescribed unreinforced concrete also for the domes in the roof. In a letter to a friend in 1902 he states:

"I feel that the old principle of construction is carried on, and that curse of modern construction and source of decay – the use of iron – has been avoided, against the consensus of opinion expressed by the engineers. This much I am proud of, for I feel

that the service to building has been ected, and that I have disproved and broken the backbone of that terrible superstition, that the use of iron is necessary to long spans".

The radial supports to the domes are 27.4 meters high and the middle dome has a diameter of 18.3 meters. The concrete thickness is 91 cm at the foot 33 cm in the middle.

There are also other examples of unreinforced concrete in the cathedral.

Some of the concrete has been damaged and then repaired. Some damages can be dated back to 1919 and are probably due to settings. Other damages originated from the bombing of London during the Second World War.

The cement and concrete chemistry technology had also made progress;

- In 1818, the French chemist L.J. Vicat found out that calcium and silicium were the most important substances.
- In 1868, the English Engineer H. Reed presented the "water-law" that all concrete technologists use even today, in a more modern way: ··· *cement, sand and course aggregate with as little water as possible is the ABC in the process of making concrete* ···
- The German chemist – Dr.W. Michaelis – studied the hydration mechanism.
- In 1883, the French chemist H. Le Chatelier made petrographic testing of cement clinker. He introduced the names alite, belite, celite, and felite for the four crystaline components of the clinker.

In September 1893, Ingeniørkaptein (Engineering Captain) Edvard Kolderup, teacher at the Military High School (Den Militære Høiskole) in Christiania (now Oslo, Norway), published his book *MONIERKONSTRUKTIONENE (The Monier - constructions)*[79]. In the preface to the book, Kolderup avers that he has been on a study trip in Germany, Austria, Italy, and Switzerland. In addition to giving a fantastic summary of the technical knowledge of his time, Kolderup gives a number of references to activities in other countries as well. This gives us a unique insight into the use of reinforced concrete just before the turn of the century. Here are some examples;

On page 5, he states that more than 1000 Monier containers have been produced in France. Some of them are up to 20 meters high, with a diameter of 9 meters.

G.A.Wayss in Berlin had taken on the responsibility of spreading the knowledge about the Monier patent and established companies in Berlin in Germany and Vienna in Austria. Responsible for the office in Berlin was the Government Building Master M.Koenen, while Wayss had moved to Vienna. The company has sub offices in Dresden, Hamburg, Hannover, Kønigsberg, Plagwitz, and Munich in Germany, Basel in Switzerland, and Copenhagen in Denmark.

- On page 7 and 8, he discusses positive durability tests and anchorage tests in Breslau (today Wrocklaw in Poland).
- On page 9 and 10, he mentions about loading tests in Budapest in Hungary and tests in Munich in Germany in 1887. Among others he also refers to the small difference in temperature elongation between iron and "Cement."
- On page 11, he discusses the positive behavior of reinforced concrete at the Helbings Spirit factory in Wandsbeck near Hamburg in Germany in 1889.
- On page 12, he refers to tests on bridge arches conducted in 1889 and 1890 at Matzleindorfer railway station near Vienna in Austria.
- On 14 November 1890, a bridge in Widegg in Switzerland, with a span of 39 meters was load tested. Other tests on bridges that are mentioned, among others, are tests on a bridge in Steyr in Austria on 18 August 1890 and a bridge in Kønigsberg in Preussen on 16–20 April 1888.
- On page 14 and further pages, he discusses various fire tests: a pipe with a diameter of 2 meters was tested in Breslau in November 1886 and two floors at Franz Clouths Rubber factory in Nippe outside Cologne in Germany were tested on 20 November 1886.
- Page 16 mentions military shooting tests on concrete structures in Germany and Austria.
- On page 20, Kolderup describes about an exhibition at Bremen in Germany in 1890, where an arch bridge was constructed in Monier-concrete, with a span of 40 meters and a height of 4.5 meters.
- On page 27, he explains that in the German colonies in Africa, Monier constructions have been built. He mentions the Royal governmental

The roof of Hellerup's Glassworks in Copenhagen.

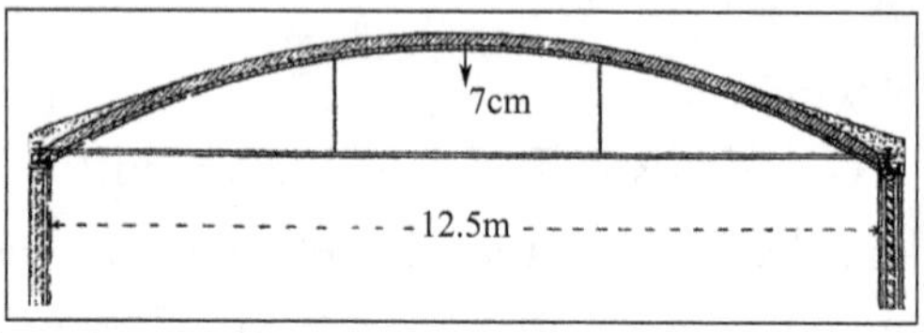

building and a hospital in Cameroon and five large buildings and 41 storage buildings in Dar-es-Salam (now the capital of Tanzania).

- On pages 36–37, he describes a Monier shell roof with tension cables over Hellerups Glassworks in Copenhagen in Denmark.
- On page 39, he mentions about building of houses using reinforced concrete in Venezuela and Brazil.
- On page 44, he refers to various road bridges near Aseod in Hungary in 1891 with spans of 10.2 meters.
- On page 45, he refers to drawings of a bridge over Peterthal in Luxemburg with "the enormous span of 80 meter".
- On page 47, Koldrup starts a chapter about the use of reinforced concrete in road and railway structures. In addition to mentioning cases from Germany, Austria, and Switzerland, he also mentions about bridges that were started to be built across a railway outside Caracas in Venezuela.

Koldrup's example of an advantageous structure for a railway tunnel.

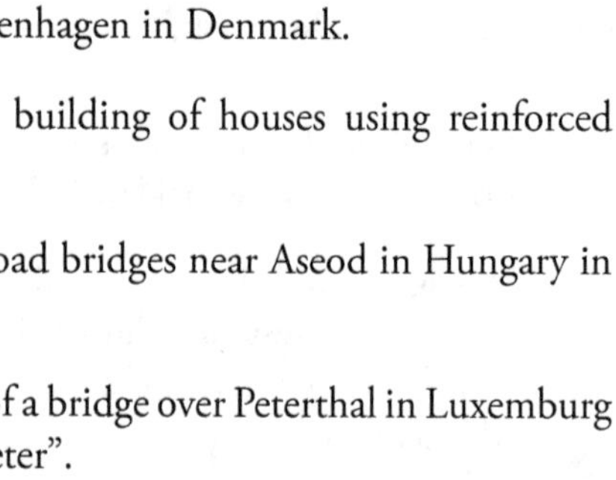

The book also contains quite a few other references to buildings, industry structures, bridges, pipe lines, water reservoirs, tunnels, and military installations, mainly from Germany, Austria, and Switzerland.

Another "military-engineer" is adding to the impression that the concrete- development, that really speeded up in England towards the end of the 1700's and the beginning of 1800's, not only has really spread to other countries, but that England as the "leading lady" has lost speed compared to USA and the French- and German – speaking Europe.

In two articles in *Concrete* published in March and April 1980 [80] & [81], Brigadier J. Hamilton-Baillie mentions about the development of concrete in the *The Royal Engineer* in the nineteenth century. He has made a survey of articles, books, and reports from meetings of the English Engineering Corp, and found about 20 references to concrete from 1834 to 1889. His survey refers to an interesting development. Here are our main impressions;

The first four articles in 1834–1835 do not mention Portland cement at all, but mention about cement/mortar of the type used by Smeaton half a century earlier. Typical is the addition of boiling water (Rangers patent) to the "lime," cement, and mixture in the proportion 1:2 ½ : 4 ½. The concrete is used for "secondary" structures such as foundations, bomb-shelters, seawalls, and similar structures. But interesting test methods were in use.

In 1838, Paisley, one of the Engineering Corp's own people, patented his own type of cement. Through the years he came up in rank and wrote a textbook about cement and concrete in 1838, with a second edition in 1841. He became Commandant of the Royal School of Military Engineering. From 1822 to 1841, he grew in ranks from Major to Major General. Even in the second edition of his book, he did not mention Portland cement. Aspdin's cement then existed only in poor quality and Paisley had not heard about it. The first time he heard about Portland cement is when he met William Aspdin at the Great Exhibition in 1851.

- The first time that Portland cement was mentioned by the Engineering Corp was by George Fredrick White in May 1851.
- In 1857, Captain H.R.D. Scott published an article about a new cement type that he had patented. He explained more about it ("Scott's Cement") in a new article in 1861. He mentioned that he had been using artificial pozzolan. Scott published two more articles in 1962. General Pasley died in 1861 and then people dared to publish something about cement and concrete which was not absolutely in accordance with the views of the old general.

Mobile and stationary mixing station, 1874

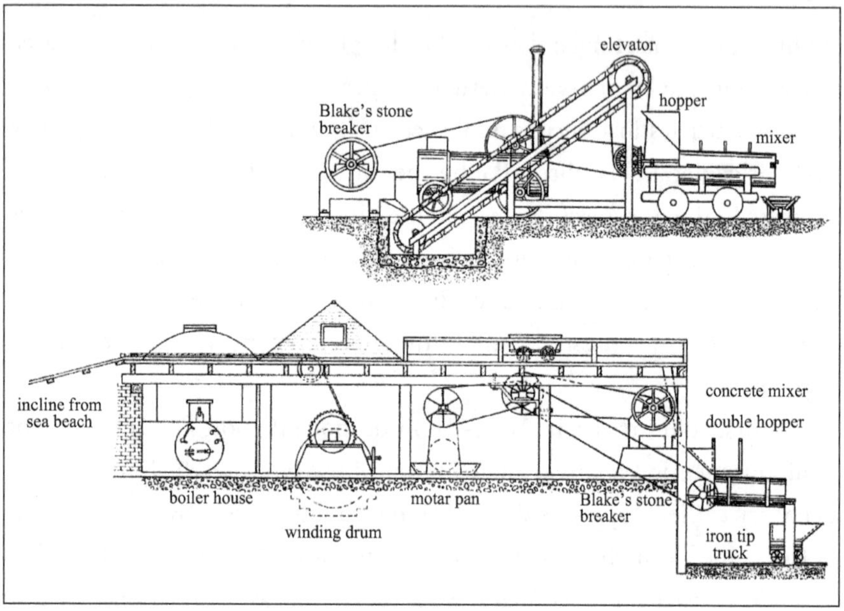

- Cement was still, 40 years after Aspdin's patent, mainly used in the mortar for masonry.

- The next article, in 1865 [66], by John Grant, discussed about the considerable use of Portland cement mortar during the masonry work for the sewage systems in South London. There are references to building of structures at Newhaven Fort in 1866, not with Portland cement, but with "Scott's cement."

- In 1870, an article that mentioned that iron–concrete can be used in fortress buildings was published.

- In 1871, an article by Grant showed that Portland cement had improved its quality and that they then had cement tests with strengths of 2.47 N/mm^2 after 7 days and 3.17 N/mm^2 after 30 days versus 1.86 N/mm^2 five years earlier. Grant then anticipated that concrete might be used in a lot of areas.

- A new article in 1871 mentions about the use of Portland cement concrete in massive structures in a railway bridge in London. The author claims that this must be one of the earliest uses of Portland cement for this purpose.

- In 1874, Major Maquay wrote an article in which he among others described a steam driven mixing and crushing machinery used during concreting at the harbor in Cork.

- Papers about concrete were presented in 1876 and 1880. At a meeting in 1880, E.A. Bernay claimed that that even if Portland cement was not mentioned in the specification of the Admiralty as late as 1867, concrete had confirmed its right as an independent building material.

- Two presentations in 1898 had interesting views. A.E. Carey claimed that England had then regained the lead in cement production from Germany. He also reported about Austrian tests with the Monier system and bridges on the continent with spans longer than 40 meters. In another presentation, R.P.Wells gave a lecture to the military engineers about *Concrete and Ferro- concrete in Arches*. He among others claimed that British engineers, both civil and military, were far too conservative, and had been too unwilling to utilize concrete. He cited examples from the Continental Europe, which had advanced much further than England. Of the examples he mentioned are the following:

    - In Germany, they had introduced three joint arches in concrete with the strength of 23 N/mm^2. A bridge in Geneva in Switzerland had a strength of 32 N/mm^2.

    - He gave illustrations from France, Germany, Netherland, Italy (bridge near Millesimo), and Spain (bridge near Bilbao).

Endre Lekve (1833–1879)

It was the Norwegian Endre Lekve who introduced concrete in **Finland** [82].

Polytechnic education was established in Helsinki in 1849. In 1858, the school got orders from the Emperor (Finland was at the time under Russia) to upgrade building technology. They got in contact with the Polytechnic High school in Hanover. The school had the reputation of being one of the best in the world. Lekve, who was born in 1833, was convinced to come to Helsinki from Hannover.

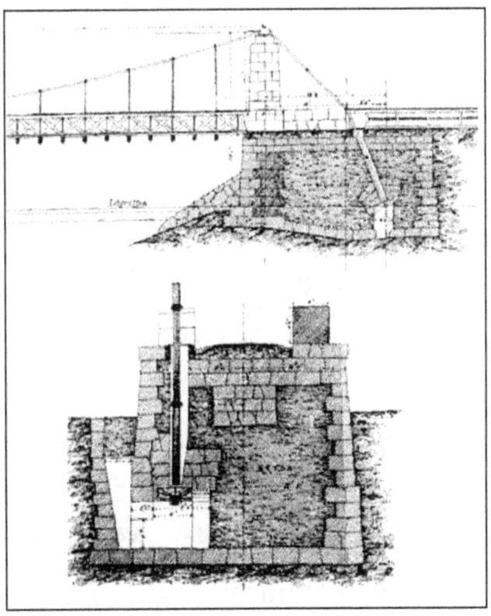

Drawing of the first suspension bridge in Finland. The bridge was planned in 1877, but the drawing is from 1889 [83]

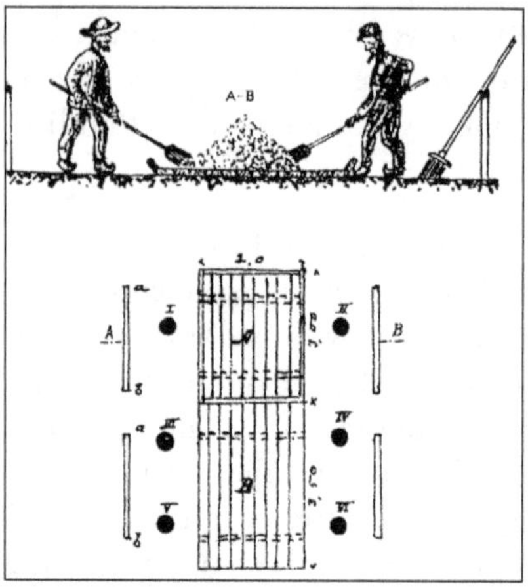

Recommended platform for concrete mixing, according to Strukel, 1895. The wooden platform is 2×3 meters.

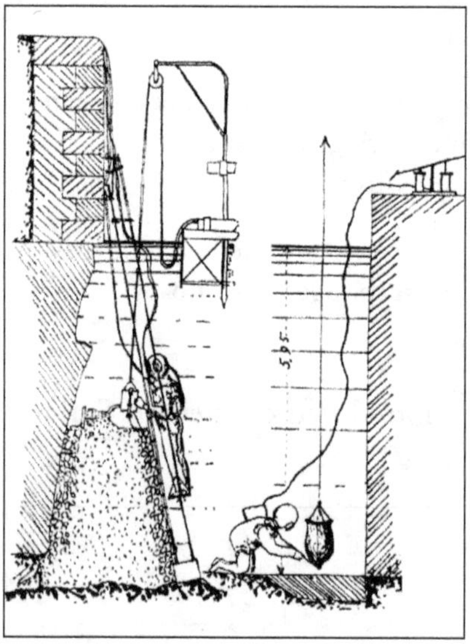

An illustration of underwater concreting – Strukel 1895

Lekve became the principal teacher of all subjects regarding building and had therefore a very wide area to lecture. He was also active as a planner of Helsinki's water system. He was the planner, controller, and contractor for the first suspension bridge in Finland. This was a chain bridge, in which the anchors were cast in concrete. The cement was imported from England. The first cement shipment had to be thrown away, as it had been damaged during the sea transport. The concrete was cast in 60 cm thick layers and the joints between the layers were cleaned for dirt and chiselled to secure proper bonding between the layers.

The proportion of the mixture of the concrete for the bridge foundations was 1 : 3 : 6, cement : sand : stone.

When Lekve died at the age of 48 in 1882, a memory plate was made for him at the bridge [82].

No publications have been found or are based on Lekve's activity, but it is a general opinion that most of his notes are included in the books of Mikael Strukel (1851–1923).

In 1879, the Polytechnic school in Helsinki had to expand their staff capacity in building education. They announced the position in various daily newspapers in Finland, one newspaper in Sweden, and four in Germany and Austria. They got eight applications for the position of assistant to Endre Lekve. The Austrian, Mikael Strukel, got the job. He had his education from the Technical High school in Vienna (1869–1874) and four years experience from bridge building in Graz in Austria. When Strukel was employed, the school in Helsinki had 33 students in the building department, out of a total of 95.

Strukel was in the beginning Lekve's assistant, but when he died, he took over the position of the principal lecturer. In the years from 1895 to 1914, he wrote three books on subjects regarding building, with a total of 1880 pages. The books were written in German and were also used at schools in Germany and Austria. In the beginning in Helsinki, Strukel was teaching in German, but later he changed to Swedish.

The concrete specifications in Strukel's books were based on the Prussian concrete norm of 1887.

In his books, Strukel mentions pozzolan, Roman cement, and Portland cement. He recommends a water/cement ratio of minimum 0.5 and maximum 0.9 (by volume).

The cement:sand ratios that he recommends for the mixture design (sand below 7 mm.) vary from1 : 2 to 1 : 5 with an average of 1 : 3 (all mixtures based on volume measurements). In blends of cement:sand:stone, he gives variations from 1 : 1 : 5.2 to 1 : 5 : 10. A typical "rich" mix is 1 : 3 : 6 and a "lean" one is 1 : 4 : 8.

As in Norway and many other countries, there were enthusiasts in the government administration, who really speeded up the use of concrete in Finland. A winter harbor in Hanko was built in 1889–1893 on the initiative of the enthusiastic Engineer Werner Lindberg. Another of Lindberg's ideas was the renovation of the lighthouse at Bogskjær. The lighthouse had been built in cast iron in 1882, but was severely damaged in 1889, and finally in 1894 they started the repair with concrete. The first Finnish standard for concrete was not established until 1929 [82].

Before 1905, concrete in Finland was mainly utilized for foundations and various uses related to "water-structures." From 1905 onwards, the use of concrete in building structures increased considerably [81].

As we move into the twentieth century, and the time of reinforced concrete, the use of concrete has spread to a number of countries, and the birth place of modern concrete, England, is no longer the "leading lady."

The first international conference on cement chemistry, however, was held in London in 1918, in the British Faraday Society.

An illustration from a Finnish publication, 1908.

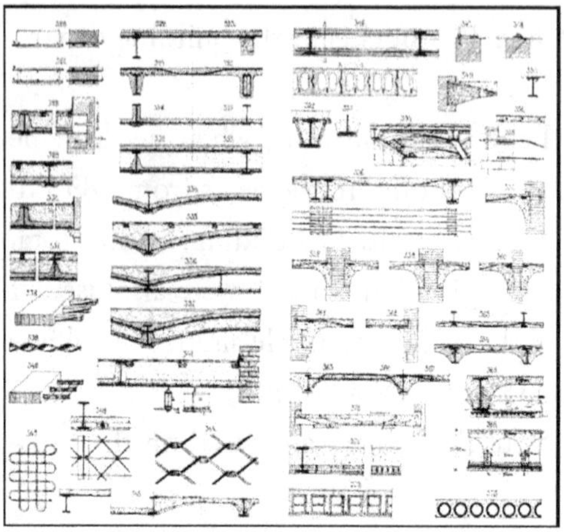

Robert Maillart from Switzerland (see Chapter 3.1.5) started building some remarkably elegant arch bridges with reinforced concrete in 1905.

In 1912, Max Berg built the fantastic Jahrhunderthalle in Wrocklaw in Poland (Breslaw). The floor area was 2 000 m^3 and the roof had a weight of 4 300 tons. (see Chapter 3.1.5)

The use of concrete in floors in commercial and apartment buildings was mainly reasoned by the fact that they were resistant to fires, and after some time, more reasonable insurance costs. A number of construction designs were developed; often only modifications of old design types that became normal at the introduction of reinforced concrete.

The illustration above, from a Finnish publication [62], shows the numerous design solutions that were used. New ideas with a more optimal use of concrete and reinforcement were presented at short intervals.

The development was astonishing not only in new design alternatives, but also with new records in length, spans, height, and volume. Also, there were developments in the production technology and new production patents were granted. We refer to a short story from USA. told by Jane Morley in *Concrete International* in 1990 [84].

Frank.B.Gilbreth was born in Maine in 1868. He probably passed the introductory exam at MIT in 1885, but instead of continuing he chose to work for a construction company in Boston, until he in 1895 decided to start his own company. He specialized in repairs of water leakages in basements in the Boston area, but moved his office to New York City in 1904, as he got projects along the whole east coast and in Canada, California, and Washington on the west coast of USA. In 1911, he had developed a number of tools to make the life at the building sites more rational. This was called “The Gilbreth System.” The system included routines for the formwork, reinforcement and concreting, transport at the site, testing etc. He also developed his own mobile concrete mixer, called the gravity mixer.

The mixer worked thus; there is a continuous outflow of ready mixed concrete, as the materials moved down the many chambers of the steel box at the same time as the many steel rods were rotated. Gilbreth also made quite a few other inventions, among others a system for concrete piles (U.S. patent 882.520 from 21 April 1908) and a pat-

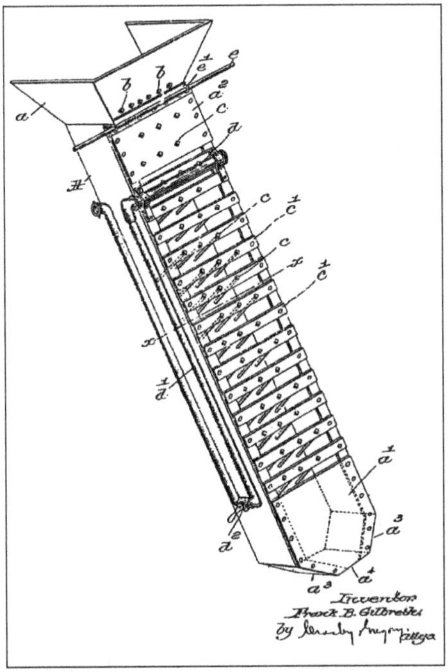

Patent drawing of Gilbreth's mobile concrete mixer

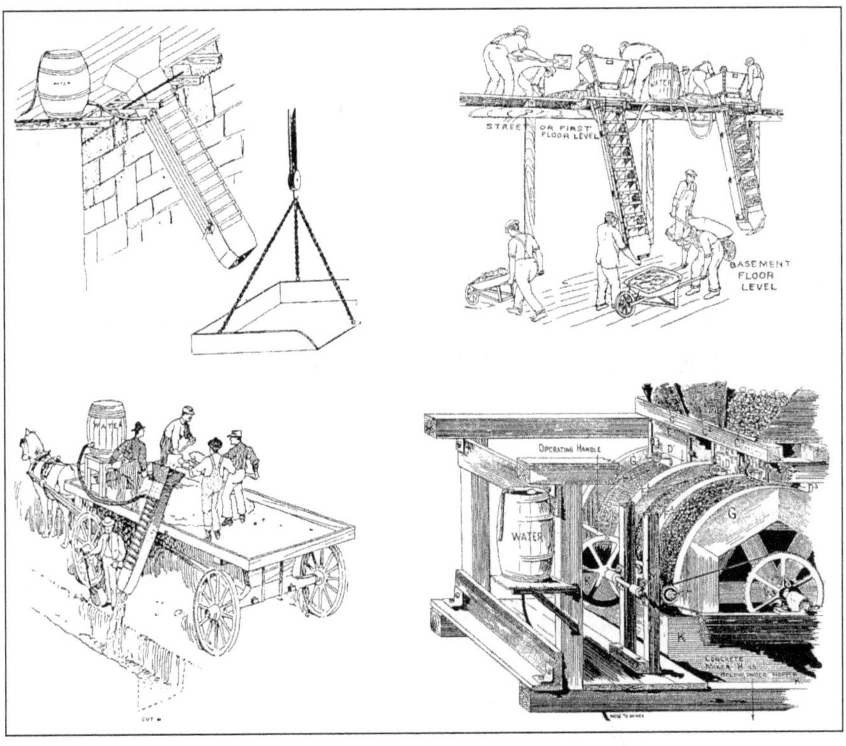

Various use of the Gilbreth's mixer.

Thomas Alva Edison, born on 11February 1847 and died on 8 October 1931, in New Jersey.

ent for ramming of concrete piles (17 March 1908).

This was really the time of inventions as far as equipments used in the production of concrete are concerned. Tools were invented and reinvented, but many of them disappeared after some time. Here are a few stories.

The "tremie"

In a paper presented at the ACI Committee 120, in the session on the History of Concrete in Baltimore in 1999, Bryant Mather [85] narrated the story about General Lee's "invention";

> *Fort Carroll, in the middle of the Baltimore harbor, was built from 1849 to 1852. Robert E. Lee had in the spring of 1849 registered that there was a stable, sound ground of 45 ft. (ca 15 meters) depth. He experimented with the placing of underwater concrete on this base by the use of a "tremie." Lee continued this work until August 1852, when he was sent to West Point as a new commander. Lee got information from General Totten on 22 June 1849 about the placement of concrete with the "tremie." Lee answered on 25 June thus; "I will experiment and test concrete with the tremie." Mather claims that this must be one of the first examples of experiments and research on concrete in USA.*

Mather states, referring to Ben Gerwick and Terry Holland, that they had found in the *Oxford English Dictionary (Oxford University Press 1971)* that the word "tremie" was first used as a tool for underwater concreting in 1905.

Mather tried to find out about this.

He refers to Holland, which means that Totten might have learned about the "tremie" through a publication from France in 1838.

Edison's invention of concrete houses

The world's greatest inventor through times, Thomas Alva Edison, tried to build houses made of standardized concrete [86].

Edison received not less than 1093 U.S. patents. Most of them were improvements of other people's or existing inventions. We know mostly about Edison from his inventions in the electricity sector. Edison could have said *we will make electricity so cheap that the rich will use candles.*

It has been said that Edison had hung a famous quote from Sir Joshua

Reynolds over his desk and several other places in the house; "*There is no refuge you will not seek, to avoid to think*".

Edison with the concrete house model

Edison is also the father of the saying or the definition for innovation; *"1% inspiration and 99% perspiration."*

Toward the end of the 1880s, Edison tried to refine iron ore using magnets and crushers. He kept on with this for some years, but lost millions of USD on the project. He tried without luck to sell the equipment. As this did not work out, he decided to use the crushers in cement production. Edison established Edison Portland Cement Company in Orange, New Jersey in 1898. Ten years later, in 1908, he introduced a much longer rotary kiln than used before. In 1905, the production at Edison's cement factory had reached a level of 715 tons a day.

Edison also received a patent on his long rotary kiln in 1909 [61]. Edison's kiln was 45 meters long, 21 meters longer than any other previous cement kilns.

In 1906, after recognizing that the cement project also had been a four-year money drain, partly because of the fact that the sale had been far lower than expected, he decided to create his own demand. During a speech at a dinner in New York in August 2006, he introduced his new baby, concrete homes. They were supposed to be fire proof, insect free, and clean. The walls would be ready painted and everything would be in monolithic structure, ready cast in only a few hours – everything – from the bath tub to the frames of the paintings. Above all, the house should only cost 1 200 USD.

Some houses were built in New Jersey, but the project was a "flop" in more than one way. He then offered contractors an opportunity to take over the equipment, but no one was interested.

But Edison did not give up concrete. In 1911, he introduced furniture

A glimpse from the Panama canal today

and other products in "foam-concrete" – everything from bedroom furniture ("more beautiful and durable than those in the palatial residences in Paris or along the Rhine" according to New York Times) to models of cats and dogs to be placed in the garden, kitchen furniture, phonograph tables etc.

Edison went bankrupt on his concrete adventure.

There were also other lapses in the process.

Brian Greene refers to the concrete dam in Austin in Pennsylvania that collapsed in 1911[87].

This is still the sixth worst dam collapse in USA. The dam was built as a massive dam, with some "saving" stones of the size 4–20 dm^3. The dam was 163 meters wide and had a height of 15 meters over the bottom of the valley at its highest. 15 000 m^3 of concrete were used. The designer had drawn a deep stoppage wall to stop water leakage at the foundation, but the owner had reduced this to 4 ft. (1.2 meter) to save money. However, bolts 7.5 meter long were drilled 1.5 to 2.4 meter into the rock foundation. The dam was completed in December 1909.

In the afternoon of Saturday 30 September 1911, the warning signal went off in the small town below the dam. Many people did not care about the warning signal. Water from the dam came down the narrow valley, with a big wave of wreckage in the front. 78 people died in the catastrophe.

The rock foundation of the dam was said to be made of sandstone. Unfortunately, this was only a layer of 2 ft. thickness. The crack appeared in the underlying slate rock.

The collapse of the dam in 1911 did not stop the development of water structures in USA. Examples of this are the great dams such as the "Grand Coulee Dam" built in 1933 and the "Hoover Dam" built in 1936 (Chapter 3.1.3).

A glimpse of the history of the Panama Canal might also be useful to get an idea about the thoughts and happenings of the period [88].

The building of the Panama Canal was a project that went on for many years and was finished in 1914 (because of the world war, the official opening was postponed to 1920). The building consumed close to 5 million m^3 of concrete. The canal had six locks with a maximum depth of 14 meters. The building had great effect on the development of the concrete technology, for example, the development in concrete trucks and standard specifications etc.

The idea to shorten the route between the Pacific Ocean and the Atlantic Ocean and to create a water way at the slimmest part between South and North America was old. Already in 1534, King Charles 1 of Spain ordered for investigations. The development that came with the finding of gold in California sharpened the interest of USA in building the canal and in 1868 the U.S. President Grant ordered for new investigations.

The development, however, did not pick up real speed until the Geographic Society in Paris discussed the case in 1876. On 20 March 1878, a French company made an agreement with Columbia (Panama was at the time a province in Columbia) on building a canal, with a back fall right to Columbia after 100 years. On 15 May 1879, a canal congress in Paris discussed the case with 14 alternatives for the canal. These were reduced to two – Nicaragua and Panama.

Work on the canal started on 1 January 1888. Originally, the plan was to build a canal without locks, but after some time came the proposal of building the canal with 10 locks. Gustav Eiffel was to design the locks. In 1894, during the French construction works, there were a maximum of 19 000 workers on the canal. It was difficult to get workers, so most of them came from the West Indian Islands and in particular from Jamaica.

In the beginning of 1889, the French canal company ran out of money and on 15 May 1889 all work stopped. The liquidation of the company took place in 1894, with a lot of consequences. The initiative taker and the leader, Charles de Lesseps, was after a number of court cases convicted and sent to prison. Most of the time was spent in hospital, where he died on 7

December 1894, 89 years old.

A new canal company was established on 20 October 1894. Also, this company was after some time short of money. The work started up again on 9 December 1894, and in 1897 the workforce counted 4 000 men.

In 1898, more than half the capital of the company was used and a petition was made to USA. On 2 December 1899, a meeting was arranged with the U.S. President William McKinley. But it took over five years before an agreement was reached.

It was President Theodore Roosevelt who took action and military forces were sent into the canal zone. An agreement was made with both the French and the freedom movement in Panama. On 3 November 1903, Panama was declared independent of Columbia. The so-called Hay–Bunau–Varilla-agreement (named after the French and the U.S. negotiators) secured U.S. sovereignty over a 10 mile (16 km) wide belt in Panama, 5 miles on each side where the canal should be built. The agreement was ratified by Panama on 2 December 1903 and by USA on 23 February 1904.

The U.S. management took over the construction on 4 May 1904.

The chief engineer was John F. Stevens, who was later named "the genius of the Panama Canal." He convinced Roosevelt that they should build six locks. He also briefed Philander Knox, who spoke to the Senate about the case on 19 June 1905. The lock alternative was approved by the Senate on 27 June with a majority of 36 against 31. Stevens' most important arguments were:

- To dig a canal that was 15 ft (45 meters) at the narrowest part without locks would lead to endless landslides.
- Without locks, the canal cannot be completed until 1924, while with the locks it could be finished by January 1914.

Stevens withdrew from the position of chief engineer on 1 April 1907 and Roosevelt appointed military leaders with authority. Roosevelt was, by the way, the first President of USA who travelled abroad during the presidential period; he visited the Panama Canal in November 1906.

The immigration of workers from other countries to the Panama Canal was enormous, and the challenge of providing food and housing was not less. On 26 March 1913, a maximum of 44 733 workers were registered, which does not include the ones that were sick or on leave. Most extreme was the migration from Barbados which was 19 900 men in 1907 or 30–40% of the male population of the small island community.

Historic picture of the Panama Canal

The workforce figures were 1 000 in May 1904, which increased to 3 500 in November the same year, and 17 000 in November the next year. As an example, they had in 1907, 5 293 workers from Spain, 1 032 from Italy, 1 101 from Greece, 6 510 from Barbados, 2 039 from Guadeloupe, 2 224 from Martinique, but only 13 from Panama. From USA there were normally about 5 000 men.

After some time, national standards, building codes, standard specifications, and directions for the production and use of concrete and reinforced concrete came into existence.

Corley [59] mentions the following about the U.S. concrete codes.

- 1907 – Drafting of the U.S. Building Code, which was in effect from 1910 and allowed cylinder strengths up to 2000 psi (13.8 MPa). The standard was in 12 pages.
- A later addition to the standard allowed strengths up to 7 500 psi (34.5 MPa).
- 1963 – The Building Code allowed strengths up to 9000 psi (62 MPa).
- 1989 – The Code allowed strengths from 2 000 to 18000 psi (124 MPa). The code has 336 pages in addition to comments.

(See also comments about codes in the introduction to Chapter 4).

The speed of the development continues to be high.

The forerunner of the vibrator we use today was a pneumatic "rammer" that came before 1907 [89].

Adams published his famous water/cement ratio law in 1918.

In 1928, Eugène Freyssinet got his patent on prestressing and in 1935 he utilized the technology for the first prestressed concrete structure in Le Havre. The harbor area was threatened by the state of the shipbuilding wharf and Freyssinet used poststressed beams to save it after some previous upjacking.

However, Freyssinet was not really the first person to work with prestressing technology [90,91]. Already in 1886, the first U.S. patent was issued to Jackson, and 2 years later, in 1888 Doehring had patented a prestressing method in Germany. None of these methods had any success in practice. The first practical utilization came with Karl Wettstein in Austria, who prestressed with piano wire. According to Anderson [91], referring to J.J.Polivka, Wettstein got a German patent on this method in 1921, based on a 1.2 mm diameter prestressed wire. This method was later adopted by Hoyer, who made a prestressed bench of a much larger scale than Wettstein in 1938.

Freysinnet's mentor, Rabut, had also made prestressed concrete corbels.

Freysinnet was, however, the first person to show that you had to use high strength steel to balance the creep in the steel. He also developed prestressing systems and other details such that we could use prestressed concrete as a solution for numerous problems. Eugène Freyssinet was an experienced and creative designer.

Marie Eugene Leon Freyssinet was born on 13 July 1879, in Objat, Corrèze in Frankrike. He worked as the director for public works in Moulin from 1905. From 1907 to 1914 he was road engineer in mid France. After a period in the French army, in 1914–1918 as road engineer, he went to Paris and worked in Ecole National des Ponts et Chaussees in Paris and for Claude Limousin until 1929. After that, he had his

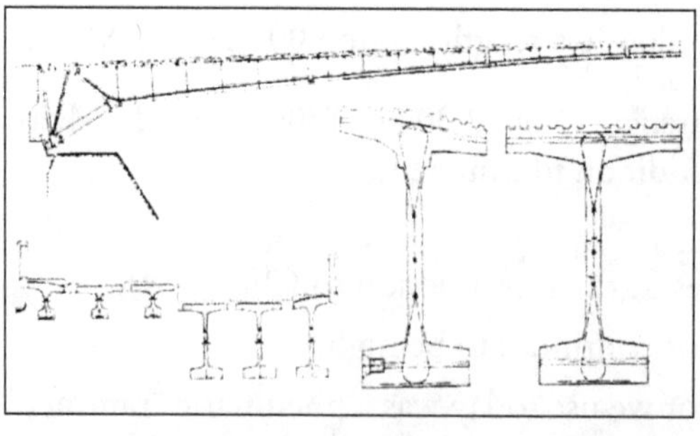

Drawing of Freisinnet's elegant bridge over Marne at Esbly [91]

own company for some years, but got into economic difficulties. After the work in Le Havre in 1935, he was employed by Campenon-Bernard, where he continued to design pressed bridges.

Freyssinet's hangar at på Orly in 1923–1924

Freyssinet designed a large number of bridges and buildings. His most famous structures are probably:

Plougastel bridge;

Pont le Veurde;

The hangar at Orly.

Eugène Freyssinet died on 8 June 1962.

Another important pioneer of prestressed concrete was the Belgian Professor Gustav Magnel (1889–1955). He was probably the one who really got prestressed concrete to take off in USA. In connection with PCI's (Prestressed Concrete Institute) 25th year anniversary in 1975, the *PCI Journal* published a special edition, describing the beginning of the development. A number of pioneers in prestressed and precast concrete from USA contributed interesting papers. Among others, Charles C. Zollman narrates the story behind Magnel's influence [92,93].

Zollmann had studied under Magnell in Ghent, but fled to USA in 1941, and he later became his unofficial U.S representative. He also translated Magnell's book about prestressed concrete to English.

The history really started in 1920, when a special relationship was developed between USA and Belgium. During the First World War, an organization with the name "Commission for Relief in Belgium" was developed. The chairman of the organization was Herbert Hoover, who later became the 31st President of USA. From the considerable funds that was available after the war, Hoover, who was a Belgium fan, established "The Belgian–American Educational Foundation." The purpose of the fund was to finance a one year study in USA for students from the universities in Ghent, Brussels, Louvain, and Liège. This system started in 1920 and greatly infleunced both the Second World War and the introduction of prestressed concrete in USA.

During the Second World War, the allied countries were considerably lacking in metals such as nickel, cobolt, pewter, zinc, and radium. Former exchange-students, who had fled to England during the war, were active in helping USA to get hold of such metals from the former Belgian Congo (later Zaire and Congo), which had considerable resources. This led to the situation in which the government of USA owed considerable debts to Belgium after the Second World War. This again led to the much faster recovery of Belgium than the other European countries after the war and might be one of the reasons why we find both the NATO and EU headquarters in Brussels.

After the war, the cooperation between USA and Belgium increased. Gustav Magnel was professor in engineering sciences at the university in Ghent and was elected to represent the Belgian building industry as well as the entire engineering profession. From April to June in 1946, he travelled around USA as "Belgian American Educational Foundation Scientist."

All the articles about Magnel discuss him as an excellent speaker and teacher.

Zollmann gives the following quote from one of his lectures:

"Gentlemen, you have now been with this University for 2 years. You must, by now, be excellent mathematicians. But do not expect me to fill this large blackboard with all kinds of complicated formulas and their derivations. It is true that I would look like a very learned man – but that is not why I am here. I must make engineers out of you – in less than 2 years- so that you can design and analyze structures rapidly and as accurately as possible – not mathematicians who will need 6 months to solve the problem based on assumptions [...]which are inaccurate anyway. By that time the building would have been constructed by someone else."

Magnel during one of his lectures in USA in 1946

The English edition of Magnel's book *Prestressed Concrete* was, however, printed in London in 1948 because the U.S. publishers did not think that there would be any interest for this material. The first edition had 6 000 copies. This was quickly sold out and 8 000 copies of the second edition that came in 1950 were printed. The U.S. publishers then bought the rights and printed the third edition in 1954.

The breakthrough for prestressed concrete structures in USA was the building of the Walnut Lane Bridge in 1949–1950. Before this, there had only been technology transfer from east to west over the Atlantic as far as this technology sector is concerned. After this event, it became more balanced. A number of practical tools, tension systems, and design solutions were then developed in USA and later used in Europe.

(For more about Magnel, see Chapter 3.1.3. Walnut Lane Bridge).

Another breakthrough in the development happened when MIT (Massachusetts Institute of Technology) in August 1951 held a 3-day conference on prestressed concrete [90]. Over 600 representatives from the industry listened to what the experts had to tell. In his closing speech, Contra Admiral I.F. Jelly, The Chief of U.S. Navy Bureau of Yards and Docks, among others said: *"This conference has given me the feeling that we are on the threshold of a new era in the history of construction technology, - about the same way as when reinforced concrete was introduced about 50 years ago"* ···

The development in USA was astonishing. T.Y. Lin came home from Belgium in 1954 and started working on his book on prestressed concrete. He proposed to arrange the first world conference on the topic. This was held at the University of California, Berkeley, in 1957 and was attended by 1200 engineers from 30 countries.

Many other pioneers in prestressed concrete could be mentioned, like Hoyer and Ables. E.Hoyer, originally from Tsjekkoslovakia, was professor at the technical university in Braunschweig. He among others experimented with thin plates of 5 cm thickness in 1935–1939 and used span wires of 2 mm diameter [91].

It was after the turn of the century that the development in the use of concrete really picked up speed.

In 2007, the cement production in the world had grown to 2.77 billion tons (see Chapter 6).

The cement kilns had developed from small vertical kilns to "wet" rotary kilns that had been increasing in length and diameter. Then came the semi-dry Leopold kilns. The dry kilns got their breakthrough after the energy crisis of 1970s, the same time when the old wet kilns were made more efficient by better dispergence of the slurry so that the water that would evaporate is reduced. Then came technological developments such as preheaters and precalcination and the kilns became shorter. The energy consumption in the burning of the clinker reduced from 10 000 kJ/kg to 3 000 kJ/kg. The kiln capacity increased from 30 to more than 10 000 tons per day. The grinding capacity increased from 10 to 200 tons per hour.

The development of the technology in the cement industry has been impressive, and the demand for expertise and heavy investments has strengthened the structural changes in the industry.

It is interesting to observe the development in the cement production across the various countries in the world. This tells, not only about the development in the use of concrete, but also the general development in infrastructure etc. in the countries. We have tried to show the development in most countries of the world during the first three-fourth of the last century here.

The numbers are in units of 1000 tons of cement production:[94]

Africa	1913	1920	1930	1938	1950	1960	1970
Algerie	16	46	68	90	324	1 061	924
Angola						161	446
Egypt	20	23	188	323	957	2 047	3 686
Etiopia					11	28	204
Camerun							40
Congo							92
Gabon				5	60	185	
Ghana							435
Kenya						341	790
Nigeria						157	604
Libya							97
Malawi							76
Marokko		20	75	165	322	580	1 403
Mozambique			23	24	50	222	394
Tanzania							168
Tunesia				69	169	405	547
Uganda						72	191

America	1913	1920	1930	1938	1950	1960	1970
Argentina	3	79	412	1 179	1 558	2 614	4 743
Bolivia			6	19	38	38	116
Brazil			87	618	1 386	4 444	9 003

America	1913	1920	1930	1938	1950	1960	1970
Canada	1 375	1 056	1 752	876	2 658	5 339	7 285
Chile	17	30	162	364	513	835	1 350
Columbia	5	6	9	142	567	1 389	2 828
Costa Rica							187
Cuba	5	77	300	110	317	813	742
Domin. Rep.					72	169	493
Equador			3	14	57	200	458
El salvador						88	167
Guatemala		4	15	10	42	117	228
Haiti						48	65
Honduras						39	141
Jamaica						213	456
Mexico	85	110	258	374	1 388	3 090	7 180
Nicaragua					17	32	127
Panama					51	110	244
Paraguay						14	68
Peru		5	26	101	331	570	1 138
Puerto Rico					549	926	1 569
Trinidad 6 Tobago						178	270
USA	15 126	16 557	27 432	17 969	38 004	53 481	64 904
Uruguay	25	56	207	151	427	462	
Venezuela		6	21	40	501	1 487	2 638

Asia	1913	1920	1930	1938	1950	1960	1970
Afganistan						35	105
Quatar							108
Saudi Arabia							675
Burma				67		45	156
China	250	300	600	1 000	1 410	12 000	26 500
Hong Kong	70	90	40	57	59	150	430
India	1	98	573	1427	2 654	7 832	13 986
Indonesia	36	80	142	160	75	379	561
Iraq					57	691	1 400
Iran				72	62	578	2 342
Israel			78	98	380	807	1 384
Jordan						165	378
Japan	636	1 341	3 277	5 925	4 451	22 425	56 543
South Korea		51	252	637	12	464	5 822
North Korea							4 008
Libanon				171	263	853	1 339
Malaysia							1 032
Pakistan					421	1 136	2 657
Philippines			100	167	301	815	2 448
Sri Lanka					12	85	326
Syria				65	68	490	968
Taiwan, China		70	150	150	332	1 185	4 541
Thailand		24	68	90	165	440	2 627

Europe	1913	1920	1930	1938	1950	1960	1970
Austria	450	120	602	650	1 289	2 828	4 961
Belgium / Luxemburg	1 420	1 238	3 504	3 130	3 677	4 598	6 971
Denmark	495	251	779	640	872	1 435	2 339
Finland		76	186	502	743	1 253	1 839
France	1 930	1500	5 437	4 121	7 208	14 173	29 325
Germany	6 833	2 250	5 671	15 265	10 877	25 845	37 482
Greece	20	20	180	313	399	1 650	4 916
Iceland						70	86
Ireland				109	427	759	867
Italy	1 360	1 050	3 426	4 587	5 035	15 817	33 128
Netherland			224	459	593	1 798	3 830
Norway	123	204	327	338	583	1 166	2 623
Portugal	12	15	99	268	572	1 198	2 329
Spain	390	281	611	993	1 945	2 806	3 998
Switzerland	390	281	611	993	1 945	2 806	3 998
Turkey	45	5	66	267	396	2 038	6 443
United Kingdom	2 900	2 333	5 111	7 838	9 906	13 500	17 583
Cypros						82	270
Albania			3	15	15	73	345
Bulgaria	15	15	174	154	587	1 586	3 667
Czchoslo-vakia	340	300	1 195	1 400	1 800	5 052	7 401
DDR					1 389	5 034	7 992
Hungary	350	100	330	343	797	1 571	2 769
Poland	650	231	1 008	1 289	2 514	5 317	12 181
Rumania	135	110	296	532	1 018	3 055	8 127
Yugoslavia	500	203	860	712	1 219	2 386	4 403
Russia	2 400	36	3 006	5 688	10 194	45 500	95 233

Oceania	1913	1920	1930	1933	1938	1950	1960	1970
Australia	230	400	552	874	1 290	2 794	4 600	
Fiji								60
New Zealand	65	80	75	204	252	618	802	

As can be seen from the table above, there are great differences in the development patterns between countries.

World's second largest producer of cement today is India. Even in 1904, a Chapter of ACI was established in India, but the first cement production in India happened only 10 years later, in 1914 [95]. The first factory was established in Porbandar in Gujarat. Prestressed concrete came to India in the early 1940s with the building of several shopping centers in Meerut. The first railway bridge in prestressed concrete was built in 1949.

In another important "concrete-country," Japan, the first reinforced concrete building was built in 1905. Five buildings designed by the Marine Engineer Kenzaburoh Mashima had reinforced concrete in columns and beams. Three years later, in 1908, Naoki Shiraishi (1857–1919) designed the first complete concrete building, Warehouse "D," for the Tokyo Storage Company in Wadamisaki in Kobe [73]. Shiraishi had, after studying at the University of Tokyo, studied in USA under Professor Barr (one of the technical experts in the building of the Panama Canal and appointed by President Roosevelt). After his stay in USA, Shiraishi also studied at the Technical University in Berlin, before he came home and became professor at Tokyo Technical University for four years. After that, he started his commercial career. Shiraishi also wrote a report about the warehouse for the Society of British Engineers.

The first reinforced concrete bridge in Japan was built in 1903, designed by Sakuroh Tanabe (1861–1944). The bridge got the name Bridge No.11, and is located near the Biwa-lake, not far from Kyoto. This little bridge is 7.28 meters long and 2 meters wide, with a plate thickness of 30 cm.

The first book on concrete in Japan was written by Tadahiko Hibi (1873–1921) and was published in January 1916 (The Theories and Applications of Reinforced Concrete). Hibi had studied in Germany from 1902 and quite a bit of the 760 pages of the book were translations from German literature. Hibi also published voluminous books on concrete in 1918 (694 pages) and in 1922 (417 pages). The publisher of the books was Maruzen, which still exists.

The development of concrete around the world picked up speed from the beginning of the 1900s. Modern communication helps in the fact that developments from other countries are more often and quickly adopted. In an article in *Nordisk Betong (Nordic Concrete)* in 1985, Arto Korhonen illustrated this regarding the development in Finland [96].

Examples of introduction dates for concrete technology innovations in Finland:	Developed	Introduction to Finland	Delay
1 Portland cement	1824-1848	1869-1914	60 years
2 Reinforced concrete	1850-1895	1905-1915	40 years
3 Precast concrete elements	1870-1910	1950-1965	60 years
4 Concrete bridges	1870	1911	40 years
5 Prestressed concrete	1900-1930	1945-1950	30 years
6 Ready mixed concrete	1925-1930	1958	30 years
7 Chemical additives	1930-1935	1945-1965	20 years
8 Fly ash	1935-1940	1965-1970	30 years
9 Silica fume	1970	1981-	10 years
10 Densit, MDF	1980	1990-1995?	

International multilingual dictionaries on concrete technology are one of the tools that have helped to increase the technical understanding across borders. As an example, we mention the two books that had been used as guides when this book was written:

- *Betongteknisk ordliste (Technical Concrete Dictionary),* published in Stockholm in 1971, with translations in Swedish, English, French, German, Finnish, and Norwegian [97], with 1310 terms and definitions.
- FIP's *Multilingual Dictionary of Concrete,* published in 1976, with closer to 1500 terms and definitions in English, French, Spanish, Dutch, and Russian [98].

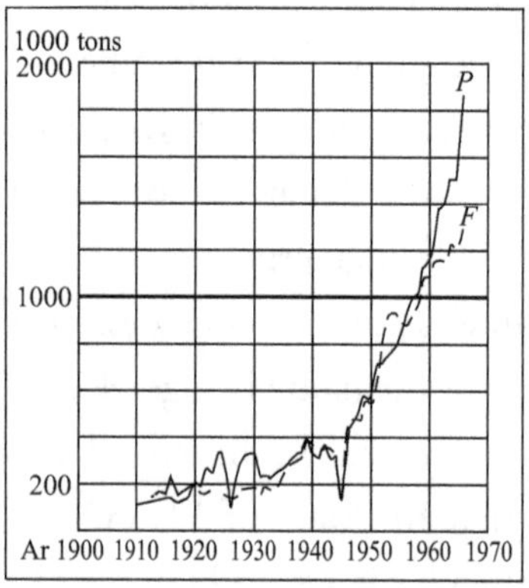

Growth in the Norwegian production (P) and consumption (F) of cement over the 50 year period from 1920 to 1970 was just as steep as that of the rest of the world [7]

In the later decades, the Internet has further increased the accelerated spread of technology, but we have to admit that only very few of the old documents from conferences and similar documents from the past are available in this media. Net-based databases and library bases, however, have improvised on this.

From the Norwegian cement curve, we can see that export of cement has always been an important activity. Today, Norway also imports close to 10% of the annual consumption. Norwegian export in 1966 was nearly half million tons, out of which 53% went to West Africa, 40% to USA, and 7% to Europe [7].

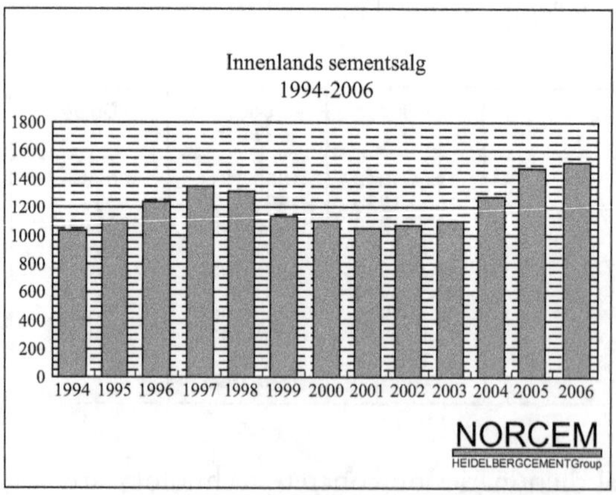

In later years, the Norwegian cement consumption has been more stable [99]

The numbers from Norway is just an example on how, over the years, cement has become an international commodity.

In the same way as in the development of cement production, we have also seen a fascinat-

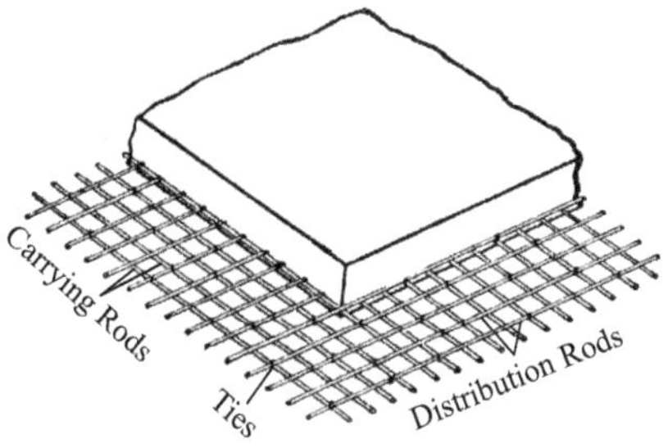

In 1865, Monier patented a system in which he had both a main and a secondary reinforcement (As Monier had not understood the importance of placing the reinforcement in the tension zone of the concrete, Loov is of the opinion that it is wrong to call him the inventor of reinforced concrete)

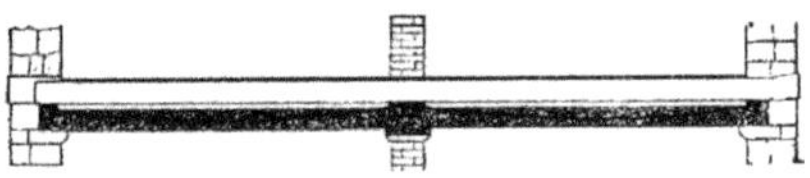

Wilkinson from Newcastle patented his system in 1855

ing development in another main

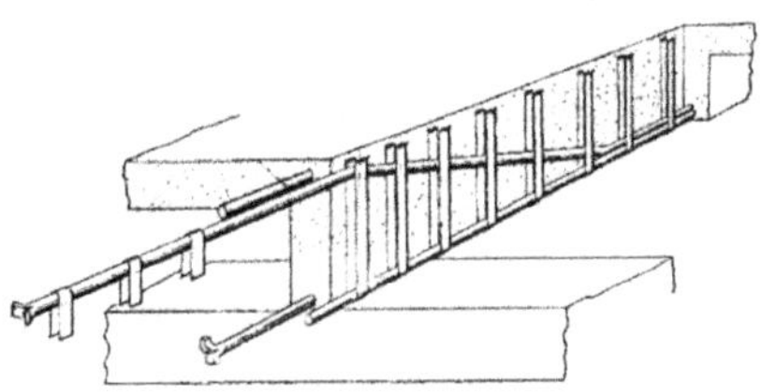

French Hennebique took out his patent in 1898. He was the first to use stirrups, and his system contributed to the development of the beam–slab systems. With reference to Mcbeth, Loov is of the opinion that in 1917 there had been 35 000 contracts that had used the Hennebiques system.

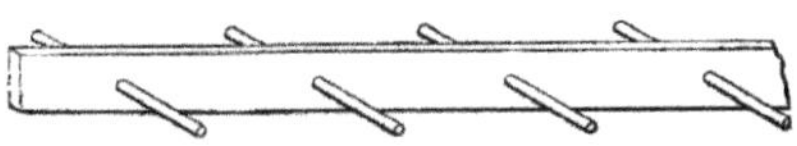

T.P.Hyatt invented the first special bars, in which the side bars were punched into a flat main bar

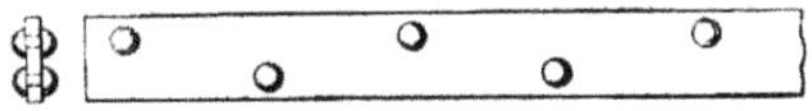

A similar system was invented by Thatcher

Klett invented a system in which he used angles bolted to the main bar

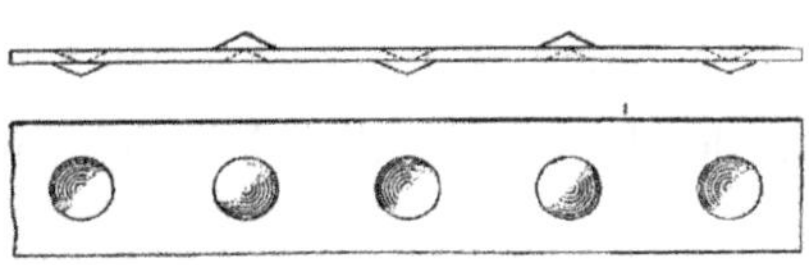

Staff used flat bars with circular punching

In 1884, Ransome patented the twisted bar. When he first presented this to a technical audience in California, he was laughed at. His workers twisted bars up to 50mm in diameter. He estimated that the cost of the operation was about 75 cent per ton.

A reinforcing bar, patented by De Man in 1898, using a crimped flat iron instead of a twisted bar

The so-called "lug bar" was twisted, with hexagonal knots, so that the bar cannot be straightened during tension

The "cup" bar has a shape that resembles modern bars a bit, but with a somewhat exaggerated corrugation

Corrugated Bar Company Inc. produced corrugated reinforcement, both with a rectangular section and a circular section

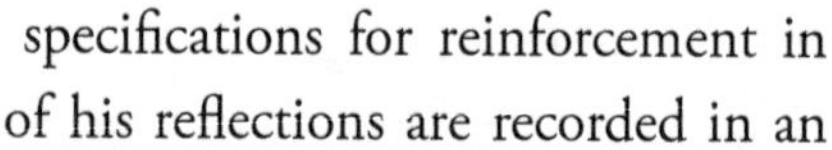

The last one has a shape that is close to modern bars

component of concrete, the reinforcement. We have tried to give a glimpse of this by citing from two references:

Gustav G. Erlemann studied the specifications for reinforcement in USA from 1911 to 1968 and some of his reflections are recorded in an article published in 1999 [100]. He points out that when the strength of an old structure is evaluated it is important to test the steel to check its quality . The yield point has roughly increased from 300 MPa during the time before 1910 till 1920 to 520 MPa for some steel qualities in 1968.

In an article in 1991, Loov refers to the development of the reinforcement about 100 years ago [89].

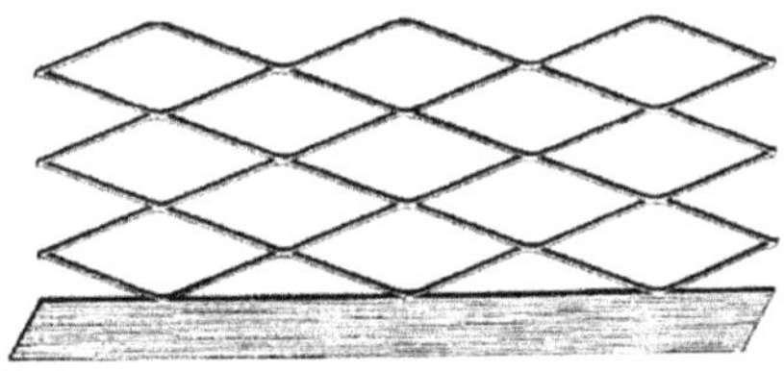

Sheet metal reinforcement sold by Expanded Metal Companies. The solution was among others tested by an institute in Chicago in 1906.

He refers to Winter who is of the opinion that 144 different reinforcing systems were in use in Europe in 1909.

Also, other types of reinforcement were developed, for example, shear reinforcement fixed to the main reinforcement. Sheet metal reinforcement also came into use.

George Hoff gives, in a paper presented at Seville in 2009 [101], the following very rough, but very interesting summary of the general technical development of concrete:

- 1913: Ready mix concrete
- 1917: Duff Abrams' water/cement ratio law
- 1920: Lightweight concrete and sprayed concrete
- 1925: The use of concrete pumps and vibrators
- 1930: Prestressed concrete
- 1938: Air entraining agents in concrete
- 1952: Silica fume
- 1955: Chemical admixtures, the breakthrough in precast components, slipforming
- 1960: Fiber reinforced concrete
- 1980: Roller compacted concrete
- 1991: Fiber reinforced polymers
- 2001: Pervious concrete
- 2003: Self-compacted concrete

The year mentioned by Hoff for the beginning of some of the technologies might be a bit inaccurate, but his summary gives a reasonable overview of the breakthrough in general in the most important markets.

The large step forward in mix-design technology came as late as in the 1970–1980s Then, as far as the cement paste is concerned, more happened over a 10-year period than during the last 50–100 years.

Akkaganse Bashi, Span 46 meters

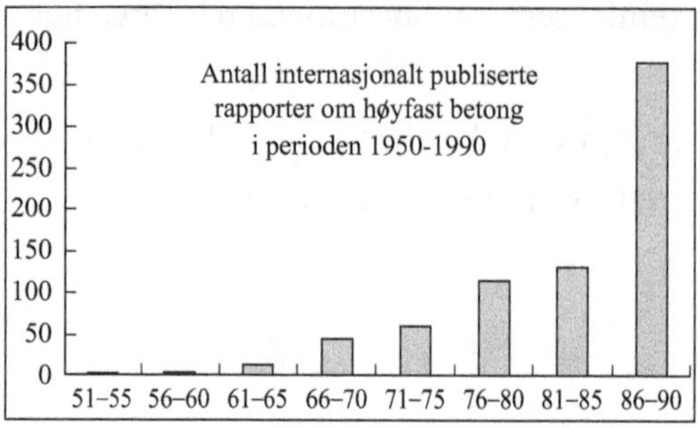

"Fra Farris til Troll" [49] shows this development in the international interest for high strength concrete by counting the number of international reports available in the period from 1950 to 1990. The development in generally accepted compressive strength of concrete in MPa.

The first super high strength concrete, with a compressive strength of over 100 MPa after 28 days, however, was produced by Yoshida in Japan as early as in 1940 [102]. Yoshida dehydrated his concrete down to a water/cement ratio of 0.23 under a pressure of 10 N/mm^2 and cured the concrete under high pressure and temperature.

The next step in Japan to produce extreme strengths was to reduce the water content in the concrete by using superplastizisers, add silicium powder (silica fume or grinded quartz) in fairly large dosages, and then autoclave the concrete.

In the 1970s, a number of structures with strengths between 90 and 100 MPa were built in Japan [103], [104]. Examples of this are Akkaganse Bashi, a railway bridge built in 1974 with a cylinder strength of 96 MPa and a slump of 15 cm, and Iwahana railway bridge, also built in 1974 with a cylinder strength of 90 MPa.

Spun piles and pylons in concrete with strengths of 60–80 MPa for nonautoclaved product and 80–100 MPa for autoclaved products were rather normal in Japan at that time.

Norway also became one of the forerunners in high strength and high performance concrete in the 1970s and 1980s. Some of this is mentioned in Chapter 4.7 "Silica fume".

In USA, the knowledge about higher strengths was, among others, utilized for foundations and columns in skyscrapers. Chicago Tower Place in 1972–1975 and Chicago Marriott Hotel in 1976 used cylinder strengths of 9 000 psi (62 MPa) and Columbia Seafirst Centre in Seattle, Washington (76 stories – 291 meters) had a strength of 10 000 psi (70 MPa) in the columns after 56 days.

The high strength concrete that is most used for practical purposes is probably what has been named RPC (Reactive powder concrete). This is really a mortar in which there is a possibility of the aggregate, sand or stones, or the contact zone between the aggregate particles becoming the weak link.

A typical RPC recipe (per cubic meter) might be:

- 705 kg cement
- 230 kg silica fume
- 210 kg crushed quartz
- 1010 kg sand
- 17 kg superplastizicer
- 140 kg steel fiber
- 185 kg water

The first RPC bridge in the world, a precast prestressed footbridge in Sherbrooke, Canada, Concrete strength of 200 MPa [99]

The concept was developed in France by Pierre Richard, former research director in the big contracting company Bouygues [8].

A similar concept has been used by Taisei Corporation in Japan. They

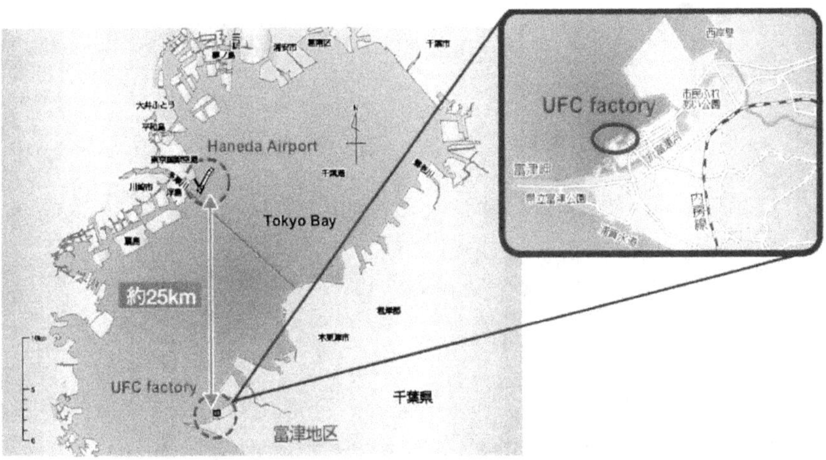

Production of UFC concrete with strength 210 MPa

The many hundred years old trees at the Confucius temple in Qufu in the Shandong province in China is repaired with concrete

call it Ductal UFC (Ultra high strength Fiber reinforced Concrete). In 2009, they built a prefabricated runway of 520 000 m^2, where 192 000 m^2 was with UFC concrete, for the Haneda airport in Tokyo. The concrete has strength of 210 MPa. The factory that produced the "super-concrete" components was situated on the other side of the Tokyo bay, 25 km from the airport [105].

The borders, frontiers, and records are still being broken in the use of concrete and it will continue to be so in the future. When Burj Dubai was officially opened on 4 January 2010, it was announced that it was the world's highest building. The comments in connection with the opening were many, and of different types, from admiration for the high engineering art to sour comments about human stupidity and vanity.

But the tower is definitely a landmark, with a total height of 600 meters and 160 stories, and with a concrete core up to a height of 575 meters. To produce such quantities of concrete in a quality 100 MPa is definitely an engineering achievement.

Some more data regarding the concrete used for the tower [101]:

45 300 m^3 concrete in the foundations.

A 3.7 meter thick foundation slab with 12 300 m^3 concrete.

650 concrete piles with 90 cm diameter and lengths up to 36 meters + 200 piles with diameter of 1.5 meters and lengths up to 50 meters.

But concrete is much more than this and most often something completely different from what is used in "Burj Dubai."

Over the years, concrete has been used in the most incredible areas. The following chapters try to give a picture of some of this.

Apart from other sectors that have been mentioned, there is a reference to the Liberty Tree in Maryland, which was repaired with concrete in 1907 and has survived at least 90 years afterward. This has been mentioned in a paper in 2004, Luke and Billie Snell [106]. The pictures above, however, are from the repair of another famous tree.

2.4 Norwegian concrete

In 1968, in connection with the 50th year anniversary of the cement factory at Dalen, Kåre Fasting and Frithjof Gartmann wrote an impressive anniversary book [107]. In the book, they also refer to other incidents that were important for the first époque in the early concrete history of Norway. Quite a few of the episodes mentioned below have been taken from this book.

Norway's first important geologist, Theodor Kjerulf, who got international acknowledgment for his thesis "Das Christiania Silurbecken" in 1855, was a keen writer about lime and possible cement production in Norway. Among others, he wrote articles in the Norwegian journal *Polyteknisk Tidskrift* in the middle and at the end of the 1800s. Kjerulf became Professor in Geology in 1858, and he was one of the founders of Norges Geologiske Undersøkelser (Norwegian Geological Investigations).

Burning of lime was a very old business in Norway, based on both fat and lean lime deposits. Kjerulf expressed strong criticism for the Norwegian way of doing it because of the too high consumption of fire wood and very inefficient kilns. Kjerulf enlisted the many places that have excellent conditions for lime burning; Aker, Asker, Eike, Sandsvær, Ringerike, Hadeland, Toten, Hedemarken, near Holmestrand, Langesund, Brevik, Porsgrunn, and Skien.

During Kjerulf's time, there were two cement factories in Norway, at Langøen in Bundefjorden and at Bagaas Brug by the Aker river (both in and near Oslo, which at time was called Christiania). Kjerulf was in particular, interested in the cement from Bundefjorden. He states that they there worked with clay-infected lime balls, the so-called "cement-balls," which were the same as the ones used at Boulogne-sur-Mer in France. Kjerulf states, among others, the following about the origin of the balls.

> ···*But, lime intimately blended with a certain amount of clay that gives cement. That is why it is often the case that lime-balls are found in rows in clay slate, over and under a continuous layer of lime, just what is called cement balls, or hydraulic limestone. This is amongst others found in England, in the Lias-formation and the there-over following Juva it is lime-concretes or balls or knots of clay blended limestone – the so- called -Cement-stones –*···

Kjerulf avers that this type of stone could be found in several islands in

Bundefjorden.

As mentioned in an earlier chapter, the Langø cement was appreciated in the World Exhibition in Paris in 1855. The jury amongst others said:

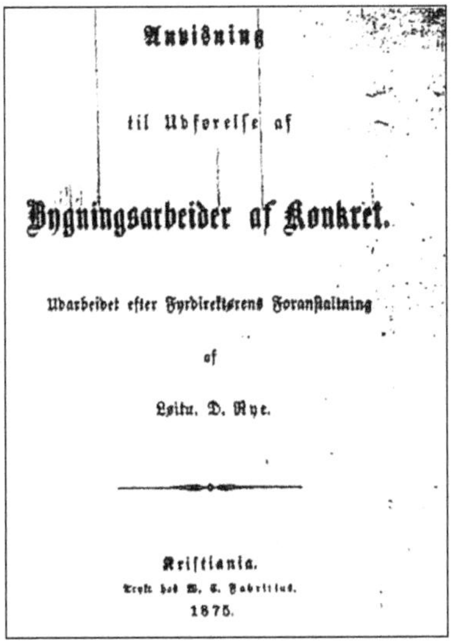
Anvisning

til Udførelse af

Bygningsarbeider af Konkret.

Udarbeidet efter Fyrdirektørens Foranstaltning

af

Lieutn. D. Rye.

Kristiania.

Trykt hos W. C. Fabritius.

1875.

Rye's specification from 1875

> … *Its Loss by Calcination was 8.3 pCt. It is powderish, of deep yellow Color. Before Calcination the Sample gave a Stone, that after setting cracked in Water after 24 Hours; this was only due to that the Powder had been in the Air, because after Calcination it was setting after 5 Minutes, and thereafter a considerable Heating took Place. This Cement is very light. According to its Constituency, this Product belongs to the Ordinary cements. It is of good quality, and the production should preferably take place in a bigger scale.*

The jury's comment about the Bagaas cement did not suggest that the production should be increased. The cement was characterized as being in Vicats class,"chaux limites," and it was said that the content was close to that of the English Portland cement.

In 1867, there was a discussion in the Polyteknisk Forening (The Polytechnic Association) about cement, with an introduction by the then Lieutenant, later Director General of NSB (Norwegian State Railways), L.H. Segelcke. The discussion concentrated on the quality of Norwegian cements compared to imported cements.

Another well spoken prophet of concrete in Norway was the lighthouse Director (Fyrdirektør) Rye, who talked about the experience in building lighthouses along the long and rough Norwegian coast [6].

Lieutenant D. Rye had already in 1875 made a specification for the use of concrete [108].

In his specification, Rye among others states:

1. *Concrete is a mixture of lime-porridge and sand and stone.*
   This mixture is filled into moulds, until it is hardened. In this way according to the circumstances, the walls are molded either by blocks that

are used as normal stones, or the complete wall, or as movable moulds are used, a complete monolithic structure is being built.

2. *As lime-porridge of ordinary lime only stiffens very slow as the water evaporates, and only after very long time gets any strength, it is not recommended as a binder for concrete. You have to use hydraulic lime. The transition of this as the water evaporates, gives a chemical process, which happen shortly or at longer time after the evaporation of the water. This time is different for different types of lime, from minutes to several days, depending on the quality of the material. For concrete works above water, a material which are binding after a few hours are to be recommended.*

3. *As well as with respect to reasonable binding time, as strength, the so-called Portland Cement is the best binder for concrete ···*

4. *Portland Cement is produced by burning of lime and silica infected clay in a particular ratio (about 22 pst. Clay, 78 Pst. Lime) ··· With stronger burning, the longer the binding time, and the stronger the cement becomes. ···*

5. *That the Cement is finely ground, is important to take into consideration, as courser grains will not be properly wetted by the added water, and therefore not behave like cement, but like sand, or even worse, react after the other mass has hardened, and thereby contribute to cracking. ···*

6. *Sand for concrete should, as other masonry sand, be sharp and clean. A mixture of coarse and fine sand is advantageous. ···*

7. *Water for concrete must be clean. Fresh water should be used; however, the strength is only mildly reduced if sea water is used. This is an advantage that Portland Cement has before other hydraulic limes ···*

8. *The strength of concrete is rather moderate just after binding, but increases over time. 7 days old concrete of good Portland cement is normally 1/3 as strong as 1 month old concrete ···*

Rye's specifications run to 15 pages with 19 recommendations.

Rye was born in 1832 and became a sea cadet in 1846 (12 years old). Thereafter, he became the first officer in the merchant navy. From 1860 to 1863, with a Norwegian scholarship, he was in the English navy and among others involved in crushing a Maori-upraise in New Zealand. When he came home, he was first a captain on one of the government's mail ships along the coast and got a very good opportunity to watch the weaknesses of the Norwegian Lighthouse system. He was also for some time involved in the fishery control navy. Before 1866, there was only one assistant to the lighthouse director and he was a marine lieutenant. From 1866, the government also provided salary for a second assistant and Rye was employed as the sea expert. In 1874, he was in England and studied about Portland cement.

As a result of this study-trip, came among others his book or recommendation.

Already in 1872, Lighthouse Director Dirik had recommended to the department that Rye, for his work, should have an increased salary.

In 1882, Rye became the director general of Norwegian Lighthouses and in this position he was responsible for a considerable upgrading of the lighthouses, the adjoining buildings, and harbors at the many rocks and islands along the coast. He was in this position until 1909, when he was 77 years old. Rye died in 1914.

Rye was an active person. He was also chairman and honorary member of The Polytechnic Association. He was honorary member of Association Internationale de la Marine and Honorable Fellow of Trinity House, and when The Rescue Association was founded in 1891, he became the first Chairman [107].

The coast was a very important market in the early period of the early new modern cement plants in Norway. Of the nearly 200 lighthouses built from Lindesnes in 1655 to Tvistein in 1908, about half of them were built during the 42 years that Rye was in the establishment.

The first lighthouse at Lindesnes (the south point of Norway) was, however, built privately by a man, who kept it running with the help of the finance from a little lighthouse tax. When in the fall of 1656, the owner of the lighthouse ran out of coal, he lit 30 large candles and kept them going for a long time. After this incident, it was closed down until it was activited again in 1720.

At the cement museum in Slemmestad, we found a picture of the eastern breakwater in the Vardø harbor, built in 1894.

The picture and its caption at the museum reveal that the end of the

The eastern breakwater, Vardø harbor, built in1894

1800s was an important and busy period for the building of communications in Norway. This was good for the new cement factory at Slemmestad; Christiania Portland Cementfabrik made a good start in its production. The harbors along the Norwegian coast had been enormously influenced by wind and weather. To overcome the technical challenges, aid of expertise from abroad was sought at the same time as when the Norwegian engineers went out to learn. In 1900, the development had proceeded so far that they wanted to show it to the rest of the world. Norsk Havnevesen (The Norwegian Harbor Authorities) took part in the World Exhibition in Paris with a number of pictures. Christiania Portland Cementfabrik (CPC) presented their cement in the same exhibition. The accounts of CPC show that they in 1896 delivered considerable quantities of cement to the Government Harbor Authorities in Vardø.

In his book *Concrete Progress - From antiquity to the third millennium,* Idorn [40] opines that the breakwater in Vardø was also part of a Nordic test project, in which test samples were set out in Vardø in Norway, Degerhamn in Sweden, and Esbjerg and Thyborøn in northwest Jutland in Denmark. A report with 27 pages and 21 attachments was presented in English, French, and German at the "International Association for Testing Materials," fifth Congress in Copenhagen in September 1909. This was a joint research project between the Nordic cement producers. The blocks that were exposed to nature's fury were not only cast with different Nordic cement types, but also with various cement substitutes such as Santorin earth, Neapolitan and Roman pozzolan, trass and slag. This must have been research and testing that were not only far ahead of their

Concrete blocks at Thyborøn at northwest Jutland in Denmark [38] 116 test blocks of 0.6 m^3

time, but were also remarkable as practical, international cooperation in testing and research even today.

The first major Norwegian *Handbook in Building art (Haandbog i Husbygningskunst)* came in 1893. The book was written by the Engineer Captain Edv. Kolderup – a lecturer at the Military Higher Academy. He wrote that concrete was going to play a major role in construction in the future. But, he probably did not envisage that concrete was going to be the most important building material in the world, with a volume that is double of all the other building materials together. (Today, we annually consume a volume of concrete that amounts to a mountain block with a base area of 1×1 km and the height of Mount Everest).

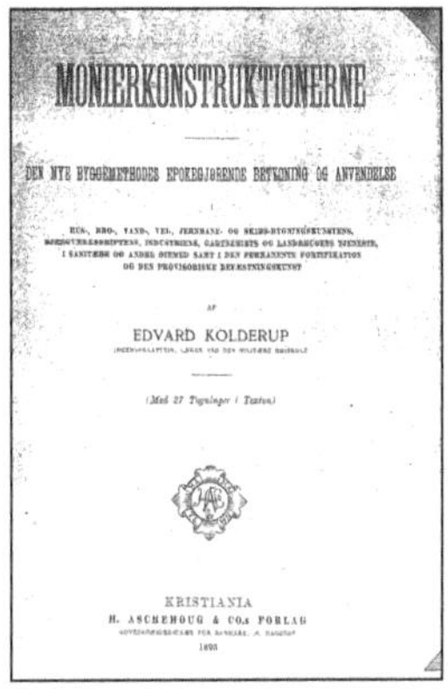
MONIERKONSTRUKTIONERNE

DEN NYE BYGGEMETHODES EPOKEGJØRENDE BETYDNING OG ANVENDELSE

AF

EDVARD KOLDERUP

(Med 27 Tegninger i Texten)

KRISTIANIA

H. ASCHEHOUG & CO.S FORLAG

Kolderup's book

We quote a few short passages from the preface to the book:

Not only in our country, but also in Sweden and Denmark, there are still very few people that are familiar with the new Building method with iron and concrete, that often are named after the Inventor and is called Monier Constructions.

These structures, however, has got a wide use abroad in the last 5 years, and will be even more popular in the future, as they have so many excellent properties, and will be of great importance in the future.

They has now by Authorities been named; "The building method of the future".

···In particular we will mention the possibilities in the art of house-building, the art of bridge-building, the art of road and railroad construction, water building art, mining, shipbuilding, in the industry, in nurseries and agriculture, in sanitary works and for other purposes, and fortifications and intermediate military structures.

···, so we can with full right, claim that Monier structures are the most genius invention, that has taken place in the art of building.

···Amongst the many I will propose to study about Monier-constructions, are; Architects, various types of engineers, contractors, building owners, industrial-

ists ··· medical doctors, officers, but most important, there should be education about this new construction technology at our technical schools.

There is no doubt that Captain Kolderup had been very enthusiastic after his study-trip to Germany, Austria, Switzerland, and Italy.

Koldrup had given a number of recommendations and comments, describing the level of the technology only a few years before we enter into the twentieth century. Here are some of them:

- Typical dimesions for the reinforcement are diameters of 0.5 to 2.5 cm.
- "Allowable stress for the reinforcement is equal in tension and in compression and can be set to 750 kg per cm^2."
- "The Cement mortar" is made of one part of Portland cement and three parts of "sharp, coarse, clean Sand" (to be sieved on a mesh width of 2 ½ cm).
- The strength of the concrete in compression is 250–270 kg per cm^2 (the tensile strength is 1/10 of this).
- The concrete shall have the consistency of clogging snow.
- The formwork should stay for 5–6 days.
- The binding thread has a diameter of 0.1 cm.
- The overlap length for the reinforcement should be 3–7 cm, with binding thread around.

It is also interesting to look at Kolderup's listing of the advantages of concrete as that of an enthusiast's, who noted it more than 100 years ago.

1. Long life
2. Fire safety
3. Great load carrying ability
4. Great resistance against shock and vibrations
5. Space saving
6. Savings in foundations and anchorage
7. Faster construction

8. Sanitary advantages
9. Low cost investment
10. No maintenance
11. Absolutely water and air tight
12. Bullet proof
13. Dryness
14. Safe against burglars
15. Great shaping possibility
16. Aesthetically advantageous

But, there were also other enthusiasts for this new material.

Two of the earliest competitors in the production of concrete were the Engineers S. Høyer-Ellefsen and J.Fr. Selmer, who started companies with concrete floors as their specialty, and left their footprints in the Norwegian concrete history by initiating the greatest construction activity in Norway which went on for over a hundred year period. The arguments they used were the ones of Kolderup.

When the Norwegian Contractors Association had their 25th year anniversary in 1985, the journal *Byggeindustrien* issued a special edition. In this, Åage Møst wrote an article about *grunderne i norsk entreprenørvirksomhet (the founders of the Norwegian contracting industry)* [109]. A major part of the following paragraphs comes from Møst's article;

In 1888–1889, Kristian Stensrud, who was from farming and craftsmen background, came to Fåberg in Gudbrandsdalen to try his luck in Kristiania (now Oslo). He had his craftsman-education from his father and a good certificate from the local secondary school, and over time he also got theoretical education from Kunst- og Håndtværkskolen (The Arts and Craftsman Academy) in Kristiania.

He came in contact with an active businessman, G.C. Boy, and got works from him. In April 1894, he started, together with his brother-in-law, Brick master L. Hansen, his own contracting business. Boy was also a key person in the cement industry and Stensrud very soon got the job with important extensions at Christiania Portland Cementfabrik at Slem-

Selmer's first regular registered project, a new packing building for Nitedals Tændstikfabrik (match factory) in 1907

mestad. Later he got more important assignments.

In 1906, 30-year-old Fredrik Selmer, came from Fåberg, came to Christiania. He had a good background, with a certificate in Civil engineering from Trondhjems Tekniske Læreanstalt (Trondheim Technical Academy) in 1896, and later two years studies in Dresden in Germany and practical experience from Germany, Austria, and USA.

Selmer marketed his specialties; floors, stairs, roofs, columns, bridges, foundations, silo and similar structures made of reinforced concrete. The first job he received was as contractors for floors and stairs at Majorstuen School in Oslo. (Majorstuen school, was the largest school in Norway after the Second World War, with nearly 2000 students. The school celebrated its 100th year anniversary in 1998).

In 1895–1896, Sigurd Høyer-Ellefsen sat in his newly established consulting company, and designed the dome on the roof of the National Theater. He was from Christiania, with civil engineering certificate from "Christiania tekniske skole" in 1899. After that he studied at the Technical high school in Charlottenburg near Berlin (The same time as Max Berg – see Chapter 3.1.5).

The well-known ski-jumper Sigurd Høyer-Ellefsen wanted to come back to Oslo (in addition to his impressive prize collection, he was known for his 25 standing jumps in the Huseby ski jumping hill on one single Sunday morning).

Sigurd Høyer-Ellefsen started his contracting company together with Wilhelm Stenersen as a 26 year old by competing for the contract of the Holmenkollen subway line from Majorstua to Besserud, and they got the job. They built it with double tracks from Majorstua to Slemdal, and single tracks further to the Holmenkollen station for the sum of 200 000 kroner (at the time 10 000 GBP).

The fourth pioneer was Haakon Eeg-Henriksen from Bergen, who was somewhat younger. Born in 1892, Eeg-Henriksen completed his study at Bergen Technical School in 1910 and the military academy in 1911–1912, before he completed his education as civil engineer at the newly started Technical University in Trondheim after three years, in 1915.

Most Norwegian civil engineers educated from the Norwegian Technical University, at least know him as the Regent for numerous years of the Civil Engineering Student Union "Århønen". During my first meeting

In 1908–1909, AS Høyer-Ellefsen built the Skjefstad waterfall Electricity station near Elverum, with among others a 415 meters long turbine pipe, with an internal diameter of 3.6 meters

with him, in 1964, he called the civil engineering department at the university at his time as "the America Department", as a great number of the students went to USA after finishing the studies.

Before he finished his studies, Eeg-Henriksen gained practical experience for a year in Poland.

Eeg-Henriksen's first job as a contractor was to transport the 265 tons "raw-Monolith" through Oslo from the sea to the Vigeland park, where it still stands as the central monument. In the silver jubilee commemorative journal *Byggeindustrien* of 1985, Eeg-Henriksen wrote an article about history and stories from the pioneer times *Fra de gode gamle dager (From the good old days)*[110]. He states that the Government Lighthouse Department had already used cement in 1870, and the Government Harbor Department in 1880, but the development in the use of concrete was very slow.

When the National Theater in Oslo was built in 1890, concrete was used at the deepest pile foundation heads, but the foundations were constructed with natural stones. About 1900, it became normal to use concrete instead of brick arches between the iron beams in fireproof floors and roofs. In 1910, they dared to use concrete in walls and reinforced concrete in the floors, but not for apartment houses. Only in 1920, it became normal to use reinforced concrete in apartment buildings. Eeg-Henriksen was of the opinion that the reason for the slow development came from the old frontiers between the strong trade unions. The brick layers held a very strong position, as well as the carpenters, while the concrete union had a low status.

Haakon Eeg-Henriksen – The Regina (the bird) and the Regent, at the abdication of the Regent from the Civil Engineering Student Union at the Technical University in Trondheim in 1965

Eeg-Henriksen always had stories from the construction field to narrate. Here is one of them;

"Tried the soil with my walk-ing-stick"

For some time, I was advisor for an insurance company, where my job was to evaluate the risk where a new building was to be constructed close to an old building. To study the archives, was part of the job. In an old document, the building inspector, gave approximately the following evaluation; Inspected the X-site where the plot had been excavated. Tried the soil with my walking stick, that easily could be injected into the soil. Ordered a double raft to be built.

Norwegian concrete technology was far behind compared to many other countries at the beginning of the 1900s. Most of the people, who later became the most important ones for the concrete development, had therefore a longer or shorter practice period abroad. As the "nick-name" of the civil engineering department at the Technical University suggests, many went to USA and many also settled abroad.

Another example of learning abroad; the same year as the Norwegian Technical University (NTH) was opened by King Haakon the VII, on 15 September 1910, Nils Christensen took over as manager of the contracting company "Trondhjem Cementstøberi og Entreprenørforretning." The next year, he went on a study trip to Germany. He learned about reinforced concrete and, in 1912, his company built the first reinforced concrete building in their area [52].

The first Norwegian production of reinforcement bars did not come until 1928, when the company Christiania Spigerverk in Oslo started to produce"monier-irons." Before this, all reinforcement had to be imported.

Testing of a reinforced concrete beam in 1904

The first deformed bars were produced in Norway from 1953.

The production of welded prefabricated nets started in 1961.

It was not until 1923 that the Norwegian Technical University in Trondheim (NTH) found out that they needed a separate professor for handling the subject Concrete. But it was not until 1930, when Anton Brandtzæg (NTH 1916) came home from USA, that the position was filled. Meanwhile, the teaching was taken care of by Edvard Harboe.

However, "research" was done even if there was no professor. The anniversary book *Fra Farris til Troll* [52] states this by citing the testing of concrete beams carried out in April 1904 as a proof. The testing of the reinforced beam had a loading of 21 tons in the middle.

When Brandtzæg's successor was Professor Inge Lyse (NTH 1923), when he returned from USA in 1936.

Sven Thaulow

Brandtzæg, who finished his education at NTH in 1923, worked for two years at A/S Jernbeton, Trondhjem, before he left for USA. After six years as a professor teaching concrete at NTH from 1930 to 1936, he got the job as managing director in A/S Høyer-Ellefsen, one of Norway's leading contracting companies, where he worked until 1939. In 1940, he started his own contracting company in Trondheim – A/S Anlegg. Then in

1952, he returned to NTH as professor in harbor constructions and civil works.

There had been many pioneers in the development of the concrete technology and activity in Norway. We like to mention two important enthusiasts.

- **Sven Thaulow** was by most people known as "The Manager of the Cement–Association" and as author of numerous publications.

During his 13 years with the Norwegian Cement Association, he among others was active in 33 committees regarding cement and concrete [6] and he was the obvious choice as the editor of the journal *Betongen I dag (Concrete today)*.

Thaulow was educated as a civil engineer at NTH in1924. After having worked as a consultant in Rio de Janeiro (1928–1931), Oslo (1931–1934), and with the Oslo Harbor Authorities (1934–1946), he came to Christiania Portland Cementfabrik at Slemmestad as building manager in 1946.

In 1955, he became the manager of the cement industry's technical information service. He later took the initiative to establish the Norwegian Concrete Technology Institute at Blindern in Oslo and was the first general manager of the institute from January 1967. Unfortunately, Thaulow died in September the same year.

In his younger days, Thaulow was also a good swimmer, with three individual Norwegian championships and three in water polo. He also participated in the Olympic Games in Paris in 1924. Thaulow managed to combine his interests for swimming and concrete. Among his books, we find *Friluftsbad og svømmehaller(Open air and indoor baths and swimming-pools)*, 1948, and *Bad og svømmeanlegg (Baths and swimming-pools),* 1965 [43][111].

Hans Henrik Thaulow looking at the original drawings for the concrete house in Konglefaret

When the Thaulow family moved from Slemmestad in September 1957, they moved into a concrete house in Konglefaret in Bærum, one of

Thaulow was one of the planners behind the bath in the Frogner Park in Oslo, designed by the architects Frode Rinnan and Olav Tveten, which opened in 1958. This was the first major and complete outdoor swimming pool in Norway and a breakthrough in this type of use of concrete in Norway [112]. Hans Henrik, the son of Sven Thaulow, said in a private conversation in January 2010 that the first plan for the project was made before the Second World War, probably during 1938–1939.

the suburbs in Oslo, built from Sven Thaulow's ideas, with good help in the planning from his wife and designed by Architect Hans Skårde. Thaulow's son Hans Henrik lives there now. Thaulow was far ahead of his time in this respect also. In the floor, over the basement, he laid copper piping for heated water. The Honnywell–Bull managing system that was installed to manage the heating must have been one of the first in a private home. This was a real concrete house, where even the bath tub was in concrete. Sven Thaulow's wife was a bit skeptical, but the bath tub was built. The present owner had considerable difficulties when, a few years ago, he installed a more modern tub. The concrete house built in 1957 is still in use, and the present owner, (Hans Henrik Thaulow) is not only very proud of the house, but also avers that only minor maintenance works have been necessary since it was built.

- **Johannes Rutle** was born in Fjærland in Sogn. He studied chemistry at NTH and started to work for Christiania Portland Cementfabrik in1930. After a crucial stint as chief chemist at the cement factory at Slemmestad, he started his own company, BETOKEM, in 1958.

It was not easy to convince the Norwegian concrete enthusiasts to begin to use admixtures in concrete, but Rutle was very well respected, had many friends in the industry, and was extremely enthusiastic. It is no doubt that he was the main factor in the development that led to Norway becoming a forerunner in the use of admixtures. I made an investigation in 1981 and found that close to 90% of all the concrete that was produced in Norway used admixtures of one kind or the other. This was, at that time, a considerably higher percentage than any other country (Ref. Chapter 4.5).

Johs. Rutle

In 1958, Rutle wrote the book *Cement-Fremstilling og egenskaper (Cement – production and properties).* A second edition came in 1980 and the phrase *andre Hjelpestoffer for betongarbeider* (*Other aiding materials for concrete works*) was added to the title of this edition. Many Norwegian concrete technologists have had great pleasure in reading the book and found it very useful. My personal, signed copy has an important place in my book shelf.

During his time at Slemmestad, Johs. Rutle also became a sort of a godfather for Norsk Fabrikkbetongforening (Norwegian Ready Mixed Concrete Association) (Ref. Chapter 4.2).

Norsk Betongforening (The Norwegian Concrete Society) was founded in 1955. In comparison to its models, NB is a young society (ACI - American Concrete Institute) and was founded in 1905 and the Swedish Concrete Society was founded in 1912.

It also took a long time before we got a separate cement and concrete research institute in Norway.

Carl Johan Bernhardt wrote an article in 1968 [113], in which he tells that "Forskningsintituttet for Cement og Betong ved NTH–FCB" (The Research Institute for Cement and Concrete at NTH–FCB") was founded on 1 January 1965. This was a extension of CAB (Cementindustriens avdeling for betongforskning ved NTH) (The Cement Industry's Department for concrete Research") that had been active for eight years.

Bernhardt, who was also a professor at the University, was the first manager of the research institute. At the beginning, the institute had seven employees.

At the 40th year anniversary of the Norwegian Concrete Society in 1995, the book *Fra Farris til Troll (From Farris to Troll)* [52] was published with Per Helge Pedersen as Editor. For more information about the Norwegian concrete history, we strongly recommend this book.

3 Structures

3.1 Historic structures

This chapter is neither a complete review nor a systematic investigation of important structures, but an attempt to show the versatility of concrete. All the examples, however, are historic landmarks and many of them have left their footprints in the history of concrete. The collection of landmarks demonstrates the fantastic manifold capability and the enduring strength of concrete. Many of the examples have interesting stories related to them.

3.1.1 NORWAY

The Namsen Bridge (Namsenbroa) (and the bridges in the Namdalen Valley)

Leir bridge in Grong – Norway's first reinforced concrete bridge, built in 1905

In 2001, the Norwegian Road Authorities department in the county of Trønderlag published an interesting book about the many bridges in the Namdalen Valley [114]. The book is fascinating to read and many of the old bridges have been built in concrete. Leir bridge in Grong was built in 1905 and is the oldest structure built using reinforced concrete in Norway.

Horka bridge, built in 1916, 24 meters long and 3.8 meters wide

Approximately around the same time as the Namsen Bridge was built, between 1912 and 1922, four other concrete bridges were built in the same district; Sandmoen, Horka, Opdal, and Skage bridges, while Bangsund bridge was built earlier in 1908.

The key person in the construction of all these bridges was Emil Bernhard Astrup. He was born in Lofoten in northern Norway and educated in Trondheim and Hannover. He completed his education in 1897. He also went on study trips to study the bridges throughout Europe in 1897, 1908, and 1912, with scholarships from the Government's Bridge Department.

Both the Leir bridge and the Namsen bridge were declared national heritage on 17 April 2008.

The Namsen bridge was an impressive structure at its time and was for many years the largest of its kind in Norway. Namsen is one of the biggest rivers in Norway and with respect to the amount of water it is just a little smaller than Norway's longest river – Glomma. The ebb and flow at the location of the bridge have a difference of about 3 meters, something that can be recognized 25 km away from the river moth. The ice sheets coming down the river in the spring can be up to 200 meter square and 40 cm thick. It is estimated that this might exert a horizontal force at the pillars, 700 tons [114].

At the location of the bridge, the width of the river is 850 meters during flood tides.

The Namsen Bridge

To reduce the length of the bridge, the river had been filled up so that the real length of the bridge became 213 meters. The bridge has four spans of 52, 52, 47, and 42 meters, and a width of 5 meters. The sailing height is 9 meters. The building of the bridge had its problems, and in particular the foundation work took a long time. The formwork for the three-hinge arches was made on land and floated into position. The floating process led to surprises and accidents, and was slowing down the construction work.

The Stavå bridge

Among the hundreds of concrete bridges in Norway, it is doubtful if the Stavå bridge can counted as one among the most historical, but it is an interesting bridge in the sense that the stories about the bridge and some of the people connected with it are interesting and this has made me include it.

The Stavå Bridge was constructed in 1941–1942 and is a road bridge at Berkåk in Sør-Trønderlag, about an hour's drive south of Trondheim. The bridge has a span of 90 meters and a height of 46 meters. Though it is not one of the largest bridges, building it would definitely have been a challenge nearly 70 years ago, a time when there was a shortage of most things.

One of the most difficult tasks of the construction was the scaffolding. In Professor Aksel Andersen's lecture notes of 1965 on wooden structures (the notes that the author used at NTH in 1963–1965), the scaffolding was particularly mentioned with drawings [115]. The scaffolding with the formwork for the concrete arches were built vertically as two separate arches at each side of the valley and afterward lowered into position by wires to form a three-hinge arch in the final position.

The Stavå bridge in October 1991

Aftenposten 07.03.1999 [116]

On 7 March 1999, the Oslo newspaper, Aftenposten [116], had a long article about the bridge with the subtitle *Hatbrev mot tyskerne murt inn i bro (Hate letter against the Germans cast into a bridge)*

During a repair of the bridge in 1999, they had found a letter, rolled into an issue of Aftenposten, dated 17 August 1942, and inserted into a liquor bottle.

Hatbrev mot tyskerne murt inn i b

Rullet inn i Aftenposten og lagt i en spritflaske, ble brobyggernes

Translated from Norwegian, the letter says;

From all the workers at Stavå bridge: We hate the German war machine and their attitude in Norway. For the time being we are raped and helpless, but we hope for better times. To them that comes after us! Remember this difficult Time, and take revenge for the injustice!

The letter has been signed by: Osvald Jahren, Ole A Økdal, Ottar Bjørnli, Alf Smørsgård, Johan Nyhus, Nils Belsvik, Hans Hansen Ole J. Bjørnli, Peder Smørsgård, Anders Horgmo, Kjartan Aspeggen og Jens Haugen.

I recognized my father's handwriting in the letter and contacted the road authorities in Sør-Trønderlag. From them I got a copy of the old yellow letter. At the same time, I made them aware of another story that my father had narrated about the construction work.

During the mining work for the foundations, they each day put aside a few pieces of dynamite from the dynamite account that was controlled by German occupation authorities. Later, probably a dark winter night, they made an explosion chamber in the foundation together with the secret resistance force (if the dynamite was actually placed there, I cannot remember from the story). The purpose was to have the possibility to demolish the bridge during a German withdrawal.

My father worked as the responsible site engineer at the bridge for the contractor Ing. Carl Pihl and I got my first lesson in holding the spade in the sand pit near the concrete mixing station at the Stavå bridge.

Bridges – new records are broken

In 1960, the Tromsø sound bridge was opened. With its 1036 meters it was at the time the longest in northern Europe.

The bridge got the award BETONGTAVLEN (The Norwegian Concrete Award) from the Norwegian Concrete Society in 1963. Like the many important bridges that were built later, it was designed by Dr.Ing. Andreas Aas-Jakobsen. In 1972, the Norwegian Engineering Association awarded him the important Eyde's Award for his work as "an extraordinary civil engineering designer" [117].

Andreas Aas-Jakobsen (1905–1980)[118]

Aas-Jakobsen was not only a remarkable designer of international standard and a first class academic in a number of sectors, but also one of the godfathers of the modern Norwegian concrete standard; he was the developer and inventor of a number of products related to concrete technology and the founder of a consulting company that has a number of important bridges in Norway and in other countries on their list.

The Tromsø sound bridge during construction

In 1992, two record-breaking bridges were opened in Norway [119]. Skarnsundet bridge in North Trønderlag was the longest in the world, when it was opened. The bridge also received the Norwegian Concrete Award and has the longest span of 524 meters.

Bergsøysund bridge in Møre and Romsdal is the first floating bridge without floating pontoons.

Skarnsundet bridge, 1992.

The Oslo Harbor Magazine (1916–1920)

The Oslo Harbor Magazine was, when it was built, the biggest concrete building in Europe. The building has 10 stories with a total floor area of

The Oslo Harbor Magazine

36 000 m^2 [120]. The building consultant was the firm Bonde and Normann, and the key person, E.A.Slettum, had many years of experience in USA. The structure received, 45 years after its completion, the Norwegian Concrete Award in 1965.

Concrete in roads

Pedersen [52]claims that the first concrete road in Norway was built in 1925 – a small stretch of 160 meters at Skøyen just outside Oslo. At this time, 600 millions of m^2 of roads had already been built in USA. We also know that between the First and the Second World War, a number of concrete roads were built in Norway.

The below picture was taken in Sandvika in Bærum, a city neighboring Oslo, in 1930. The cement factory at Slemmestad had supplied cement in bags for the construction of the road, and the picture was used to promote the cement. The concrete was transported by horse wagons to the site, so no wonder that the car on the picture looks a bit lonely. The cement factory at Dalen, further south, was very active in their marketing of their

Concrete roads outside Sandvika, 1930

cement for concrete roads and by 1936 a total of 33 km of the so-called "Holter-type" concrete roads had been recorded. The General Manager Alfred Holter was one of the founders of the factory at Dalen and was an active promoter of concrete roads. (See Chapter 5). These 33 km of concrete roads must be viewed against Norway's 450 km of "solid-surface-roads" and a total of 40 000 km of main roads in 1936.

In a lecture at the Annual Concrete Day in 1972, the Director of the Norwegian Road Directorate, T. Borchgrevink, described the development that took place until then [121]. He referred to a paper of Gunnar Trevland at the Concrete Day in 1970, which states that from 1930 until 1970 Norway had in total made 110 km of concrete roads. This is a very small number compared to other countries. For example, Sweden cast 80 km of concrete roads in Skåne in the 1960s, and in Europe as a whole, 200 km of concrete roads were cast in 1970. A paper by Ole Kr. Sylte presented at the Concrete Day in 1974 [122] confirms the low numbers as well as the fact that there was little progress in this sector during this period.

The Concrete Dams

It is very difficult to judge which one of the many elegant, fascinating, and large concrete dams in Norway is the most impressive.

In total, there are about 2 600 dams in Norway. In the old days, the dams were built to provide water for everything; from ice production to grain-mills, saw-mills, iron works, and similar needs. In the last century, the dams were being built to provide Norway with more electricity, which has rendered Norway unique as far as electricity production is concerned. While the rest of the world gets

Zakarias dam in Tafjord in Møre and Romsdal

The Førrevass dam in Rogaland

more than half of its electricity supply from coal fired power plants, Norway has been fortunate to get more than 90% of its electricity from hydroelectric power.

Most of the dams are either rockfill dams or concrete dams. Using concrete for dams has been a fascinating and exciting part of the development of concrete technology with numerous challenges in its design, material technology, and organization of construction.

Several Norwegian consultants and contractors have brought their experience in building dams in Norway to other countries across continents.

The first concrete arch dam in Norway was designed by Chr. F. Grøner in 1929. That was the Lake Storli dam with a height of 31 meters and a crown length of 60 meters. Later he constructed Lower Skjerka Lake Dam in West Agder with a height 18 meters and a length of 190 meters.

The arch dams are probably the most spectacular to look at.

Here are some examples of Norwegian dams; (dam height has been measured from the lowest point of the foundation of the dam to the top)

- The Zakarias dam in Tafjord was built by Høyer-Ellefsen in 1967–1969, and being 100 meters tall, it was the highest concrete dam in Europe.
- The Svarte-lake dam in Sira-Kvina has a height of 125 meters.
- Alta dam in Finnmark has a height of 145 meters and a length of 140 meters.
- The Førrevass dam in Rogaland is 95 meters high and 180 meters long.

The Concrete Off-shore Platforms

The concrete oil platforms in the North Sea have already made their way into the history of concrete. When someone writes a book about the history of concrete, hundred years from now, they will probably be a definite part of the book, keeping in line with Roman concrete and Smeaton's tower. They will stand as landmarks of Norwegian concrete technology in the world history of concrete.

Much has been written about the platforms in the last 30–40 years. We will only give a glimpse into the history.

Oil was found on the continental shelf outside Netherland in the beginning of the 1960s. In 1963, Norway demanded ownership for the Norwegian continental shelf, according to the part line principle. In 1965, the first concession for the investigations was announced and, in 1966, the first drilling started. The Norwegian Oil Directorate was established in 1972.

Already during the first concession round, the Norwegian contractors Høyer-Ellefsen and F. Selmer entered the playfield and came out with proposals for artificial islands in concrete, but did not get any acceptance for this.

In a paper at the Norwegian Concrete Day in 1974, Sigvard Hafskjold [123] among others stated:

> ... *When the need for bottom -fixed structures for the difficult ocean areas came up towards the end of the 60s, the time was ripe for solutions that were completely different from the pile based steel frameworks. This type of structures had been normal on depths of more than 10 meters, since the first had been built outside Louisiana, USA, in November 1947. This design method had reached its limits.*
>
> *For Arctic Seas and for the northern part of The North Sea, it was necessary with new types of design, and the generation of steel structures was also pioneer projects. Delays, redesigns and cost excess were proofs of this.*
>
> *At the same time, this was the breakthrough for the use of concrete, and it was the need for very large constructions that gave concrete its chance. This possibility was used to show all skeptical people in the oil industry that big-size production of concrete could take place within narrow limits of quality specifications, as for other materials.*
>
> *From a total of 429 tests on the Ekkofisk-platform, with 15×30 cm cylinders, we received at 28 days a compressive strength of 437 kPa/cm² with a standard devia-*

tion of 30 kPa/cm².

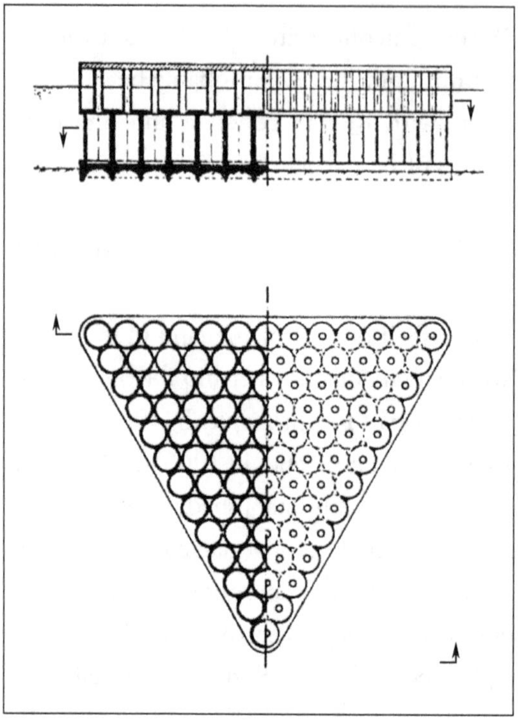

The design proposed by Ingeniør F. Selmer A/S for the Norwegian Industry Department in August 1970 [123]

Then the French company C.G. Doris got a contract from Phillips Petroleum for building the large storage tank at the Ekofisk field. The contracting company Ingeniør F. Selmer A/S got the building contract along with Høyer-Ellefsen. In 1971, started what was to become an interesting story. At the most, there were 600 workers on the tank.

The Ekofisk tank had the height of a 30 story building and the weight was over 200 000 tons. The towboats had a total force of 40 000 hp and undertook a 270 km trip to the Ekofisk field. The tank was placed on 70 meters depth of water. The storage capacity was 160 million liters of oil [124].

A drawing of the Echofisk tank under construction

Stavanger Spennbetong A/S produced 3000 "reels" at their factory at Vaulen, near Stavanger. The components were 190×190cm and170×170 cm.

The wave breaker wall became an interesting challenge to the Norwegian precast concrete industry. At Stavanger Spennbetong and O.C.Østraadt, the so-called reels were produced. They were formwork components that formed the characteristic wave breaker holes in the tank.

The Ekofisk tank is towed out from Jåttavågen in Stavanger in June 1973

The idea for the Condeep platform came in 1971 from Senior Engineer Olav Moe in Høyer-Ellefsen. He introduced the idea of building large storage tanks in concrete as foundations for a platform structure.

The platform building activity influenced the work done by the Norwegian concrete enthusiasts and led to a number of spin-off activities. During the Norwegian Concrete Day in 1978, P.S. Hafskjold from Ingeniør F. Selmer A/S presented a paper about the research project "MARBET" [125]. In his introduction, he among others states:

> ···*The development in the North Sea has gone very fast and raised questions that needed thorough investigations and decisions before answers could be given.*
>
> *First and foremost, the size of the project, was far beyond any previous experience. How would these structures behave during real loadings? Would the very extensive and costly calculations that were required be relevant and give the right results? Or would simpler calculation methods be better? The questions had to be asked, as the big computers were basing their programs on linear elastic relationship that was not completely in accordance with the real life of concrete. The curing development and temperature smoothening because of the heat of hydration complicated the picture, as much as the construction technique that gave considerable stops in the casting of the tank walls.*

He then discusses the cooperation with FCB (the Research Institute for Cement and Concrete) in Trondheim and the measuring program to verify the real strength of the structures, which was frontier-breaking news as far as structures of such sizes are concerned.The research and development carried out by the contractors and FCB was extended to new sectors and new institutions came into the cluster, The contractors were Thor Furuholmen A/S, Høyer-Ellefsen A/S , and Ingeniør F. Selmer A/S. This led to a new contract from 1 January 1977 between FCB and the new contractor company that was established – Norwegian Contractors.

An illustration of the development in the North Sea

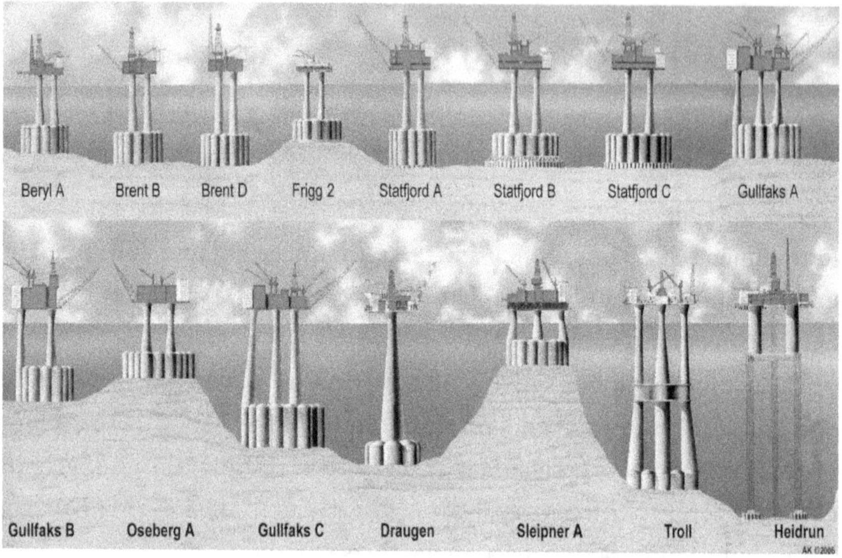

In 1985, 15 years after the adventure in the North Sea started Bernt Jacobsen from Norwegian Contractors presented a paper at the Norwegian Concrete Day in Trondheim [126], summarizing the development till then and indicating further development. At this point in time, 18 large structures had been built or were under construction in deep waters in the North Sea (15 in place and 3 under construction), from 70 to 153 meters water depths, with altogether more than 2 million m^3 of high performance concrete. All the structures were gravity based. At the same time, new solutions for deeper waters were ready on the drawing board. The Gullfaks C- platform for a water depth of 216 meters was ready for bidding, and Condeep SP and Troll designed for water depths of more than 300 meters were in the preliminary stage. The Condeep T300 for water depths of about 240 to 375 meters was also ready.

There were several "off-shore-lectures" on the Norwegian Concrete Day in 1985. Ove T. Gudmestad from Statoil presented a paper on "Concrete in Off-shore-structures" [127]. Today it is interesting to note that he in his lecture labeled water depths up to 150 meters as "relatively shallow waters." He also opined that:

> *The technology to build concrete platforms is developed in Norway and Norwegian industry is today the leading lady in the world for these kind of structures. "Another interesting remark" … The inspection reports from the concrete structures in the North Sea are very encouraging (Fjell and Røland 1982). Few*

damages are observed and the repair has been very satisfactory. However, it should be remarked that none of the concrete structures in the North Sea is older than 12 years, so we have to follow the aging effects very closely.

Many of the platforms and the people involved have received international acclamations and awards, for example [124]

- 1975: American Concrete Institute Medal awarded to Knut Tovshus.
- 1980: FIP medal awarded to dr.techn. Olav Olsen.
- 1988: Distinguished Achievement Award of Offshore Technology Conference (OTC) awarded to Norwegian Contractors.
- 1990: FIP Award for Outstanding Structures awarded to the Gullfaks C platform.
- 1994: Golden Medal of Gustave Magnel awarded to dr.techn. Olav Olsen.
- 1995: The Norwegian Concrete Award given to the Troll A platform

In 1995, FIP's President Jan Moksnes congratulated his colleague Steinar Helland with the FIP Medal. Jan Moksnes was the Managing Director of Norwegian Contractors and a key person, both on the offshore-sector and in the Norwegian concrete field in general for a number of years [124]

The building of the platforms has been very important for Norwegian companies and technologists in an international environment.

The Troll platform has been consid-

The fantastic Troll A platform is towed out in the summer of 1995

Troll A

ered as a fairytale by international authors. In a "development-paper" at a conference in Seville in Spain in October 2009, Geoge Hoff, a specialist from the oil industry for many years and former president of ACI names it the eighth wonder of the world and mentions that it is the largest structure ever moved by human power. It has a height of 474 meters and stands on a water depth of 305 meters. In total, 245 000 m^3 of concrete was used [101].

3.1.2 England

In 1970s, the English journal *Concrete* published a number of examples under the title "Historic concrete." Several of the following examples have been taken from the series.

Brunel's Thames tunnel, 1825–1843

The work on Brunel's Thames tunnel started in 1825 and pedestrians walked through the tunnel for the first time in 1843. This was not only an épooch breaking tunnel and an engineering achievement, but also the first really big concrete work in modern times. Shani Wallis wrote an article in 1997 about the construction and the renovation work in 1995–1997

[65] and called the tunnel the eighth wonder of the world. (*over time, there have been many eighth wonders of the world).*

The Thames tunnel was repaired in 1995–1997 and got, thanks to the English Department of the Environment, a look that is similar to Brunel's original structure

The inventor and builder was the French-born Engineer Marc Isambard Brunel. Later, his son Isambard Kingdom Brunel also joined the company. The work started in 1825, less than a year after that Joseph Aspdin took out his patent on Portland cement. There seems to be various opinions on whether "Roman" cement or Portland cement was used for the tunnel. Wallis gives a reasonable explanation for this:

- Brunel planned to use "Roman" cement because this had a shorter setting time.
- There are no papers that indicate that he changed to Portland cement or about the company Jos. Aspdin & Sons in Brunel's large archive.
- In a brochure from William Aspdin in 1854, he refers to the collapse of the tunnel in 1828, which partly stopped the work for seven years, and goes on to mention that *" When the Thames broke through the tunnel[...]and filled it with water, a large quantity of this (Portland cement) was thrown into the river which eventually stopped the cavity and enabled the contractor to pump out the water ··· soon afterwards the work resumed its wonted appearance subsequently obstructed for want of funds"*

Tall's house in Bexleyheath in Kent, 1866 [66]

Albion Villas, No 65 and 67 Woolwich Road, Bexleyheath, Kent, built in 1865

Chapter 2.2, "Portland cement," refers to the enthusiastic inventor and contractor Joseph Tall. In an article in *Concrete* in 1976, R.J.Barfoot mentions about one

Drawing of the house from 1866

of the first houses that Tall built and gives an interesting account of it to throw light on Tall's activity. Most of the following is taken from Barfoot's article.

The house in Kent was, in 1976, probably the only surviving building built by the pioneer Tall. The outhouse was also built in concrete, but some dynamite managed to pull it down in the beginning of the 1900s. The house originally had a flat roof.

Originally, the cost of the house was 100 GBP and it was meant to be built for farm workers. *The Civil Engineering and Architect's Journal* dated 1 September 1866, writes among others *can··· be built to a cost that satisfies the low wages that the farm workers gets, and the owner gets a good return on his investment. The use of concrete for inexpensive construction of walls has recently been tested by the building of a two family house in Woolwich Road, Bexley Heath, Kent. All the walls and floor in the first floor and the roof are in concrete···*

The article also states that the concrete mix design for the house was 30 yard^3 of gravel that passed a mesh of ¾ inches, 3 yard^3 of Portland Cement, and 3 yard^3 of river sand, and that 30 yard^3 of concrete was needed to build this house, which means there might be a misprint somewhere,

Campbell's farm, 1869 [128]

In 1859, Robert Campbell came home from Australia. There he had made

Campbell's farm

a fortune and he settled in Buscot in Oxfordshire. He had a lot of enthusiasm and started several local projects. He also built a large barn on the property. Only the latest in technology was used on the property. The barn was probably built about 1869 and was among others used in a gigantic research project for cows.

Axemouth Bridge, 1877 [129,130]

In 1877, Sir Walter Trevelyan initiated the construction of a concrete bridge between Axemouth and Seaton in South Devon. A local report dated 1885 states that the bridge had various concrete mix designs and also used "cellular concrete" to reduce the stresses in the arches. Instead of the original proposal to build a bridge with one span, an arch bridge with three spans of 10, 16, and 10 meters respectively was built. The bridge was a toll bridge until 1907, when it was bought free for a sum of 5000 GBP and through which Sanders Stephens, Lord of Mayor, had raised 2 200 GBP. A write up by the local tourist organization [131] mentions that this was very well celebrated at the time.

In an article in *Concrete* in 1990, Waine and Long [131] refer to the testing of the concrete in the bridge. They claim that this is the oldest surviving concrete bridge in England. The first bridge built in England was the

Axemouth Bridge [129]

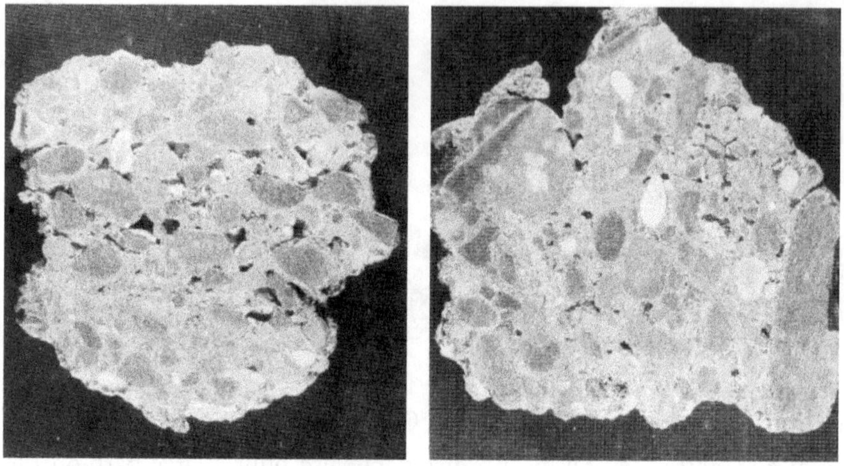

Sections through the cores of the old concrete [131]

railway bridge between Earls Court and Gloucester Road, built in 1867, which was demolished eight years later. The second was in Homersfield in Suffolk and that has also been demolished.

The historical significance of the Axemouth Bridge was acknowledged in 1956 when it was named a "Historical Monument".

Axemoth Bridge had had a number of damages. Among others, settlement of up to 600 mm had been registered in the western foundation. The opinion is that this must have come from the removal of the scaffolding. The bridge was strengthened in 1956. In 1990, cores were drilled out from the old concrete.

In addition to visual evaluation of the samples, a number of microscopic tests were done (RLM, SEM, and XRD). The tests showed;

Large variations in the size of the aggregate and the cement paste distribution.

Very large cement particles, probably originating from a shaft kiln.

Bad aggregate distribution and numerous pores

Probable mixture between sand and coarse aggregate: 1 : 1 : 6

Remnants of coal ash which might have given some pozzolanic reactions and might also be the reason for this "bad" concrete to have survived over time.

In the conclusion of the article, the authors feel that this tells a lot about concrete because such a bad concrete, badly mixed, with very coarse cement of bad quality, has endured for over 100 years. We can only speculate the possible life span of a quality concrete produced with modern cement.

Glenfinnan Viaduct

Glenfinnan Viaduct, 1888

In an article in *Concrete* in October 1975, Barfoot [132] wrote a fascinating story about the bridge structure, which he had visited earlier. He calls the viaduct *a monument to concrete.*

The viaduct was built between July 1897 and October 1898 and the cost over 100 years ago was 18 904 GBP. The bridge is 100 ft (30 meters) high, and with its 21 spans, it is 1000 ft (300 meters) long. The rock in the area is mainly quarts and gneiss, very well suited for aggregate, but found not suitable for a stone construction. The conclusion was therefore to build the viaduct in massive concrete. Barfoot refers to documentation from Walter Stuart Wilson in Proceedings from Institution of Civil Engineers in 1906–1907, which claims that the concrete mix design varied from 5 : 1 to 4 : 1.

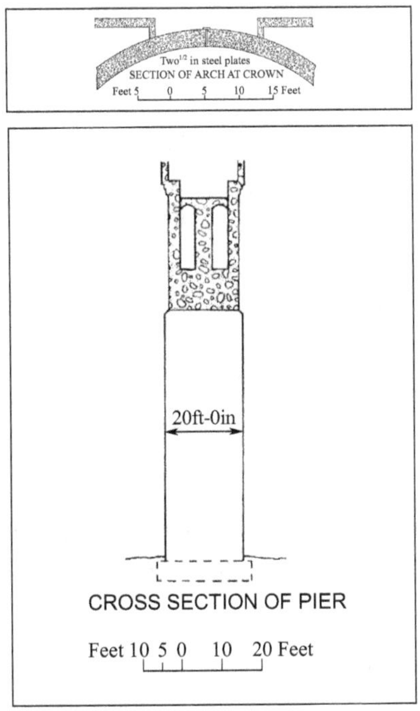

Section and length section of the bridge

The contractor of the construction was Mr. Robert McAlpine or "Concrete Bob" as he was called had 20 years experience in concrete constructions. Steam driven crushers were put at each end of the bridge and hundreds of carpenters produced the formwork. The wages varied from 4 ½ d. to 5d. per hour, plus bonus. According to the history of the building, a horse fell off the scaffolding and down into one of the 30 meter deep pillars. As it was very difficult to save the horse and the equipment, it was decided that the easiest solution was to bury it in the pillar.

Water tower at Newton-le-Willows – 1904

The demolition of the tower in 1979

Water tower – Newton-le-Willows, 1904 [133]

When the water tower in Newton-le-Willows in Lancashire was built in 1904, it was the second reinforced concrete water tower in England and probably the largest in the world.

The tower had a capacity of 300 000 gallon (1.36 million liters) and a height of 117 ft. (12 meters). The tower has many times been inspected to calm worried neighbors. During an inspection in 1925, rust was found on the concrete, but the following additional comment was found in the report; *the lasting qualities of ferroconcrete cannot be disputed when good workmanship exists.*

The article published in 1975 states that the tower had still been in good shape and could be seen on the left side of the road when travelling M6 between exit 22 and 23.

However, an Internet check in 2009 revealed that the tower was demolished in 1979 [134].

Tuckton Bridge, 1905 [135]

Tuckton Bridge [135]

From the load testing ("tractor" with wagons carrying 65 tons of load) of Stanford bridge in March 1906

Tuckton Bridge over Avon, which is on the main road between Southbourne and Christchurch, was opened for traffic in 1905. The bridge is 350 ft (105 meters) long, and in reinforced concrete, resting on 42 reinforced concrete piles that have been driven 90 ft (27 meters) down into the ground. The main span is 41ft 2 inches (12.35 meters)

Standford bridge- photo anno 2009

Stanford bridge, 1906, over the river Teme [136]
The bridge has later been replaced by a more modern construction, but has been remodeled as a footbridge. The design is by L.G. Mouchel and Partners, using Hennebique's system.

Cerrigdrudion Dam, 1910 [137] (and the Dam busters)
Cerrigdrudion is in the Snowdon area (highest peak in Wales) between Bangor and Llangollen. In 1910, a gravitational dam was built downstream in the 5 km long Alwen reservoir. The dam is 100ft (30 meters) high and

The Snowdon area

450 ft (135 meters long).

To match the environment as well as possible it is clad in precast concrete blocks with exposed granite aggregate. The blocks are 5 ft (150cm) long, 2 ft (30 cm) high, and 3 ft (45 cm) wide. Another interesting point with the dam is that it was used as a training object during the Second World War, in connection with the so-called "Dam Busters" that bombed the German dams, Mohne and Eder, in a spectacular air raid.

Andrew Budek discussed in the ACI Committee 120, Concrete History Session, "The Dam Busters" and related issues [138]. During the Second World War, the allies wanted to blow up the German electricity production unit in the Ruhr area. The targets were the dams at Mohne, Eder, and Sorpe. The problem was that the reinforced concrete dams were regarded as being impossible to bomb from the outside with existing technology. For example, the dam at Mohne was an arch dam, over 30 meters thick at the foot and 7.5 meters at the crown.

An experienced airplane designer, Barnes Wallis, found a solution. During piling at the Waterloo Bridge in London, numerous cracks had been found in the pile heads due to the rebounce from the rock. The other observation was that the stones thrown by small boys bounced on the water. These two observations created the solution for the "Dam-busters" but many tests and evaluations were behind the final plan.

Avro Lancaster

The solution was to drop a sphere-shaped bomb from low altitude, let it bounce on the water in front of the dam, over the torpedo nets, and then sink down toward the bottom so that the explosion

The dam at Eder after the collapse

absorbs the rebounce force from the water.

There were long training and testing sessions before the final raid. A new bomb squadron, "617," was formed and the bomb was made with a weight of 3000 kg in a casing with a diameter of 1.2 meters and with Torpex explosive. Leader for the squadron was Wing Commander Guy Gibson, 25 year old then. The bomb had to be dropped from low altitude, 20 meters, and with an accurate speed of 250 m/h. The bomb raid started early in the morning on 17 May 1943. The attack force comprised 19 planes of the type Avro Lancaster. Four planes were shot down over the Netherlands and one had to return due to damages from low flying over the North Sea.

The first target was Mohn. The planes attacked one by one. Gibson tried first. The bomb was delivered as planned and the bomb exploded against the dam wall with a huge water pylon, but the dam held. Plane number two got a hit during the flying in and the bomb that was dropped jumped over the crown of the dam and exploded on the roof of the power station downstream. Three more bombs were dropped and the dam collapsed.

The second dam was Eder. Upstream from the dam was a narrow valley, making Eder a very difficult target. The first plane took the wrong course and had to abort the attempt. While the first plane circled back, the second plane came in. The bomb hit the crown of the dam and exploded as the plane was passing, and the plane got lost. The first attacker came back, made a perfect drop, and Eder collapsed, sending 200 million liters of water down the narrow valley.

Different circumstances, among others fog, allowed only one plane, flown by an American volunteer of RAF to attack Sorpe. The bomb was correctly delivered, but the concrete dam in Sorpe held.

10 of the original 19 planes returned to England.

The plane and bomb designer Barnes Wallis was active long after the war. He was knighted in 1969 and died in 1979.

Guy Gibson was killed in the Netherlands, the year after the raid.

3.1.3 USA

Erie Channel

In an article in *Concrete International* in December 1996 [139], Luke M. and Billie G. Snell narrate a fascinating story about the first really big concrete project in USA.

(Luke, Professor at Southern Illinois University, and his wife Billie have written a number of articles about the history of concrete and have been very active in the ACI Committee 120, History of Concrete).

In the beginning of the 1800s, the challenges of transport across USA were considerable, and the cost of crossing the Appalachia was a great economic burden. An eight-span wagon train took 15 to 45 days on the trip from Buffalo to Albany and cost 100 USD per ton.

No doubt that official representatives from the New York State got interested in a channel project in the Hudson river at Albany to Buffalo at the Erie lake – the Erie channel. When the idea was put forward to President Jefferson, however, it was blankly rejected, with the reasoning that this was madness. On 4 July 1817, the state of New York granted 20 000 USD for a detailed study. The channel from Albany to Buffalo became 584 km long with 83 locks.

In 1816, a self educated engineer by the name Canvass White was employed as an assistant. Later he became a key person in the building of the channel. The Governor of New York, DeWitt Clinton, demanded a study trip to England to study the existing channels and locks and to buy the most advanced surveying instruments that could be found. White made the trip and used his time well.

It was decided to construct the locks in natural stone, with lime mortar in the joints, with the exception of the more complicated parts. In such places they wanted to use hydraulic cement. Hydraulic cement had to be imported from Europe and was used in the beginning. White had seen cement production on his study trip to England. He found suitable limestone near Chittenango, south east of Syracuse, that could be used to produce hydraulic cement, when it was burnt in an acceptable way. The "White-cement" was used in the channel construction, as early as in 1818, and the production went on until 1840. Even though White patented his simple cement, he did not trust that it had a superior technology. *Cement must be both a cheap and good commodity* – White claimed.

Upper New York State

Wards House 1871–1876

In an article in *Concrete* in May 1979 [70], John Fiedler enlisted the pioneers in concrete in America and claimed that the *first reinforced* concrete house in USA was built in Port Chester in New York State in 1871. This was, however, not the *first ever concrete* construction in USA. The oldest is of Spanish origin and is found in St. Augustine in Florida. The floor at Castello de San Marcos was cast in 1656 and consists of primitive concrete based on sand shell and lime.

Corley [59] claims in an article in 1992 that the first house in concrete based on Portland cement in USA was Milton House in Milton, Wisconsin, built in 1844. The house stood for well over 100 years. The reason behind the building of the house was better fire protection. The cement was imported from England.

Dabney [72] states in an article that the first reinforced concrete house in USA was built in 1885–1888. That was Ponce de Leon Hotel, built for the railway tycoon Henry Flagler in St. Augustine in Florida (see next page).

William E. Ward's house in Port Chester New York State, built in 1871– the first reinforced concrete house in USA [67]

Thaddeus Hyatt (1861–1901) got his patent on reinforced concrete in 1878. Among others, he showed that reinforced bars were more economical than iron profiles. An "artificial stone" house was built in 1835 by Obodiah Parker in 107, Maos Street, in New York. As a consequence of this, several houses of the same type were built upstate New York in the 1850s.

Fidler [70] states that the house in Port Chester was built with 4000 barrels of imported Portland cement and 8000 barrels of beach aggregate, and that the reinforcement was according to the French Coignet system. The house is today registered as a "Historical Landmark."

Ponckhockie Union Chapel (PUC), 1870

In an article in *Concrete International* dated October 1993, Heun [71] opined that PUC was the first reinforced concrete house in USA. Heun also discussed the first development of cement technology in USA. According to him, White found natural cement stone near Chittenango in 1818; similar deposits were found in 1825 in Ulster County, New York, and in 1828, a mill was built in Rosendale. This was developed further and in 1896, 15 cement factories were found in the Rosendale-Roundout district. This made up more than half of the U.S. cement production at that time.

The chapel, built in 1870, is claimed to be the oldest reinforced concrete structure in the USA [68]

According to Heun, all together, 300 million barrels of Rosendale-cement were produced and famous structures such as the Brooklyn Bridge and the Statue of Liberty in New York and the U.S. Capitol, the U.S. Patent office, and the Department of Finance in Washington DC., and many more were cast with this cement. Whether the buildings came as a consequence of the cement factories or vice versa could probably be discussed. At least, a number of concrete structures were built in and around the Hudson River.

PUC was built in 1870. The high concrete spire was demolished in 1965 as the quality of the concrete started to pose a danger to the environment. The chapel was built by and for the workers at Newark Lime and Cement Company.

Hotel Ponce de Leon (1887) and Tampa Bay hotel (1891) in Florida [140]
Henry Bradley Plant, born on 27 October1819, grew up in simple surroundings in New York and Connecticut. He did very well in shipping, until he moved to Augusta Georgia in 1854 to work for the Adams Express Company. In 1861, he started his own company – The Southern Express Corporation. In the 1870s, Plant bought up bankrupt railway companies and connected cities like Savannah, Jacksonville, and Tampa together in a railway network.

Henry Morrison Flagler, born on 2 February1830, grew up in humble surroundings in New York, until he left home at the age of 14. He first worked in his uncle's business in Ohio, before he, during the civil war, got into nitric-mining, which was the base for gunpowder production. Later, he together with his friend and partner, John D. Rockefeller, got into the

University of Tampa – 2004

field of oil and started the Standard Oil Company. Both of them became multimillionaires.

In 1877, Flagler's wife Mary fell sick and the doctor recommended that they should spend the winters in Florida. In 1881, Mary died and Flagler came back to Florida in 1881, where he lived with his new wife Ida Alice. They left Florida again in March 1884, but Flagler had planned to come back.

Flagler College – 2004

Flagler and Plant knew each other. Plant became a railway tycoon and Flagler a Standard Oil baron. They started an investment company together.

When Flagler returned to Florida in 1885, he started working on his plans to build a magnificent hotel in St. Augustine, USA's oldest city, and he called it Ponce de Leon. In the summer of 1885, he started the construction work by moving tons of sand into the swampland of Florida. Barrels of salt were spread out to protect the workers against malaria. The cement came by steam ships from New York and Germany. The strong mixture proportions in the 1.2 meter wide foundations were 1 : 1 : 2, while the mix for the walls was 1 : 2 : 4. They tried to use reinforced concrete in the floors, but it did not get further than the testing. Timber was used in the final structure. Ponce de Leon was opened in 1887.

While Flagler worked with his hotel, Plant bought land outside Tampa, the last station along one of his railway lines. Plant built Tampa Bay Hotel. In this case also the cement came from New York (Rosendale Brooklyn Brand (1.75 USD per barrel) and Germany (3.00 USD per barrel). The concrete floors were precast on the ground and hoisted into position afterward and placed on iron beams. The concrete was made of one part of cement and eight parts of crushed bricks, oyster shell-sand, and white sand. The hotel was opened for the first guest on 5 February1891.

Both buildings are in use today:

In 1933, the Tampa Bay Hotel was closed. It is today a part of the University of Tampa.

In April 1967, Hotel de Leon closed its doors after 80 seasons. 18 months later it was reopened as Flagler College.

USA's First Reinforced Concrete Bridge, Alvord Lake Bridge –1888
[101,141]

Alvor Lake Bridge spans over a foot path in San Francisco's Golden Gate Park. The bridge was designed by Ernest L.Ransome, an inventor and a key person in the development of a number of concrete sectors in his time. Probably, Ransome had used his own patented reinforcement system on the bridge. The concrete surface of the bridge has been chiseled to look like natural stone. The bridge was one of the few structures that survived the earthquake in San Francisco in 1906.

Alvord Lake Bridge – 6.1 meters span

USA's First Concrete Road–1891 Bellefontaine

The concrete was cast with a hard aggregate on the top to increase wearing resistance and in the shape of squares of 10 cm so that the horses would not slip.

The contractor had to deposit a sum of 5 000 USD as a guarantee for 5 years. The guarantee did not have to be paid.

The streets on the three other side of the town square in Bellefontaine were laid with concrete in 1893.

USA's first concrete road: 1891 Bellefontaine

Detroit – The First Mile–1918

Woodward Avenue, Detroit – the first mile in Portland cement concrete

The Worlds First Skyscraper in Reinforced Concrete–1902

The first skyscraper in reinforced concrete was built in Cincinnati in Ohio in 1902.

The construction was made by the Ferro Concrete Construction Company and this is the first building with more than six floors built with reinforced concrete. The building had 16 floors, and is still in use [142]. Three larger buildings in reinforced concrete were built approximately at the same time. This must be considered as the real breakthrough for reinforced concrete in USA.

- The Ingall's Building in Cincinnati, Ohio; 1902–1903.
- Terminal Station, Atlanta, Georgia; 1903–1904.
- Marlborough Hotel, Atlantic City, New Jersey; 1905–1906 [68].

The world's first skyscraper in reinforced concrete – the Ingall's Building in Cincinnati

The most impressive was the Ingall's Building, the world's first skyscraper in reinforced concrete. They used Ransom's reinforcing system in the large monolithic beams and slabs with tensile reinforcement. The two-way slabs used the Monier and Wayss systems, while the bent stirrups came from Hennebiqeu's systems. In addition, they also used the reinforcement ideas of Hyatt and Considere.

The Large Dams–The Hoover Dam (1931–1936)

The Hoover Dam

The Hoover Dam, also known as the Boulder Dam, is an arch-gravity dam in Black Canyon that dams up the Colorado River between the states of Nevada and Arizona. When it was ready in 1936, it was the largest concrete structure in world until then and the base for the world's largest hydroelectric power plant. In both the areas, it was surpassed by the Grand Coulee Dam built in 1942 [143].

The dam is located 48 km south east of Las Vegas and is still a popular sightseeing spot.

The dam is the end of a large 640 square km water reservoir called Lake Mead.

The building of the dam started in 1931 and was finished in 1936, two years ahead of the plan. The dam has a height of 221 meters and a length of 379 meters. It has a thickness of 200 meters in the bottom and 14 meters at the crown.

After long discussions, a federal commission was established in 1922 to clarify as to what should be done with the Colorado River, which with its many flood catastrophes had made serious damages. The river affects many states; Arizona, California, Colorado, Nevada, New Mexico, Utah, and Wyoming. The Governors of these states met with the federal representative Herbert Hoover in January 1922. Hoover was at that time Secretary of Commerce under President Warren Harding. The official commission reached an agreement, on 24 November 1922, that among others split the Colorado River into an upper and a lower part. It also made the foundation for what was later called the Boulder Dam. The agreement was named "The Hoover Compromise."

Already in 1922, the first proposal to get the building of the dam accepted by the Congress was made. However, it was not until December

1928 that both the House of Representatives and the Senate approved the proposal. The president signed the bill on 21 December 1928. At that time, Calvin Coolidge was the president. Preparation for the construction work started in July 1930.

The building was far from being without challenges. An important argument in favor of the building and the forced start was the provision of employment opportunities after the economic crash. A new town was built for the workers – Boulder City – but this city was not ready when the first batch of workers arrived early in 1931. Disapproval of the living conditions was also the main reason for a strike on 8 August 1931. The work with Boulder City was speeded up and all the transfer from the intermediate city, "Ragtown," happened in the spring of 1932. Gambling, alcohol, and prostitution became forbidden in the Boulder City. Even today, Boulder City is one of the two places in the state of Nevada that does not allow gambling. The sale of alcohol was permitted in 1969.

Officially, 96 workers died during the construction of the dam, many because of the contaminated air in the tunnels. The unofficial numbers point to about 112 deaths. One popular story narrates the first death, the surveyor J.G.Tierney, who drowned during work at the start of the project. His son, P.W. Tierney, is said to be the last to die during the dam construction [143].

The concrete work for the dam started on 6 June 1933 and posed a lot of challenges, in particular, in the temperature sector. A number of technologies that were developed at that time have later become the standard for building large concrete structures. Among others, 1 inch (25mm) cooling pipes were installed in concrete. After concreting, they were cut and injected with mortar. All concreting took place in trapezoid sections. In total 3.33 million m^3 of concrete was used in the dam, enough to build a 2 lane concrete road from San Francisco to New York.

Herbert Hoover played an important role in the building of the dam. That is why the dam has taken his name. Originally, the dam was named the Boulder Dam. Hoover became the president and on 17 September 1930, his Secretary of the Interior, Ray Lyman Wilbur, announced that the dam should be renamed Hoover Dam. At that time, it was not unusual for large dams to be named after the president. The name was officially approved by the Congress on 14 February 1931. In 1932, Hoover lost the presidential election to Franklin Delano Roosevelt. Roosevelt's Secretary of Interior, Harold Ickes, sent out a memorandum on 8 May 1933 stating that the dam should be given its original name, Boulder Dam. Roosevelt's presidential period ended in 1945 and Ickes retired in 1946. On 4 March 1947, a Congress man from California made the proposal to again rename the dam as Hoover Dam. This was approved by both chambers on 23 and 30 April 1947.

Here is a small story about the dam, narrated in the September issue of *Fortune Magazine* in 1933 [144].

Arthur Powell Davis was the name of the engineer who showed interest in the dam project as early as in 1902. His uncle was John Wesley Powell, who did the first investigation on Grand Canyon as early as in the 1860s–1870s. Arthur Powel got his education as a civil engineer from the Columbian University (now George Washington University) in 1902. For 14 years, he was fighting for the construction of the dam during his tenure at the potentially responsible builder, the U.S. Bureau of Reclamation. The difference of opinion between Davis and his superiors, however, got so serious that he in 1923 tendered his resignation as director and chief designer of the Bureau. Disillusioned, he returned to California, where he worked with local water projects. Later he also worked in Turkistan as the Soviet Union chief consultant for irrigation projects.

The work on the Boulder Dam continued for 10 years without Davis. Davis lived in California and his health started deteriorating. Then, in July 1933, Davis was happy and satisfied to work as the chief consultant for the dam. Davis was 72 years and his health did not allow him to take up a heavy work load.

The Grand Coulee Dam (1933–1942)

The Grand Coulee Dam is a gravity dam, in the state of Washington, that dams up the Columbia River. The water reservoir above has got the name Franklin Delano Roosevelt Lake. The lake contains 85 km^2 of water and is 243 km long.

The dam is 168 meters high and has a length of 1592 meters. To build the dam, a total of 9,16 million m^3 of concrete was used. This makes the

Grand Cooley Dam

dam the largest concrete structure in USA and in 2008, the electricity production from the dam was still the fifth largest in the world.

The building of the dam was approved in 1930 after 13 years of debate and President Roosevelt passed the bill for the construction in 1933 and the foundation work started on 16 July 1933.

The dam was extended between 1966 and 1974 [181].

Walnut Lane Bridge (1949–1950)

Walnut Lane Bridge was not the first prestressed concrete structure in USA, but became a symbol of the technology breakthrough in the country and rippled out to other countries.

Here are some previous incidents.

- 1896: Henry Jackson produced precast concrete window sill, in which the reinforcement was prestessed. It was a success in the start, but they cracked after 2–3 years due to the stress relaxation in the steel [89]
- 1910: One of USA's concrete pioneers, Ernest L. Ransome, tried to produce prestressed beams with wire ropes. Unfortunately, the ropes twisted during heavy load and the experiment was a failure.
  (An interesting background story goes like this; for some time, Ransome's chief engineer was Ben C. Gerwick Sr. Among others, he was involved in development of precast bridge deck components in North California in 1915. He was also involved in the establishment of a factory for precast piles in Petuluma in California. After the Second World War, his son, Ben C. Gerwick Jr. (later, a very well-known concrete technologist, professor, and designer) joined the company. One of his tasks was to look for new products. This led to the production of prestressed piles in 1959.)
- 1939: Prestressed piles used by Raymond Concrete Pile Corporation in New York harbor [145].
- 1946: Prestressed slabs were produced by Roebling Engineers for their warehouse in Chicago.

- 1947: Lumberville suspension bridge finished, with a prestressed bridge deck as specified by Roebling
- 1949: Rio Paz suspension bridge between Guatemala and El Salvador was opened, with a prestressed bridge deck.

The planning of Walnut Lane Bridge started in 1948 [145]. Charles Zollman was employed by Preload Corporation. Preload persuaded the authorities in Philadelphia to build the bridge and Professor Magnel and his staff in Belgium designed it, while the detailing work took place in USA. Late in the summer of 1948, the contract documents were ready. Included in the contract was also a full scale test of a 49 meter long bridge beam. The bidding documents were sent out on 19 January 1949. The Preload–Zollman–Magnel group, however, had a problem; a competing offer was quoted by a

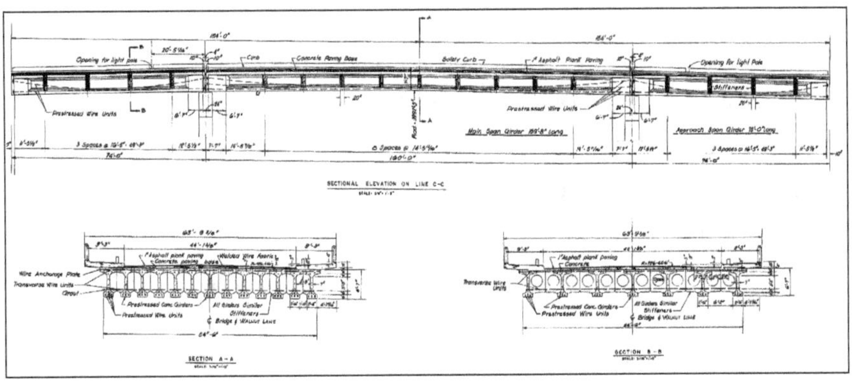

Part of the construction drawings of the Walnut Lane Bridge

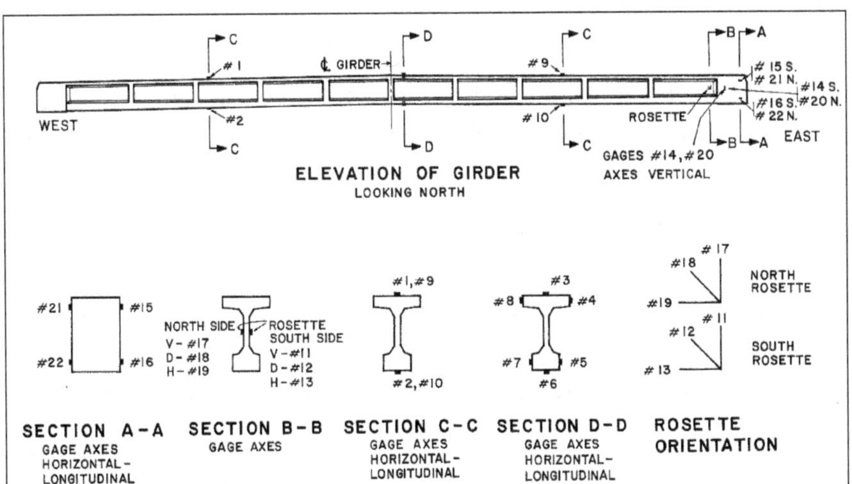

Plan and section of the test beam with the placement of the instrumentation [146]

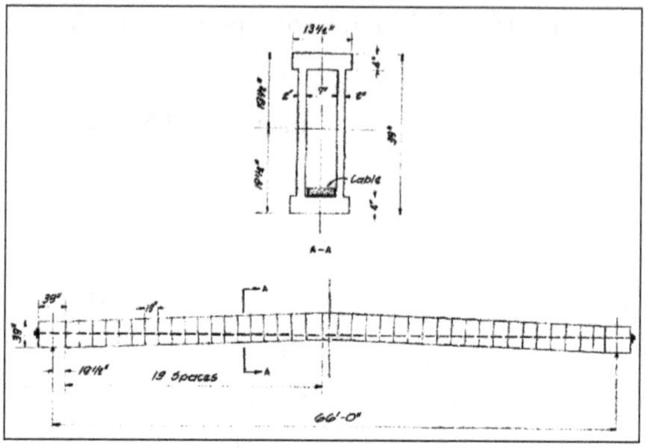

Precast beam design with concrete blocks, Terneuzen Canal, 1944

joint venture between Corbetta Construction Company and Raymond Concrete Pile Inc., based on Freyssinet's design. In the end, in the spring of 1949, the contract was awarded to Henry W. Horst Company, while Preload got the subcontract to produce the bridge beams, which were produced on the site.

Testing of the bridge beams took place on 25 October 1949, with 300 engineers from 17 states in USA and engineers from England, Cuba, Mexico, Canada, and Belgium were also present.

The instrumentation for the testing of the bridge beam was done by a later very famous concrete pioneer, Arthur Anderson. (see Chapter 4.1 and 5.9). At the time, he operated a consulting company in Springdale, Connecticut, and had specialized in instrumentation, with equipments of his own invention.

He got the contract to instrument the 49 meter long test beam in the city of Philadelphia. The test was very successful.

Zollman [92] narrates the following story about another bridge, in prestressed concrete, that Magnel designed in 1944. Toward the end of the Second World War, in the beginning of 1944, Gestapo gave the order to arrest Magnel. He was not in Ghent on the day he was supposed to have been arrested and with the help from his secretary he kept himself hidden, first in Brussels and later in other places.

Later, in 1944, the allies marched into Belgium. The engineering soldier, Brigadier Jean Paul Carrière, from the Canadian forces was to build a bridge over the Terneuzen Canal in Ghent to carry fuel pipes. He asked for Magnel's assistance and he promised to design the bridge in one day. Long before this, the Germans had ordered Magnel to produce some special concrete blocks. However, he had delayed the delivery and the blocks for a prestressed bridge were still lying at the producer's plant.

3.1.4 The Nordic countries

The concrete runway at Lindholm

- **The first runway in concrete, Denmark – 1906**

The world's first runway was built at the island of Lindholm at South Sjelland in Denmark in 1906 [38]. A self-educated mechanical talent, J.C. Ellehammer (1871–1946), had quite by himself designed an airplane, with which he flew 42 meters, half a meter above the ground in 1906. He also designed the first runway in concrete.

NK in Stockholm built in 1912

- **NK – Nordiska Kompaniet's warehouse in Stockholm, Sweden – 1912**

As one of the first reinforced concrete buildings in Sweden, the NK's shopping center in Stockholm was built in 1912. The building was designed by Ferdinand Boberg [147].

Reinforced concrete came to Sweden in 1900. At the start, very few people knew about this new material. But, long before the other Nordic countries, Sweden established their Concrete Society – Svenska Betongföreningen on 15 January 1912. In 1915, the Society constituted a committee to establish a national concrete standard.

The Alvettula museum bridge

- **Bridge with precast concrete components in Finland – 1916.**

Alvettula Museum Bridge in Pælkæne was built in 1916. The components of the bridge were precast after a patented method called the "Tapani method" [148].

Pohjankuru Bridge in 1899

The first reinforced concrete bridge in Finland was the Pohjankuru bridge over a railway track, built in 1899. The bridge was reinforced with curved I-beams. The first nonreinforced concrete bridge in Finland was the Mannerheimintie bridge at the center of Helsinki, built in 1894.

- **The main railway station, Helsinki – 1919**

Eliel Saarinen's design of the railway station in Helsinki [82]

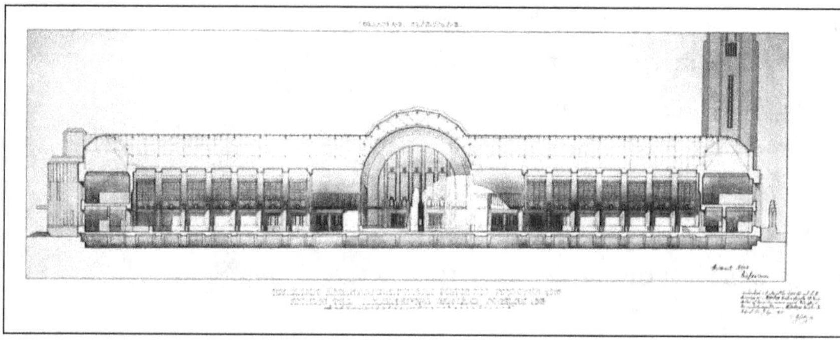

In 1904, the Finnish government announced a competition for the design of the new railway station in Helsinki. The proposals resulted in a major public debate. The competition was in the end won by Eliel Saarinen, who proposed a reinforced concrete structure. The station was not only a major breakthrough for this material in Finland, but is still today a landmark that people in Helsinki are proud of.

The first building in Finland built completely with reinforced concrete was Suvilahti Power station in Helsinki, began to be built in 1907 and completed in 1909. The architect was Selim A. Lindquist and the constructor was Jalmar Castren [82].

- **Olympia stadium in Helsinki, Aalto, and the Finnish architectural trend[149]**

In 1952, the summer Olympics was conducted in Helsinki. For the next ten years or so, concrete designers went on a pilgrimage to Finland to study the new trend in Finnish design. The Olympic stadium became the symbol of an architecture in which Finnish architects had succeeded in creating a new unique Finnish design.

Finland got their independence in 1917. Stockholm, Sweden, conducted the Olympic Games in 1912 and the very sports-oriented Finland had both,

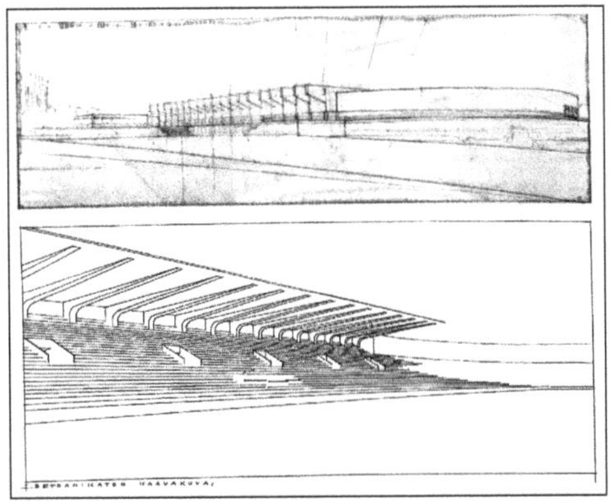

Alvor Aalto's proposal for the grandstand at the stadium

Olympic stadium in Helsinki designed by Yrjö Lindgren and Toivo Jänti, built in 1934–1940

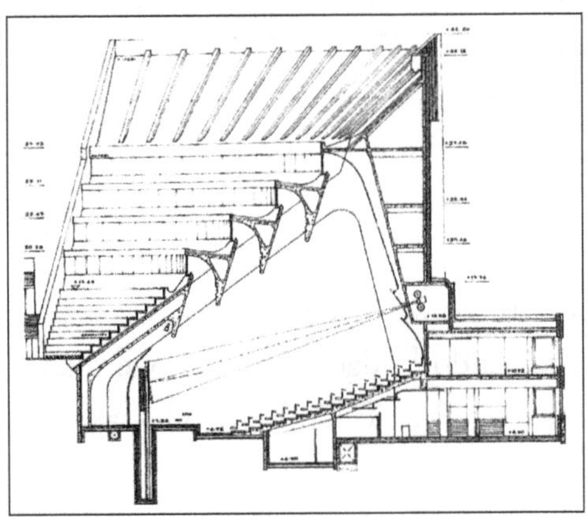

Section of the Auditorium in Otaniemi, designed by Alvar Aalto, and completed in 1964

a marking need and a strong desire to show their sports and cultural ability.

Originally, the Olympic Games should have been held in Helsinki in 1940, but the Second World War delayed the games till 1952. The competition for building the stadium started in 1933 and the various proposals led to strong debates. Alvor Aalto did not get acceptance for his proposal for the grandstand.

The proposal that won came from Yrjö Lindgren and Toivo Jänti. The construction started on 12 February 1934 and the opening was on 12 June 1938. The spectator capacity during the games, from 19 July to 3 August 1952, was 70 000 (record 70 435). The capacity was later reduced to 40 000. The stadium is 243 meters long and 159 meters wide. The landmark, the Olympic tower, is 72 meters high [150] or as high as the length of a good Finnish javelin.

However, the Olympic stadium was only the symbol, and Alvar Aalto became a sort of head priest.

Alvar Aalto was born in 1898 and got his certificate from Jyvæskylæ Lyceum in 1916 and from the Helsinki Institute of Technology in 1921. Aalto died on 11 May 1976 [151].

To left:
Aalto's Voaksen-niska church in Imatra in 1956–1958

To right:
The TWA's air terminal at JFK airport in New York designed by Eero Saarinen

Of the many impressive building structures designed by Aalto, we in particular mention the main building and the auditorium of the new Technical University in Otaniemi, Espoo, a suburb of Helsinki. This is new inventive concrete design at its best.

We will not try to judge who was, among the impressive group of Finnish architects, the greatest. We only mention Aalto as a symbol of the new trend.

Another example from this impressive group was Eero Saarinen.

- **The Sandö bridge – 1943**

The Sandö bridge over the Ångerman River in Kramfors in Sweden was the longest of its kind, when it was built in 1938–1943.

The Sandö bridge with a span of 264 meters

From May 2003 to November 2003, the contractor PEAB made comprehensive repair work on the bridge for a contract sum that was close to 100 million SEK [152].

- **The Kaknas tower – 1967**

The Kaknas tower was the highest concrete structure in Scandinavia, when it was opened on 12 May 1967 [153].

The tower is 155 meters high and was slip formed with a slip forming speed of 3.65 m per day. It has been anchored in the bedrock with 72 tension cables and has a ground area of 10×10 meters.

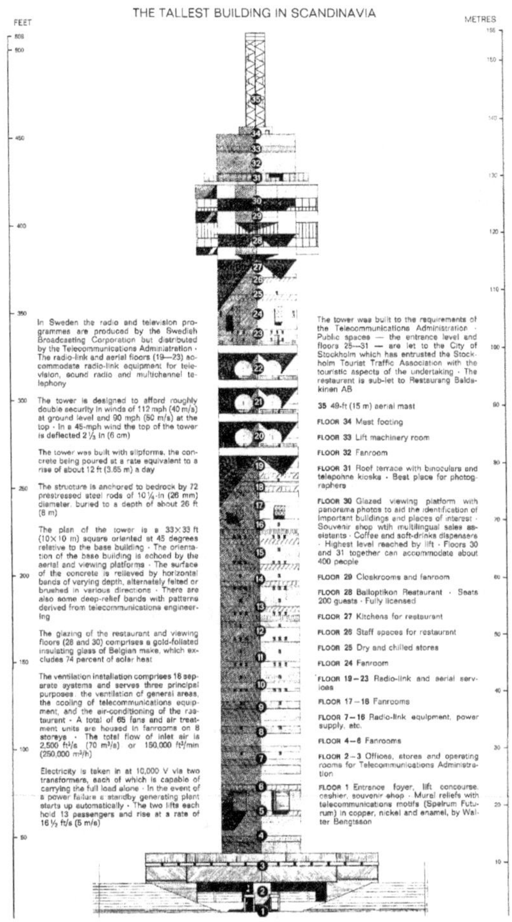

The Kaknas tower, built in 1963–67: Architects: Hans Borgström and Bengt Lindroos. Consulting engineer: Professor S.O. Asplund. Main contractor: Yngve Kullenberg Byggnads AB.

- **The Limfjord tunnel – 1969**

In 1950, a committee was formed to look at the traffic problems in crossing the Limfjorden in Aalborg, Denmark. The tunnel was completed in 1969 with six lanes. The tunnel is a part of Nordic building art that has also made its footprints in other parts of the world. The tunnel was constructed with the help of five tube shaped tunnel segments, each 105 meters long, cast in a dry dock, and floated to the construction site and then placed in their position. The lowest level for the driving lanes is 18.32 meters and the highest sailing depth is 12.2 meters. At the shore line, there is 9.5 meters of water above the components [154].

Crossing the Limfjord was one of Denmark's worst traffic challenges and when the bill regarding building of the tunnel was finally accepted in 1963, its cost budget was 143 million DAK. The original budget in 1961 was 81 million kroner. The building committee made an agreement with the

Section of the Limfjord Tunnel

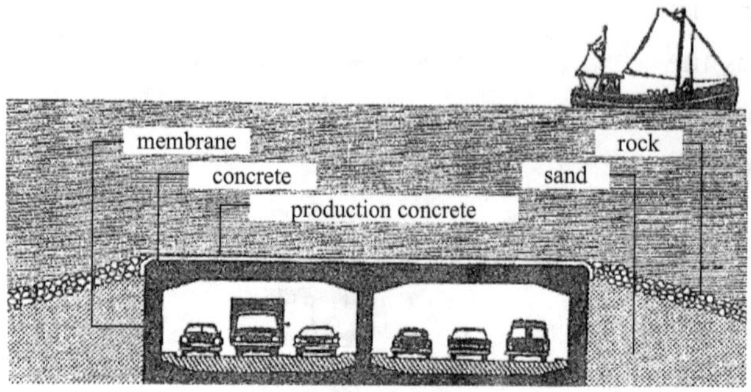

contracting company Christiani & Nielsen that had designed similar structures in the Netherland and in Canada. The company had among others done the engineering work for the tunnel under Maas in the Netherland and it was the same technology that was utilized in the tunnel under Limfjorden. The tunnel was constructed 2.5 km east of the existing bridge over Limfjorden. The construction work started in 1964. The five tunnels with 105 meter long components were produced in a dry dock in Romdrup, east of Aalborg, in an area that later became part of the Aalborg harbor. The components were placed on pillars in the fjord. Originally the tunnel was planned with four lanes but a later analysis regarding the traffic between Aalborg and Nørresundby showed that they had to build six, with a total width of 28 meters [154,155].

In the summer of 1902, a dynamic lecturer at the technical University in Copenhagen proposed that a young engineer, Christiani, who had just finished his examination, should make a visit to Hennebique's company [156]. Consequently, he worked in the Hennebique's company for two years. After his return to Denmark, Christiani, together with his friend Captain Aa. Nilsen, established the contracting company – Christiani & Nielsen A/S. In the later years, they got a number of very interesting assignments and contracts, among others the circus building in Copenhagen in 1914. The reinforced roof dome has a diameter of 40 meters.

3.1.5 Other countries

- **Jahrhunderthalle, Wroclaw, Poland – 1913**

  (when built, this was Breslau in Prussia)

In 1904, a large art exhibition was arranged at Breslau for 900 visitors. During the two and a half month period the exhibition had 625 000 visitors. Consequently, the city gained considerable reputation as an exhibition city. King FredrickWilhelm of Prussia wished, in 1910, to celebrate the 100th year anniversary of the freedom from Napoleon, and a building site at the east side of Breslau was chosen [157].

The architect and town planner Max Berg was born in Stettin in the Prussia's Pomerania (today Szczecin in Poland) on 17 April 1870. In 1909, he was employed as town architect, the highest position in the field of construction in Breslau [158].

Max Berg's father, Nicolaus Eduard Heinrich Berg, was a teacher at the local gymnasium. His mother was Emilie Luise Theresis, born Radeloff. After studying in a local school, Max Berg studied at the Technical University in Berlin, Charlotte Burg, under Carl Schafer during the years 1889–1893. During one semester, he was also a student at the Technical University in Munich.

After his studies, he climbed up the ladder in administrative jobs, from town architect to expert, Regierungsbaumeister. In Frankfurt he was a Stadsbauinpector and he stayed there until 1909. During this period, he developed his special architecture and designed several buildings, among others an airplane hangar. He was also married to his wife, the pianist Edelgarde Gerlach, in Frankfurt. On 17 December 1908, he was employed as town architect (Stadtbaurat) in Wrocklaw [159].

He became the designer of the "hundred- year-hall," that later inspired the design of a number of new roof structures.

Opening of the Jahrhunderthalle in 1913

The construction period was from 1911 to 1913.

The construction was ingenious, innovative, and very big for its time.

The structure has a column-free room of 65 meters, more than three times as big as the St. Peter Church in Rome, but it weighed only half [160].

Berg's construction started a new epoch and is by many regarded as one of the foundations for Modernism. The Jahrhunderthalle is one of several buildings in the exhibition area in Wroclaw, in which other architects have also been involved. Berg also designed other structures in the area. Some people opine that the Jahrhunderthalle is known for ferroconcrete and that it is the forerunner of Nervi's ferrocement. However, we should be careful not to mix up the two expressions. We should not forget that Berg's structure came at a time when reinforced concrete of any size was very rare and

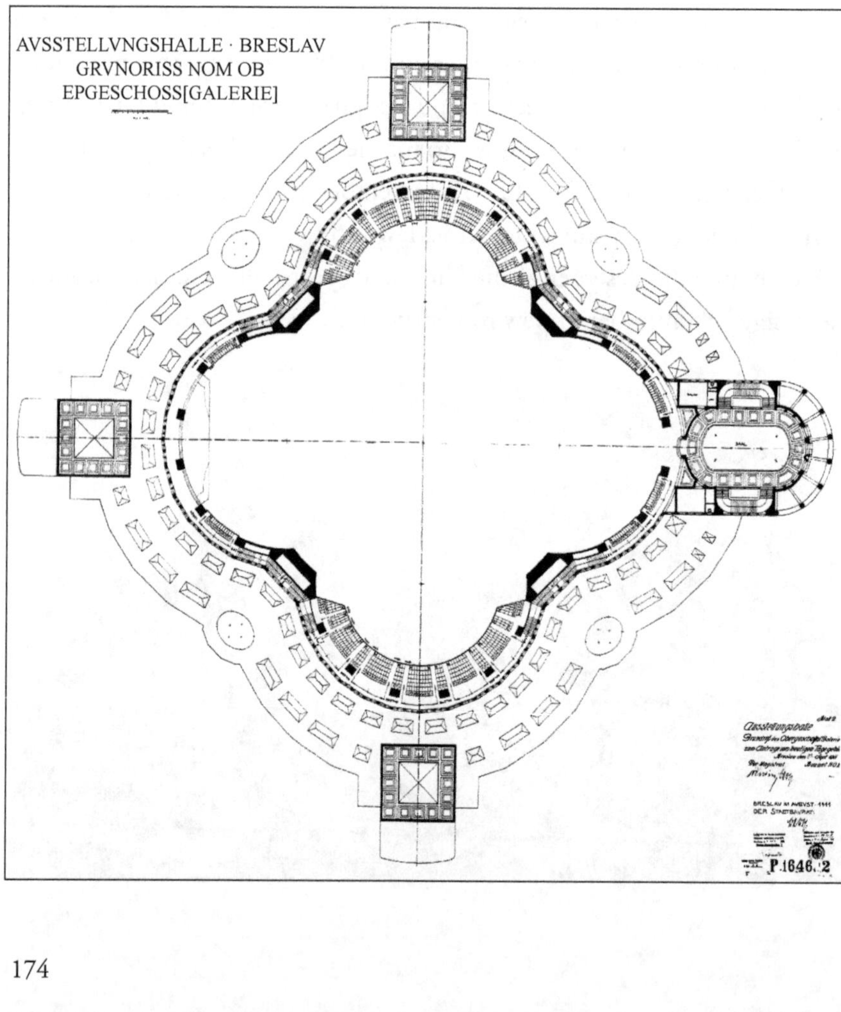

Berg had many editions of his drawings of the hall. This is from the final version that was approved by the city council on 25 September 1911

The hall, as it looks today

something completely new. Ferroconcrete was the normal way of referring to reinforced concrete at that time.

The structure had an intricate geometry of circles and sphere forms at the center and out toward all four directions.

Breaking barriers posed many challenges. Dyckerhof & Widmann were contractors for the construction. During the construction period, the formwork for a warehouse in the Franken factory in Nuremburg collapsed. Dyckerhof & Widmann were contractors for that project also. In May 1912, Berg had to write an article for the local press and state that the formwork for the Jahrhunderthalle was safe. One of the arguments was that this was a wooden formwork, while the one in Nürnberg was steel. Berg had his own architectural philosophy. After the First World War he said:

> *When you look at the pictures of Kandinsky, Klee and, for example our Leistikow, you will find that this art has its origin in experience, that is, it is influenced by the deep musical way in which you arrange planes and colors that create harmony and disharmony and give us an original music experience against the well known and experienced reality..... Architecture must deny all ornamentation and fraud in the same way as the painting when the focus is taken away from the object*

Berg moved to Berlin in 1925 and died in Baden-Baden in January1947 [158].

In 2006, the hall was included in the prestigious list of UNESCO as one of the World Heritage Sites.

The Hall today accomodates 7 000 visitors under one roof and has a multifunctional use.

The hall has been used for many important occasions, for example,

1948: Peace Congress.

1964: Marlene Dietrich fascinates a packed audience.

1997: Pope John Paul II speaks to the Eucharistic Congress.

Salganitobel Bridge, 1929–1930

One of the first academic studies of reinforced concrete was done by the Swiss Professor K.W. Ritter at ETH Zurich [161]. One of his pupils was Robert Maillart. Many have called him ' concrete-virtuous,' and one of the fathers of modern concrete bridges. Robert Maillart was born in Bern in 1872 and died in 1940. After having finished his studies as civil engineer at the Federal Polytechnic Institute in Zürich in 1894, he worked for several Swiss engineering companies. He established his own company in 1902. In 1912, he moved his company to Russia [162]. This caused problems during the Russian revolution in 1917 and he moved back to Switzerland. He then started to work with Lucien Meissner and Ernst Stettler as consulting engineers. Between 1910 and 1912, he participated in five different bridge competitions and he was allowed to build three of them, based on the jury's evaluation of quality and price. Maillart built some of the most beautiful concrete bridges in twentieth century. Statically determinate, three hinged arches became his trademark.

When Maillart was 58 years, in 1930, he had the pleasure of watching the opening of the Salginatobel bridge. With a span of 90 meters, this beautiful three-hinge arch bridge spans a steep valley and at the same time it, in an elegant way, blends naturally into the landscape. The bridge design was the lowest bid among 19 competitors [161].

Salginatobel bridge, near Schiers in Switzerland, built in 1929–1930

In an "embracing-article" in *Concrete International* in 1993 [163], Professor Jørg Schlaich from the University in Stuttgart in Germany, discusses the bridge artist Mailllart and his bridges, and hails him as the greatest master in his area of his time. He compares him with a number of personalities, associated with concrete, of that time, and Schlaich claims that no-one managed to transform his/her technical talent and knowledge into such reality, in terms of quality and quantity, as much as Maillart managed with his bridges.

He cites some of Maillart's own sayings. We cite some of them here;

From "Designing with Reinforced Concrete," *Schweizerische Bauzeitung, Jan. 1, 1938:*

> *···The structures of Hennebique and his successors mimic the customary design of earlier steel and timber structures. Reinforced concrete does not grow as wood grows, neither is it rolled as are steel bars, nor assembled as masonry is. Perhaps it could best be compared with cast iron, a material also poured into forms, and we may possibly learn something from the shapes of cast iron deriving from years of experience with a fluidity avoiding any rigid harsh transition from functional element to the other. It is the holistic view which emerges from this fine liquidity. The designers of automobiles and airplanes always know how to achieve this to a high degree, while only a few reinforced concrete structures manage to do so.*
>
> *But it is not aesthetic desire alone which raises the wish to maintain the overall view of the total prior to that of its individual members. This overall view of the total will also always result in economical advantages.*
>
> *···these heterogeneous structures, piecemealed shape borrowed from the design vocabulary of former building materials, cannot possibly satisfy aesthetically. They are also, by far, less economical than a structure which is conceived as a single unit between and above its abutments and duly constructed accordingly. Only by this a clear structure with nearly total utilization of the structural material will be achievable.*
>
> *May therefore the structural engineer break away from the existing tradition of forms given by older building materials and with this freedom and in observation of the total picture obtain the most expedient use of his material. Perhaps*

we will then also reach, as in airplane and automobile production a new style, truthful to the material and beautiful. And perhaps then it may also happen that the public taste will clarify and the traditionally shaped reinforced concrete bridges will be judged like last century automobiles, modeled from the horse-drawn-carriage.

Important words that probably should be studied by both designers and marketing people of concrete even today.

With the increased tendency toward standardization at the background, we also repeat the following from *L'Ingenieur et les Autorites,* Vie, Art et Cité, Jan–Feb. 1940*:*

Many engineers are content studying the codes: That is the safest method to avoid responsibility, and to be on good terms with academic and administrative officials. Those who leave this environment live dangerously, since professors seldom admit that their teaching does not always correspond with the latest stage of the arts, and administrative offices out of habit (as a rule) are scared of anything new.

(Then further on, following the proof to the load capacity of his new mushroom-head flat plate by experiments). Built structures in Switzerland, in France, Spain and Russia have therefore proved themselves excellently from the economical and practical viewpoint. But Building officials refused to recognize the system and in Germany, thanks to the opposition of academic honoraries, it is forbidden altogether.

- **Habitat 67**

Some months before the World Fair in Montreal in 1967, the housing concept, Habitat 67, was presented. The structure is beautifully situated at the shores of the St. Lawrence River and was designed by Moshe Safti for his Master thesis at Mc Gill University. The structure became a part of

Habitat 67

Cliff-houses in Mesa Verde National park, Colorado, USA [165]

Expo 67 as an example of houses of the future. During the World Fair, the apartments were used as intermediate houses for important guests. They were later sold, based on an idea that these would be low-cost houses. However, the market demand is such that these houses today are both popular and expensive [164].

In a lecture at the ACI's fall conference in Seattle, Washington, in 1987, titled "Back to the future," Bob Philleo compared Habitat 67 with former Indian houses in cliffs [165].

The roofs of the lower houses in "Habitat 67" had gardens of the houses above. Every apartment had a garden and the largest houses had two. The project originally consisted of 158 houses, built up of 354 standard components. There were 15 types of houses. All the units had an underground parking spot. The area around had been embellished with parks and playgrounds. The box-shaped concrete components were 5.25×11.35 meters in ground area and 3 meters high. Every unit weighed 70–90 tons. The components were cast

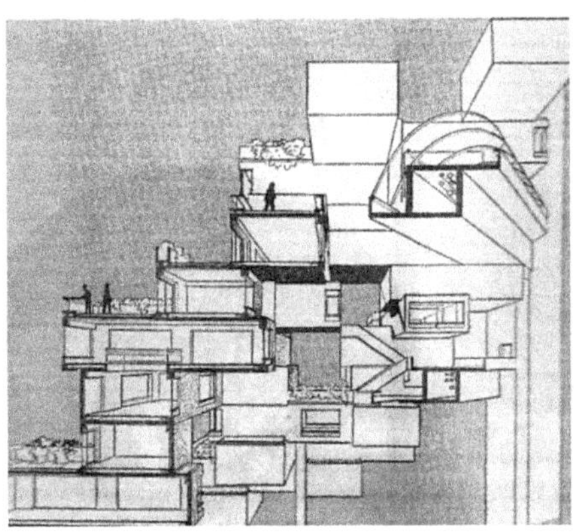

Section showing the box-units and the terraces, streets, and pipelines of the structure.

in steel molds and were transported to the building site with a 100 ton s "travel-lift."

The building was finished in November 1966 [166].

Habitat 67 has later inspired architects in many continents.

Sydney Opera House, 1974

As windy sails, the 12 white, 60 meter high concrete shells are placed on a peninsula in the Sydney harbor. This has become not only for Sydney but also for the country, Australia, and the continent a central illustration and symbol in presentations on the continent and the country.

The Danish architect Jørn Utzon won the international architect competition for the opera house in 1956. Though he started the detailing work in 1959, the structure was not completed until 1974. There might have been several reasons for this. The building was to be financed by a national lottery, and the solutions that Utzon suggested went beyond all budgets by several hundred percent. A lot has been told and written about Utzon's lack of interest for detailing drawings. Experts opine that the building would not have been possible without the consulting assistance of the London-based, Danish born, consultant Ove Arup, and several Australian architects who made modifications to Utzon's original drawings.

People who are interested in concrete and visit Sydney Opera House will be just as impressed with the quality and the workmanship of the concrete, in particular the exposed aggregate in stairs and parapets around the building, as the structure itself. Later, many fascinating structures using Utzon's ideas have come up, but it is hard to find similar workmanship.

Sydney Opera House

Beautiful concrete detailing 1990

Some data about the architect, Jørn Utzon [167]:

- Born in Copenhagen, Denmark on 9 April 1918.
- Studied at Copenhagen Art Academy between 1937 and 1942.
- Travelled a lot and studied with a number of famous architects in the years 1942–1948 (Gunnar Asplund in Stockholm, Alvar Aalto in Helsinki, Frank Lloyd Wright in Arizona).
- Started his own office in Copenhagen in 1950.
- Refrained from further work at the Sydney Opera House and closed his offices in Sydney, USA, and Switzerland in 1966.
- Designed a number of buildings in Copenhagen and Kuwait in the years 1971–1986.

Sagrada Familia (1884 -?)

Antoni Gaudí started the work with the Sagrada Familia in 1884, and probably no one knows when it will be finished. This is to really carry forward some of the traditions from ancient times, when it comes to the time taken for construction.

Gaudí was born in Reus, just outside Barcelona on 25 June 1852. After being educated as a blacksmith, he took his exam as an architect from

Sagrada Famila, the unfinished church in Barcelona.

Escola Superior d'Architectura in Barcelona in 1882 [167].

Gaudí's architecture is a landmark of Barcelona and it is "different." Any visitor to Barcelona can hardly avoid a visit to the church.

It is not difficult to find critics of Gaudí's style and designs, but it is unique. Gaudi made his own architectural thesis and his works have also become symbols of the Catalonian independence. The detailing is overwhelming, and more than three-fourth of a century after his death, new architects have worked on new specialties and details in Gaudi's spirit. To make concrete mix designs that can satisfy the designers' wishes and fulfill their needs has also become a challenge to the concrete technologists in Barcelona.

Gaudí was killed in a collision with a tram, on 10 June 1926 and was buried in a crypt in the church [167].

Brasilia, 1960

In 1955, Juscelino Kubitschek was elected as president of Brazil. He gave the job as director to Oscar Niemeyer to build Brazil's new capital.

In the first Brazilian Constitution in 1891, it was decided that the capital should be built where Brasilia is, that is in Distrito Federal [158].

Oscar Niemeyer was born in Rio de Janeiro on 15 December 1907 and started his architectural education in Rio de Janeiro in 1930. In 1936, he was employed by the Brazilian education department and among others got the opportunity to work with Corbusier.

The Congress Building in Brazil's capital Brasilia (1958–1960)

Before he got the Brasilia assignment, he had

already designed several important buildings in Rio de Janeiro and Sao Paulo.

Lucio Costa designed the main plans for Brasilia, which looks like the body of an airplane with wings, while Niemeyer filled the city with adventurous, modernistic concrete structures. The best known is probably the Congress building that is characteristic for its huge flat roof with a bowl-shaped structure on top.

Brasilia has a historic design in more than one sense – a whole city built using a completely new design philosophy, which is impressive, setting a new development trend in an underdeveloped country – something not found in any country where we have seen concrete development previously.

In 1960, the Congress Building was completed, but Niemeyer had to escape from the country. His lifelong membership in the communist party was no longer acceptable. Niemeyer designed, in the 1960s, several important buildings in Milano, Paris, and Alger [167].

Brasilia has today about half a million inhabitants, with 2.2 million in the satellite towns around it [158].

Oosterscheldekering, 1986

An English proverb goes like this: *God created the world, but the Dutch created Holland.*

American Society of Civil Engineers has nominated the dams in the Netherlands as one of the modern wonders of the world.

The storm barrier Oosterscheldekering (kering-barrier) was officially opened by Queen Beatrix on 4 October 1964. The cost was 2.5 Billon Euro, and it is claimed that the project has reduced the possibility of flood in the country to once in every 4000 years [168].

Netherland has at all times been fighting to tame the water forces, and the big spring flood in 1953 gave the signal to begin the work on a barrier and water regulation project ahead of any

Oosterschelde (East part of the river Schelde) is located in Zeeland in the southwest part of Netherland.

other earlier project. The largest spring floods comes when extreme flood (high tide) comes at the same time as storms.

One end of the Oosterscheldekering starts at the artificial island Neeltje Jans. Here is found a placate that says: *Hier gaan over het tij, de wind, de maan en wij (Here the tide is conquered by the wind, the moon and us)* [169].

Originally it was the planned that the link between Oosterschelde and the North Sea should be shut by a dam. The work started at the end of the 1960s by constructing several artificial islands, among others, Roggenplaat (1969), Neeltje Jans (1970), and Noordland (1971).

Toward the end of 1973, the dam was 5 kilometer long, but environmental protests stopped the work in July 1974. In 1976, it was decided that the remaining 4 kilometers should be equipped with locks. Normally these are not closed, but open, so that the tidal water can run. The idea is to keep the salt water isotope inside intact.

The work with the most difficult and costly part of the delta project started again in April 1976. A four square kilometer dry dock was constructed at Neeltje Jans, where the bottom was 15.2 meters below sea level. Here, they cast the gigantic concrete components that should form the pillars for the lock gates. Every pillar was between 30.25 and 38.75 meters high and weighed up to 18 000 tons. This means that 7 000 m^3 of concrete was used in each pillar. The construction time for each pillar was 1.5 years, and starting up the work of a new pillar took place every third week. The work went on 24 hours a day and up to 30 pillars were worked upon at the same time. All together 65 pillars were cast, plus 2 reserve pillars. When the casting work was finished, the water was let into the dock, and the pillars were transported out with specially designed ships. The two reserve pillars can today be found at Neeltje Jans.

The pillars are the foundations for 62–42 meters wide lock gates. The last gate was lifted into its position on 26 June 1986 and the opening took place on 4 October of the same year [169].

The road across the dam was opened by Princess Juliana on 10 May 1987.

The annual costs for running the gigantic project are estimated to be 17 million Euro [169].

Oosterscheldekering

The 62 gates are closed during storms (wind higher than 9 Beaufort) or during extreme high tide, so that the high tide does not reach Oosterschelde. If there are no people at the command central at Neeltje Jans, the gates close automatically if the high tide exceeds 3 meters.

Oosterscheldekering is the largest of the 13 dams and barriers in the delta project.

On 8 May 2002, the Oosterschelde was declared to be part of the national park in Netherland. The very special underwater fauna is a popular place to visit for both national and international divers.

The world's highest bridge: The Millau bridge, France, 2004

This is a steel bridge in southern France built across the River Tarn on the motorway between Paris and the Mediterranean, designed by the British architect Lord Norman Forster [170]. The seven high towers of the cable bridge is in reinforced concrete. The bridge was opened in December 2004 by President Jacques Chirac, who called it a monument of French engineering and art. The cost of the bridge was 394 million Euro and it is financed privately.

Height of the highest pillar (tower) – 342 meters

3.2 Design

The old Greeks forbade people to show objects and structures that were ugly or unpleasant because this might have a negative psychological effect. They tried to develop their aesthetical thinking in literature, art, sculptures, and architecture. We have taken lessons from them in many ways. Some of their design principles, which are based on what the human eye observes, is partly valid today, e.g.

- The eye observes a straight horizontal line to be sunken at the middle. A weakly upward bent line seems lighter and more elegant.
- A column that bends out slightly in the middle gives the illusion of being concave.
- By leaning the outer column a bit inwards, the structure does not give the appearance of leaning outwards.

The "golden cut" of the Greek, which is valid even today, is the base for a number of formats and not least, most of the papers we write on or read. Even most modern laptops often have a golden cut.

(The proportion between the short and the long side is equal to the proportion between the longest side and the sum of the sides, $l/L= L/l+L$, i.e. $L = 1.618\ l$).

Later designers have given us new models and ideas. In the following section, some of the artists who have played an important role in giving concrete design to the state that we experience today has been discussed.

The discussion that follows is far from an analysis of the architecture of the designers we have picked out, but a discussion on small groups of founders and idea-givers who have contributed to many of the aesthetic expressions we see in concrete. When one wishes to take a glance at the history of architecture, particularly in concrete, one comes across several important personalities and ends up in a confusion as to whom to include and whom not to One feels that there are so many that should have been included. The small number of designers who have been discussed here have all left their footprints in the field of concrete, but probably other authors might have chosen other examples.

3.2.1 Frank Lloyd Wright (1869–1959)

F.L. Wright was born in 1869 and lived a very creative life as architect, thinker, teacher, and speaker for more than 60 years. When he was 19 years old, he got his first architectural assignment in the progressive Chicago Company of Adler and Sullivan. When he was 24 years old, he started his own architectural practice. Wright died in 1959 [171]. His work as architect and his creative use of concrete, has inspired, and still inspire, designers all over the world.

Wright's early works are characterized by strict modular rectangles. Only in the 1930s came the diamonds, triangles, and hexagonal patterns into his planning systems. After the Second World War, we also find the circular and the spiral shapes.

He had during some periods been very popular and productive. In the years just after the turn of the century, he often designed 4–5 houses a year. Just afterwards, he got very fascinated by Japanese culture and art and from 1916 and some years later he often visited Japan. He designed the Imperial Hotel in Tokyo and became a legend in Japan as the structure was the only one in the neighborhood that was left after the heavy earthquake in September 1923. In the new Imperial Hotel in Tokyo that can be found in Ginza near the Emperor Palace, you still find Frank Lloyd Jones' bar on the second level.

A small historical sidestep in this history:

The Hardy Cross lecturer at the ACI Convention in New Orleans in 2009 was Professor Shunsuke Otani [172]. He opined that the Kanto earthquake in 1923 forced Japan to become the first country in the world in 1924 to set the standard for earthquake designs, that is, the Japan Urban Building Law.

However, it was Italy that set the rules first. These came in 1916 because of the large Messina quake in 1908, when more than 80 000 of the 130 000 inhabitants of Messina were killed.

(Hardy Cross (1885 – 1959) was the professor, who in 1930 developed the Cross-method for an approximate design method for moments in static indeterminate structures, with reasonable accuracy).

Frank Lloyd Wright's "Falling water" is probably the one project that

Frank Lloyd Wright's masterpiece "Falling water," built in natural stone and in-situ concrete It was built in 1936–1937 and is located in Bear Run outside Pittsburgh, Pennsylvania, and it is now a museum, opened to the public.

Under: From the Guggenheim museum in New York

is most often mentioned as an example of "organic architecture," with its strict horizontal lines and its cantilevers in reinforced concrete. The house was designed for Edgar J. Kaufmann, who was a well-to-do man, whose son worked for Wright. The building has had some durability problems and several experts in concrete who spoke about it at a seminar, arranged by ACI's Committee "Concrete History" in Baltimore in 1999, were of the opinion that Wright's capabilities as a concrete material technologist was in no way a match to the standard he maintained as a masterly form-giver and sculptor.

The Kaufmann family owned a department store, and lived in Pittsburgh, and asked Wright to design a vacation house. They had showed him the waterfall that was their favorite spot, and hoped that he would design a house close to this. Their astonishment and joy knew no bounds when they saw the drawings. With the large balcony cantilevered out over the fall, he produced an organic design that blended into the landscape. The house was finished in 1937, with a dent in the budget of 50 000 USD.

Edgar Kaufmann Jr. later became the curator of architecture at the New York's museum of Modern Art [167].

In 1943, he started his work with the Guggenheim Museum in New York. This is not only one of his most famous and most popular designs, but probably the work that took him the longest to get realized. The drawings were approved in 1945, but the building did not start until 1956, and it was opened in December 1959 (6 months after Wright's death). The building is a landmark in New York. It is also a landmark in the sense that it is an artistic and rational presentation of art. The idea for the spiral ramp originated from V.C. Morris Gift Shop in San Francisco that Wright designed in 1948 [173]. The ones who have not already planned to walk slowly down the ramps in Guggenheim and enjoy the pictures have something to look forward to.

The Guggenheim is exceptional in many ways. To study books about Wright's works is in many ways a walk along the artistic horizontal and rectangular forms and shapes that he had created. However, Guggenheim is not just the only project in which he showed how to master curves; the Annunciation Greek Orthodox Church in Wauwatosa in Wisconsin (1957) and the house for Norman Lykes in Phenix, Arizona (1959), are other examples.

In addition to his more well-known structures, Wright also designed projects for a number of close and distant members of his family [174].

His first project was in 1887 and it was a school designed for his aunts Nell (Ellen) and Jane Lloyd Jones, the Hillside Home School in Spring Green, Wisconsin. The school was extended in 1903 and the original parts were demolished in the 1950s.

In 1903, Wright designed Abraham Lincoln Centre in Chicago for his uncle Reverend Jenkins Lloyd Jones. The building was renovated in 1978.

In Oklahoma, we find three Wright-designed projects. Two were built in his later architectural period; the house for Harold Price Jr. (1953) and The Price Tower (1952) [174].

However, it is the third project that is the background for the following, a bit personal, story.

Wright's cousin, Richard Lloyd Jones, was the founder and editor of "The Tulsa Tribune." Even if Richard and Frank had been fighting as boys, Richard supported his cousin, also financially, during a period when he had few assignments. In 1928, he asked his cousin to design a new home for him. (the final drawings were registered in 1929) [174].

My father, H. Oswald T.J. Jahren worked as a site supervisor in USA from 1924 to 1934. During the period 1929–1931, he worked for the Chicago-based contractor Paul F. Mueller & Son. From September 1930 to April 1931 he was asked to go to Tulsa to be the building master for the construction of Richard Lloyd Jones house. My father took a number

Richard Lloyd Jones' house – Westhope in Tulsa, Oklahoma. The house had a floor area of 1000 m²

Photos and paper clips from by H.O.T.J.Jahren [175]

of photographs during the construction period. At a meeting in ACI's (American Concrete Institute) Committee 120, History of Concrete, I showed some of the pictures. This resulted in a presentation that I made about the construction at the ACI fall conference in 1999 [175].

One of the remarkable features with the building is the precast modules that were cast at the sites. Every main module is 6 ft (150 cm). Also, the columns were precast segments stacked on top of one another.

The following is one of the stories that my father narrated about the construction;

One day, Frank Lloyd Wright came on a site visit, accompanied by his cousin and his wife. Wright was not satisfied with the stair to the top floor and demanded that it should be removed. Mrs. Jones protested because she liked the stairway. Wright told her very clearly and convincingly that the house was his until it was finished and asked her to leave the building site.

Wright then made some sketches for my father, probably on the back of his lunch packet. At night they made a full scale model of the stairs in cardboard. The next morning Wright showed up at the site and approved the stairway. That was how it was built.

It is interesting to register the following comments in one of the many books about Frank Lloyd Wright that describes the house in Tulsa; "The stairway to the top floor was added during construction and is not on Wright's original plans".

Whether F.L.Wright looked upon Corbusier as an "idea-competitor" or not, we can only speculate. The following citation taken from *World Unity* in September 1928 can give a certain impression of this [171].

> *America and Le Corbusier. All Le Corbusier says or means was at home here in architecture in America in the work of Louis Sullivan and myself, more than 25 years ago, and is fully on record in both building and writing here and abroad.*
>
> *The Third Dimension. In this matter of art, the Frenchman has seldom got inside. He has usually discovered the surface "effects" best suited to the time, the place, and the hour. It is no small virtue in him and the trait has heaped honors upon his nation for centuries.*
>
> *But that "flair" is no longer good enough. In this architectural matter France may for once find herself behind. America's minority report, already handed in a Quarter – century earlier, goes deeper and the French movement may soon lose its two dimensions ("surface and mass") within the three that characterize American work. The Third dimension we already have, to be added to the two in France, is depth.*
>
> *It is this quality of depth that alone can give life to architecture [171].*

Frank Lloyd Wright has left his footprints. When he died, he left 400 buildings and projects he had put his name on [160]. Due to lack of funding, Wright never got any university education, but he taught himself.

Wright practiced the theory that the building shall be in harmony with the construction and the construction shall be in harmony with the materials. No wonder concrete became his favorite material!

We do not know if he was an inspiration for Jørn Utzon's design of the Sydney Opera House, but Utzon was a student at Wright's studio in Taljesin West in Arizona in 1947–1948 [167].

Two days after Frank Lloyd Wright's death in 1959, Odd Hølaas published an article in the Norwegian newspaper *Dagbladet* with the title *Jeg skal bli verdens største arkitekt! -sa Frank Lloyd Wright – og ble det(I am going to be the greatest architect in the world, Frank Lloyd Write, said, - and became so)* [176]. In the article, there is not a word about Wright's death!

The article starts like this (from Norwegian):

My strongest impression of Frank Lloyd Wright was formed when he was standing at the podium in San Francisco Art Museum and demonstrated the enormous butterfly bridge he had planned over at Oakland. This is now 4–5 years ago. I had a strong feeling for him, and if anybody had asked me to mention the most important personalities that have been on this globe at our time, I think I would have mentioned Frank Lloyd Wright as one of them.

It was something like an old man's arrogance over this small tight body. He was already in his 80s, but that could not be seen. The projectors were shining in his silvery hair and the large silk handkerchief that hang out of his breast pocket was shining in his silver lorgnettes between his fingers, and he moved back and forward over the scene with a strong degree of self-respect. He was extremely well spoken, safe, pointed and rude. It was natural that those that not regarded him as a genius would look upon him as one of the biggest poseur that has ever lived. But, he was a genius. Early in my life, he once said, I had the choice between arrogance and humiliating hypocorism. I have chosen honest arrogance, and I cannot see that there are any reasons to change this ···

I cite a few more paragraphs from Hølaas' article:

The Japanese government gave him the assignment to build the Imperial Hotel, something that was

···much more than an ordinary hotel, also a monument over the New Japan. The continuous earthquakes in Japan were a problem, and Frank Lloyd Wright

thought through the task in his own way. It must have been wrong to build on rock. His big building was therefore rested on a raft that should give more resistance and movement-possibility. Then came the great quake in 1924. Half of Tokyo was leveled with the ground, but the building smoothly came back to its original position. Since then, Japanese always sought shelter in the Imperial Hotel when these nightmares of earthquakes came to the city. But, during the Second World War it was hit by a bomb and destroyed. ···

As mentioned earlier, Frank Lloyd Wright's bar in the new Imperial Hotel, a place where normally there are not so many people and it might feel a bit gloomy, but it is elegant.

A visit is recommended!

The last paragraph from Hølaas' article about Wrights many marriage scandals;

···*He was limitless in his anger over the puritanical America because he had been going in and out of marriages in a way that was unknown at the time. A dreadful tragedy came over his house. After he had been in a sort of exile because of his marriage-stories, he came back to Wisconsin in 1911, and built his famous house, Taljesin. He lived there together with Mrs. Cheney and her two children. One day, when he was away, his negro butler got insane and threw a burning torch into the house. The unlucky that managed to get out, he killed with an axe. Six people got killed, including Mrs. Cheney. There were no lack of people that meant that this was God's punishment. But for Frank Lloyd Wright, this horrible reaction that bread over America was the savior. As a demonstration, he built up the house again*···

The stories about Frank Lloyd Wright were many. I grew up with them because my father had worked for him several times. I got Wright's "On Architecture" as a birthday gift more than 50 years ago. After that, it has been many interesting hours with chronologies and articles and books, but as this chapter is not only about Wright, we have to end here.

3.2.2 Pier Luigi Nervi (21 June 1891–09 January 1979)

Nervi was born in Sandoria in Italy in 1891 and got his certificate as civil engineer from the University in Bologna in 1913. In 1920, he established his own company together with Nebbiosi in Rome. In 1932, he again established a company in Rome with his cousin. He designed a number of buildings in Italy and other countries. Among the best known are probably his fantastic new shapes for the four different structures for the Olympic games in Rome in 1960; Stadio Flaminio, Corsa Francia-viaduct, Palazzo dello Sport, and Palazzetto dello Sport.

Palazzetto, the small stadium, is 78 meters in outer diameter, with Y-shaped outer columns, an intricate, symmetrical, stress-rational, and beautiful roof of precast concrete. The components were during erection laid on a steel scaffolding and then cast together at the top. The stadium, or rather, the floor, is sunk into the ground. The hall has an inner diameter of 52 meters and space for 5 000 spectators.

The big palace had the largest spherical shell at its time, with a diameter of 96 meters.

This is concrete at its best!

Nervi was a real concrete enthusiast and shall among others have said: "*reinforced concrete is the most beautiful load-bearing system that the world has discovered until today*" [167].

Palazzetto dello Sport – interior

Before Nervi designed the Olympic structures, he had designed a number of large projects. Some are of the opinion that his elegant stadium in Firenze, built in 1930–1935, was more important as far as the trend in concrete architec-

Palazzetto della Sport – exterior

ture is concerned than the structures in Rome in 1960.

Palazzo del Lavoro, Torino 1961

While he was working with the Palazzetto, he was experimenting with what he called ferrocement. Some people are of the opinion that this technology today is passé, but the technology met with significant reception, not least, in the boat sector (see Chapter 4.1). From an engineering viewpoint, the ferrocement technology is very interesting. With rising labor costs and the need for optimum use of material resources, sometimes in the future, it is not unlikely that ferrocement again will get its renaissance.

Nervi might be called both engineer and architect, but most of all, he was a form-designer. He utilized his constructive knowledge and spontaneous ideas to create new forms and shapes. The statical analysis that his intuition was not off the track came afterwards. Palazzao del Lavoro in Torino is another example of Nervi's stress-optimal structures.

Ernest Hauser wrote an article in the journal *Byggmesteren* in 1964 after meeting Nervi in Rome [177]. The author did not manage to hide how impressed he was after having met the maestro. On a day in January in 1960, Nervi took him up on a hill outside Rome, from where they can see the Olympic structures, lit up by myriads of internal lights, "Perbacco" (by Baccus!) Nervi mumbled and he seemed very satisfied with his work.

Le Corbusier was also satisfied with Nervi's creations. He shall have said "*To see Nervi create concrete skeletons to a structure is a wonderful experience. He never creates anything that is clumsy. What an elegance! He does not call himself an architect, but he is better than most of us.*"

Hauser describes Nervi in the following manner: *Even if he is overloaded with new assignments and honor awards, he is and stays the same simple building master as he has been all the time. He has little time for partying, but for most of the time he is sitting bent over his drawing board in a over-filled office just across the hall of his private apartment in Rome. Three of his four sons work with him as architects or engineers.*

Nervi-hangar in Orieto 1935–1936

Nervi is of medium height, slim and strong, and has the same ideal shape and elegance and strength that characterize his structures.

Nervi had been seemingly proud of his first engineering work after his moving to Rome. This was a cinema in Naples, "Augusteo." In the basement was a station for a cable car line and the thin 30 meter wide concrete roof was the foundation for an office floor. The structure did not get much public attention because the structures were hidden "*It was a daring structure, and I would have been proud to make something similar now,*" Nervi said.

Nervi was preoccupied by the thought that concrete creates its own shapes in accordance with the laws of static and dynamics, "*A structure is good architecture when it is constructed the right way,*" he said.

One of Nervi's largest works before the Second World War was a number of concrete hangars, built for the Italian air force. Four of them were thin-skinned colossus 100 x 40 meters that were supported by six sloped columns. No one had built anything like this before. Some inspiration might have come from Max Berg's Jahrhunderthalle in Wrocklaw (See Chapter 3.1.5).

During the war, all the hangars were bombed when the Germans withdrew. "*One single American sergeant might have saved them, if he had been there in time*" Nervi said.

3.2.3 Le Corbusier (1887–1965)

Hardly any other architect in the twentieth century is better known than Le Corbusier. The artist's pseudonym, which he took later, was Le Corbusier but at birth at la Chaux-de-Fonds in Switzerland in 1887 he was christened Charles Édouard Jeanneret-Gris.

Even at the age of nineteen, in 1906, he designed his first house in his home town.

Architect Jeanneret-Gris was seemingly a restless soul, and in his early working years, he travelled and worked in a number of offices of architects in Switzerland, Austria, Italy, France, Germany, and Greece. His inspiration was a number of sketches and his reading of several books, such as *Urbanism* in 1925.

Possibly, his most important contribution to the concrete history in his early years was the sketch of his domino-system (probably"Dom-ino" [178]. This drawing, based on a simple principle for a monolithic, static indeterminate, beamless system with cantilevers has become an illustration in engineering and architectural teaching books all over the world. Le Corbusier tried, without luck, to patent the system in 1915.

The autodidact architect Jeanneret-Gris, with great self respect, came to Paris in 1917. A great deal of his works in the years 1915–1925 was to create minimalistic building systems for houses, in which some of them were based on prefabrication, and he tried to patent some of them.

In 1920, he took the pseudonym "Le Corbusier." About the same time he started the collaboration with his third cousin Pierre Jeanneret.

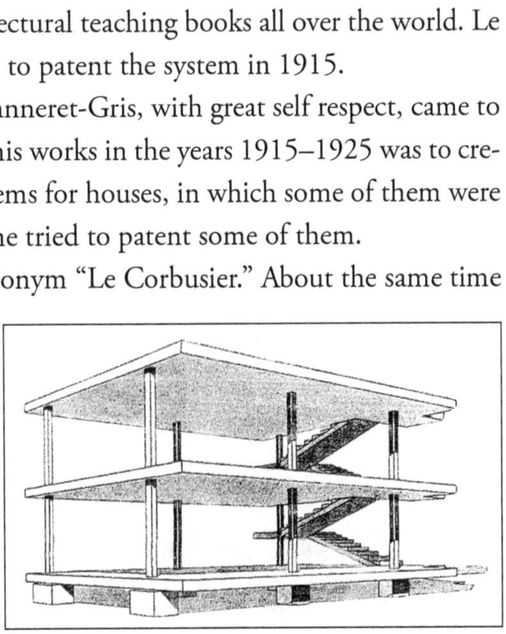

Le Corbusier's Domino-system from 1915

By the end of the 1920s, Le Corbusier was an architectural icon, who in addition to design, wrote and gave lectures about

Le Corbusier and Pierre Jeanneret: Soviet palace in Moscow, model 1931, rejected by the authorities!

Notre Dame-du-Haut

his ideas in a number of countries, from Buenos Aires to Moscow.

The structure by Le Corbusier that most often found a place in publications on concrete is probably Notre Dame-du-Haut in Ronchamp.

Le Corbusier had full freedom as far as the design is concerned and he has utilized the special freeform abilities of concrete in a path breaking way. When he got his assignment in 1950, he was very skeptical, but he fell for the possibilities when he saw the landscape. His brutal concrete, convex shapes, cantilevered shell-roof, window and light settings in various shapes, formats, and colors, have later directly and indirectly inspired buildings in many countries.

Le Corbusier was very productive and impressively variable in his shapes and new creations. It is impossible to give a fair outline of the architect in a few pages, A trip to the library to study a few books is recommended for those who would want to study his works in depth, or are interested in more variation. Here we only include one more excellent example – the Philips pavilion at the World exhibition in Brussels in 1958.

Philips Pavilion, the World Exhibition, Brussels, 1958

The builder of this building was

the Dutch radio, Philips. The structure probably resembles a church more than an exhibition building. The building was equipped with the latest in electronics, both for sound experience and light effects. The intricate double curved shell-roof was prefabricated against sand molds and bound together by posttensioned cables.

In the morning of 27 August 1965, Le Corbusier walked into the sea at Roquebrune, where he died of heart attack, possibly to fulfill his own words that "everything returns to the sea" [178].

3.2.4 Kenzo Tange (04 September 1913–22 March 2005)

Kenzo Tange was born in Shikoku in 1913. He studied architecture at the University of Tokyo in 1935–1938 and he took up his doctoral research in 1942–1944. His thoughts and theories on architecture were to a great degree shaped when he was a professor at the University of Tokyo in 1946 and 1974. He is probably best known as the forerunner of the modern Japanese town planning, but his ideas on form shaping in concrete has also been followed by many after him.

He was visiting professor at MIT (Massachusetts Institute of Technology) in USA in 1959–1960. Among the many awards he has received is the Gold Medal from the Royal Institute of British Architects [167].

After having won the architect competition, he was asked to design the Hiroshima Peace Centre. A visit to this museum will for several reasons make strong impression on any "concrete-person."

On 6 August 1945, the first atomic bomb was dropped at the center of Hiroshima. Three days later, the second bomb was dropped over Nagasaki. In 1949, it was decided to establish a peace park where the bomb fell. Kenzo Tange's museum is situated near the southern end of the park. The memorial for the dead is in the center of the park in the shape of a double curved arch structure. The museum building floats on columns from the ground.

During a first visit about 25 years ago, I found that the elegant and genial concrete structure was not the only one that made an impression. Inside the museum, they among others showed various objects that had been affected by the bomb. Some of them were concrete products that

The peace center in Hiroshima built in 1949–1950

1960–1964 – Tokyo's Olympic halls

had been hit by the radiation. Concrete surfaces were directly hit by radiation and the parts that were shielded by a steel handrail, which had in areas reduced the damage, had made "pictures" on the surface. In view of my experience and knowledge about the durability of concrete in general, the intensity of the damage effected by radiation was almost shocking. Together with the other presentations in the museum, there are presentations on atomic weapons and probably not many who leave the museum would believe in the use of atomic weapons, at least people with a minimum knowledge in concrete technology.

In 1964, the two halls for the Olympic Games in Tokyo were ready. Here the clean horizontal line from the peace museum had changed to roof constructions with images from the old Japanese Shinto temples. In 1965, Tange together with Noboru Kawazoe wrote a book about the temple in Ise. The first temple in Ise was built by the first Emperor (Mikado) who ruled over Japan, Temmu 678–687. It was rebuilt by Empress Jito in 692 and the building we see today is a copy of this. Tange was fascinated by the shape of the roof and he recreated similar forms in Tokyo.

Town Hall Tokyo

The roof of the Olympic halls is not concrete, but steel, but the cantilevered grandstand shows that he could also master the art of molding concrete into forms. When it was built, the roof was the greatest indoor structure of its time.

The Tokyo halls were probably also an inspiration for the Olympic structures in

Munich, which were designed in 1972.

While the first two examples have a somewhat horizontal expression, this is not the case with the town hall in Tokyo, built in 1986–1991.

Kenzo Tange was not only a master of giant dimensions but he also had a fabulous sense of details.

By standardizing and carefully designing the precast details, he used connection details to break the monotony and gave life to the facade as well as giving the viewer an impression of the detailing inside the structure [179].

Tange – detailing in precast concrete

3.2.5 Some Norwegian designers

Many could have been mentioned. We have chosen five designers who have had their influence on Norwegian concrete design and mention them in alphabetical order

Jarle Berg (1908 –1972)

Jarle Berg, who was educated at NTH (the Norwegian Technical University in Trondheim) in 1935 [180], designed among others the Thiis building, Oslo, and the parking garage at Hjortneskaia, Oslo (1959), and a number of other buildings with precast concrete in the façade.

Berg's play with light and shadow in the components at Hjortneskaia was followed by a number of architects afterwards. He was a pioneer in the use of precast concrete during the early years of precast concrete in Norway.

The façade components in the ground floor at Hjortneskaia was shaped by the artist Ørnulf Bast [181].

Sverre Fehn (1924–2009)

Sverre Fehn was an exponent of what might be called "poetical modernism," and had to his merit a number of structures that have received "the Norwegian Concrete Award";

The car house at Hjortneskaia (Hjotnes quay) was demolished toward the end of the 1980s. The building got "the Norwegian Concrete Award" in 1962. The building was a transit-garage for 500 cars and reminds one of Frank Lloyd Wright's Guggenheim Museum in New York.

- 1976: Museum building, Domkirkeodden (cathedral peninsula), Hamar
- 1992: The Norwegian Glacier Museum, Fjærland
- 2000: The Ivar Aasen structure
- 2008: The Gyldendal House, Oslo
- 2009: The National Museum, Oslo

In addition to a number of Norwegian Awards, Fehn also received awards for his designs in Sweden, Denmark, Finland, France, and Scotland [180].

Fehn was a professor at the Architectural High school in Oslo from 1971.

Fehn's poetic modernism is also expressed in a report to the journal *Byggekunst* in 1952, after his visit to Morocco (translated by C.Norberg-Schulz [182]).

The Norwegian Glacier Museum in Fjærland lies like a deported stone in the field where the glacier was once ruling the landscape.

I discover and I am in what I discover. Today, when one visits French Morocco to study primitive mural architecture, it is not a voyage to discover new things. One recognizes. This is how Frank Lloyd Wright's house in Talesian must seem – an entity broken up in parts and with the same coarseness of the materials. Mies van der Rohe's walls must be like this. The same character of infinity. Here, too, is Le Corbusier's poetry of the terraces and the roof. Architecture works to perfection because it works in a timeless space. Its signature is Anonymous, for it is nature itself.

The Ivar Aasen structure

Jan Inge Hovig (1920–1977)[183]

Hovig was educated at NTH in 1946, and after his studies, he established an office in Narvik in northern Norway. He became town architect in Narvik in 1950 and after he moved to Oslo, he established an office there [180].

Jan Inge Hovig passed away early, but he left his footprint in, among others, the Tromsdalen church.

Tromsdalen church was also called The Polar Sea Cathedral. The large roof slabs were made in lightweight concrete aggregate. Together with the Tromsø bridge in the background, it is an important part of Norwegian concrete history. The consulting engineer for both structures was Dr. Ing. Andreas Aas-Jakobsen

Lund and Slåtto

Slaatto and Lund – 1985 [184]

Kjell Lund, born 1927, took his exam from NTH in 1950 and Nils Slaatto (1926–2001) in 1947 [180].

Details ftrom the Kredittkassen building

Lund and Slaatto have had the fantastic ability and impressive knowledge to transform most building materials into fascinating and practical structures and received numerous awards for this. By being the architects who had received the highest number of Norwegian Concrete Award, "Betongtavlen," awarded by the Norwegian Concrete Society and Norwegian Architects Association, they have set their deep footprints in Norwegian concrete history. We let the awards narrate the story;

- Betongtavlen 1977: The Veritas Center, Høvik
- Betongtavlen 1988: "The Silver Building" – Stavanger Cultural House
- Betongtavlen 1989: St. Magnus Church, Lillestrøm
- Betongtavlen 1996: Grønland Bridge, Oslo
- Betongtavlen 2006: The new Svinesund bridge between Norway and Sweden

Office building and laboratories for The Norwegian Veritas (1972–1976). Columns, beams, and U-shaped floor components in precast concrete. Offices 43 800 m^2 and parking garage 9 000m^3 [184]
Betongtavlen 1987: Kredittkassen's (bank) new head office, Oslo

The interior of the "The Silver Building"

The St. Magnus Catholic Church at Lillestrøm has also received a number of other awards,such as the Norwegian Precast Concrete Award 1998, Citation Award IFRAA USA 1992, and A.C. Houens Fonds Diploma in 1991.

The new Svinesund bridge over the Iddefjord between Norway and Sweden was opened in June 2007

Erling Viksjø (1910–1971)

Erling Viksjø took his exam from NTH in 1935, participated in the creation of the "Naturbetong" (Nature-concrete) époque, with buildings such as the government quarter and a number of other important buildings in Oslo to his credit[48].

Erling Viksjø won the competition for the building of structures for the Government-building in 1939, but the war delayed the construction until 1950–1958. His design conquered new frontiers in many ways. The structure negated Oslo's building-size code at the time and was forward-looking in its idea. The façade was sandblasted, and with a decorative pattern in sandblasted concrete it was something that was never attempted before. Erling Viksjø developed the methods along with the artist Carl Nesjar. Several

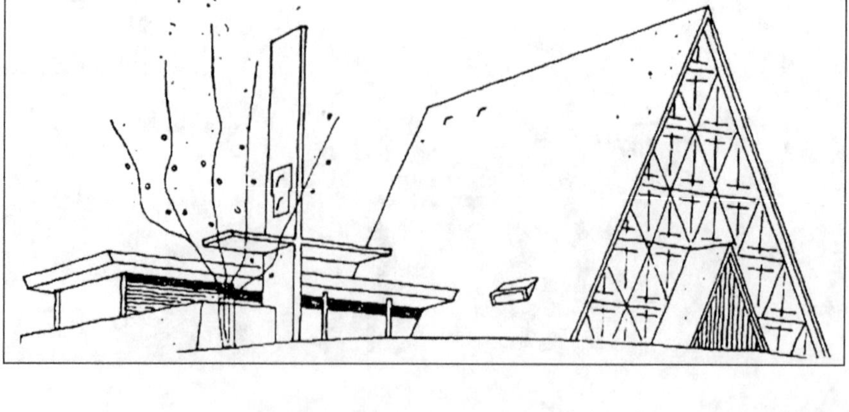

Bakkehaugen church

artists were represented in the decoration of the building; Kai Fjell, Tore Haaland, Carl Nesjar, Pablo Picasso, Inger Sitter, and Odd Tandberg.

Viksjø also designed Bakkehaugen church that received the Norwegian Concrete award in 1961. Also in this case the decorations were done by the artist Kai Fjell.

The Government building, Oslo anno 1960 [185]

3.2.6 Others

There is no doubt that many examples of historic concrete designers could be mentioned, however, some more are also mentioned in other chapters. But still many have not been mentioned.

- Among them could be some idea-creators like Santiago Calatrava, who with his thrilling shell structures has and will create a school in the future. But Calatrava is too young (born in Valencia 28, July 1951), and he will probably be there when the history is written in 50 years or so.
- Another architect who has not been mentioned is Candela, who was a forerunner of double curved shells.
- But the idea behind this chapter has not been to give an overview of excellent architects. For that purpose, we rather recommend a trip to the library to view and read some of the many fascinating books on this topic. The intensions behind this chapter have been to remind about some of the people who have created and inspired in creating the many shapes and possibilities we see in concrete today, and that still makes concrete the most important building material in the world.

The designers we have chosen have exploited the possibilities in designing concrete and created some of the best traditions in concrete designs. But people do not always see this. They only see the grey mastodons in concrete in which shape and design have been given by reasons other than architecture and technology, namely that concrete is an inexpensive material, which can quickly and simply provide help when there is an urgent need for buildings in the society. A lot of the "grey" concrete, we are less proud of today, has its background in the need for quick, effective, industrialized, and cheap construction of apartments after the Second World war. At that time, the architects had to go behind the economists, the stop-watch-specialists and the engineers, when decisions were made.

In Norway, a separate "Housing Directorate" (*Boligdirektorat*) was formed to speed up the development. We have chosen to cite a bit from a more than 50 years old article from a Memo from the Directorate (BD *orientering, Melding fra Boligdirektoratet*) Nr. 3, 1959 [186].

The article was written by Carl Anonsen and has the title; - *Notater fra en samtale med statsminister Krusjtsjov om russisk boligbygging. (Notes from a conversation with the Prime Minister Khrusjtsjov regarding the Russian building of houses.)*

UN's economic commission for Europe had its own housing committee. After meeting in Geneva, the committee went on excursions to the member countries for 10 to 14 days. After the spring session in 1959, the 30-man strong delegation went to the Soviet Union. On Saturday, 4 July, they were about to meet representatives from Gosstry (the organization in charge of building activity in the Soviet) and Vice Prime Minister Koutcherenko. The day before, they also got the message that between 11 and 13 hrs. they shall meet Khrusjtsjov. At the same time, they were told that they should not go soft on their criticism of the Russian way of building houses and apartments as the Russians liked to learn.

The dialogue with Khrusjtsjov is interesting: it describes a situation in which the whole Europe wanted to build quickly at a cheap cost, to fill the gap in housing demand.

A few paragraphs related to design are cited here;

- As an introduction, Krusjtov says,
  "The finish" in our apartments are miserable, but the construction is good. The houses will last for 500 years··· The capitalistic countries might have some advantage in their army of jobless people, we do not have such an army, we have to recruit the labor force for our increased industrial and building activity from agriculture, where rationalization gives us a surplus. The building activity must therefore to a higher degree be transferred to the industry.
- As an answer to a comment from a representative from England, in which he has criticized the architecture and the town planning, Khrusjtsjov stated
  I agree that the architecture is not so good, but that relates to what have been

built so far. This will change now. He talks about his impressions from London, and he was not impressed with the architecture when it relates to common people, long monotonous rows of brick buildings. He gets personal, and talks about his childhoods experience, and that he in London gets the same depressing feelings as at that time. He gets more rational again and gets into the demand about simplicity in architecture versus the need for a certain "festivities" and variation. The houses shall be comfortable to look at. He is teasing about the Russian failures in trying to create a type of new monumentality in the architecture. He calls these buildings for "Turkish baths." The apartments shall look like apartments, but for public houses and other official buildings there might be room for some decorations in the architecture I have seen apartment building in Finland. That is good architecture!

- The following was given by way of an answer to the comments from the representative from Luxemburg;
  He tells that he 2 years ago was in the Don- area and looked at apartments for people from the mining industry. One of the engineers was very proud when he showed one of the "Turkish Baths." They had wasted materials and labor force so that I could cry. The man did not understand why I criticized the buildings. He might have been a very good engineer, but he understood nothing about architecture. When even the engineers does not understand how ugly it looks, then it will take time to change the production.He then starts to answer an English delegate about the one-sided use of 5-stories buildings. He refers to a British investigation that has shown that this is the cheapest way to build. The investigation was used as a model in Soviet, where they concluded with the same results. There is much more expensive to build higher buildings. It is the land speculators that have created the sky-scrapers in USA. This is not a consequence of economical analysis. He tells that he gets letters from people that want to move down from higher floors, but never the opposite. Even in a 5-story house there should not live families in the top floor. The architects themselves never live in houses of more than two stories, - if they can avoid it.

- The answer to the representative from Netherland who had taken up questions about living habitats and town planning on one side related to the long life expectancy of concrete structures on the other hand;
  This will not mean that the development goes towards larger apartments. Larger apartments mean more work for the people that are living there. We will look closer into this question towards the end of the 7-years plan. However, the family life will be different when every family gets their own apartment. We cannot handle this problem now. We must have more wealth first. It has taken 800 years to build Moscow. In this period it has been built between 11and 12 million m^2. This year we build 2.7 million m^2, next year 3 million m^2. If we continue at the same speed, we will have a new Moscow in 3 ½ years.

4 The many aspects of the concrete industry

What might be called "the concrete industry" covers a wide range of technologies, uses, and trades. In this chapter, We have tried to cite some historical examples and provide an overview of some of these sectors and at the same time remind the readers about the many challenges and ideas that concrete represents. Most importantly, the chapter also briefs about the many enthusiasts who have played characteristic roles in the industry and set their footprints in the history of concrete and have spread technology, experience, and enthusiasm among their colleagues in the concrete industry and the society as a whole.

Of the areas we have chosen not to mention too much about is, for example, concrete research. Research is and will continue to be in the future an integral part of all the various sectors we have mentioned, but research is also a separate sector in itself that covers knowledge that is common for all the various industry segments. We have probably been a bit unfair to the research sector as we only briefly mention it in a few sentences here and there. However, a thorough historical analysis of the concrete research sectors would probably demand a book with several volumes of what has been the intension in this book.

Old concrete testing machine. Today it is found in the hallway at SINTEF Byggforsk (Norwegian Building Research Institute) at Blindern in Oslo.

Another example of the many subtechnologies in concrete that *we have not* covered properly is Fiber concrete[187]. George Hoff from USA stated during a "historical conversation" in a conference on concrete in Seville in Spain in 2009 that he for many years collected material about the topic and that he made a reference list of the literature in the area. In 1984, this

Description of the testing machine on the last page. A bit of Norwegian concrete history in itself

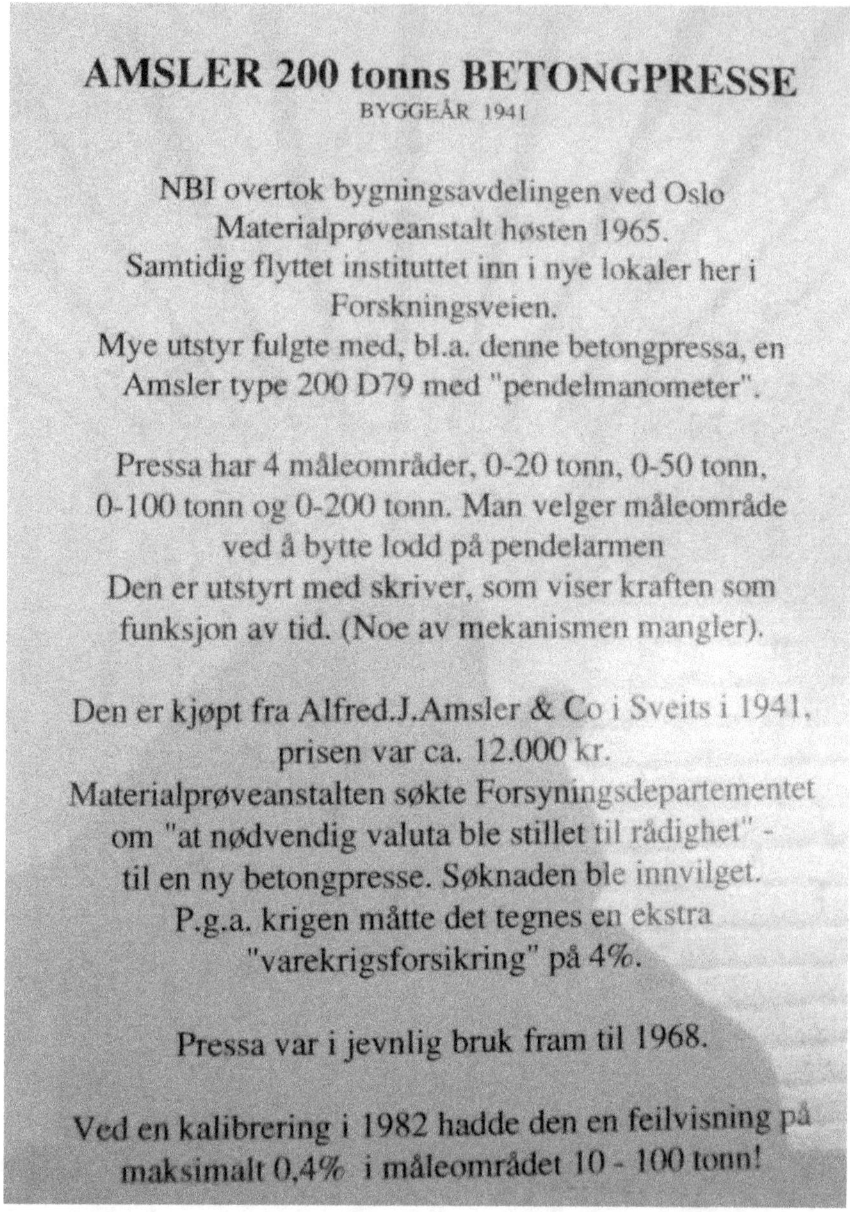

had come up to 3 700 references and then he gave up on this work. He commented that a great deal of the material was references regarding glass-fiber reinforcement and little about plastic fiber. He was convinced that if someone made a survey today, the main content of the literature would be about steel fiber and plastic fibers.

In a lecture at the annual Norwegian Concrete Day in 2009, Standal mentioned that the use of fiber-reinforced polymers in concrete was first demonstrated in the 1950s and that the interest was renewed in the 80s [188].

Yet another subarea that is not covered is standardization. In a paper at the annual Norwegian Concrete Day on 12 October 2009 [189], Steinar Leivestad states that the first concrete specification in Norway was Koldrup's textbook published in 1893. (Leivestad is probably correct when it comes to reinforced concrete. Rye's specification of 1875 was for unreinforced concrete (see Chapter 2.4).

With reference to Leivestad and Vik in 1995, Steinar Helland presented a paper at the Norwegian Concrete Day in 2009, showing the following development in the maximum allowable concrete strength in the Norwegian Standards [190].

The Norwegian Standard for the maximum allowable concrete strength to carry a load of 100 tons and minimum area of a short column:

Norsk Standard	Høyeste betongfasthet	Nødvendig tverrsnitt for at en kort søyle skal bære en last på 100 tonn
1923	C - 18	51 cm × 51 cm
1939	C - 29	40 cm × 40 cm
1963	C - 45	28 cm × 28 cm
1973	C - 65	23 cm × 23 cm
1989	C - 105	15 cm × 15 cm

Development in the maximum allowable concrete strength in Norwegian Standards

Two days after Helland's lecture in Trondheim, Norway, George Hoff delivered a lecture in Seville, Spain [101], in which he gave an overview of the early international standards for concrete;

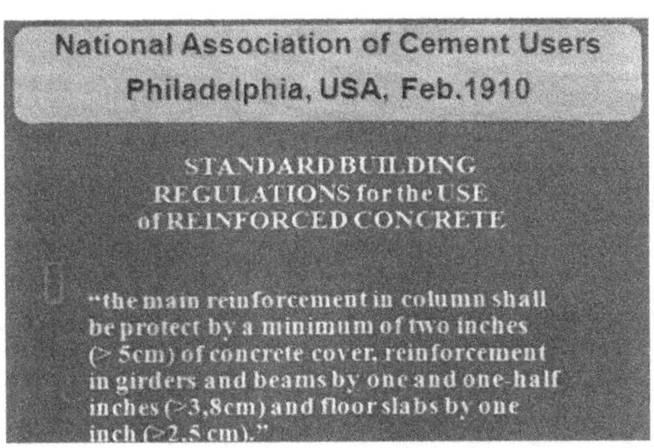

Summary of the early developments in concrete technology in USA according to Hoff [101]

Was it the load or dimension or concrete cover standard or the corrosion-resistance rules that were not followed?

- 1903: Switzerland
- 1903: Germany
- 1906: France
- 1907: England
- 1910: USA

We have in earlier chapters also referred to concrete regulations;

January 1886; Metropolitan Board, London – regulations for concrete (Chapter 2.2).

Prussian concrete norm in 1887 (Chapter 2.2).

The standards have many functions; one of them being the use of experience to secure a reasonable "standard" from the construction work that takes place. Increased knowledge in the field has resulted in a ten-fold increase in the number of pages written about the standards and in what may be called the standard regime.

4.1 Concrete boats and ships

Concrete has a more than 150 year old history as a material for making boats and it might seem like this use has over time come in instalments.

The first known concrete boat is Jean-Louis Lambot's (France) rowboat of 1846. The wall thickness was 30–40 mm and the boat had a length of 3.6 meters. The boat sank in an accident, but had later been recovered from the mud and been exhibited 100 years later.

In his book *Seagoing and Other Concrete Ships* published in 1922 [191], Nic. Fougner draws a parallel between concrete boats and those made of iron. Through thousands of years, boats and ship had been built in wood until a barge was built in iron in Yorkshire in 1777. The first "iron steamer" was Aaron Manby, built in London in 1821. Two historical incidents speeded up the changeover to iron, and later steel. The coastal ship Garry Owen was built in Birkenhead in 1834 and together with several wooden ships it was stranded on a reef during a storm on its virgin voyage. The wooden boats were wrecked, but Garry Owen was nearly unhurt. The first Atlantic ship in iron was Great Britain. She was built in Bristol in 1843, designed by I.K.Brunel, and was the largest ship of her time. After one year of use, she stranded outside the coast of Ireland. Because of inclement weather conditions, she was left there through the whole winter, exposed to the forces of the nature. When she was collected, or rather saved, they found rather minimal damages.

An important argument against the iron ships was that they were too heavy and would sink. (the first steel ship was registered in 1878). Similar arguments have later been used against concrete boats and ships.

When Fougner wanted to design and build ships in reinforced concrete, he got little support from the shipping department. It was accepted that concrete could be used for barges, but not many accepted the use of concrete in seagoing ships. It is interesting to read Fougner's arguments for using concrete contra steel (from his book in a 1922) [191];

> *Considering, therefore, a concrete ship and a steel ship of equal deadweight carrying capacity, it is obvious that the former will have a greater total displacement because of her greater weight of hull; and the greater displacement can, in general, only be obtained by the adoption of larger dimensions. Hence the concrete ship has a greater cubic capacity, and a larger space available for deck cargo, than the steel ship. This is a distinct advantage in trades where the cargoes commonly carried are light and bulky, measuring about 70 cubic feet, or more, per ton weight. For example, by far the greater proportion of the war supplies transported from the United States to France were so light in relation to their bulk that the ships carrying them were completely filled before they had on board sufficient weight to sink them to their designed load draughts, Of course, the increased capacity involves the disadvantage of increased harbor dues, since these are assessed on a cubic capacity bases.*

The first seagoing concrete boat was the Norwegian ship **Namsenfjord**, with a displacement capacity of 400 tons and a length of 26 meters, built in 1917, based on Fougner's drawings. However, Fougner had already built a concrete barge in Manila in the Philippines during his work as a civil engineer in Asia. Some authors say that the building of Namsenfjord came as a result of the World War. This is not completely correct. The planning had already started in 1910–1911, and Fougner's first patents on the subject are dated January 1912. Lack of funding resulted in the delay in the completion of the building.

In January 1917, Fougner got a contract from Mr. Anzjøn from Namsos for a total cost of 90 000 kroner. Anzjøn was to use the boat for timber transport along the Norwegian coast and across the North Sea. The length of the boat was 25.2 meters, the width 6 meters, and the depth 3.5 meters. The loading capacity was 200 tons. It was designed for a crew of seven persons. Namsenfjord was launched into the sea on the morning of 2 August 1917 with the help of a crane from the Norwegian navy. On 5 August, the engine was placed in position, and the ship was towed to a wharf in Moss. The official test voyage was on 8 August, with inspectors from the

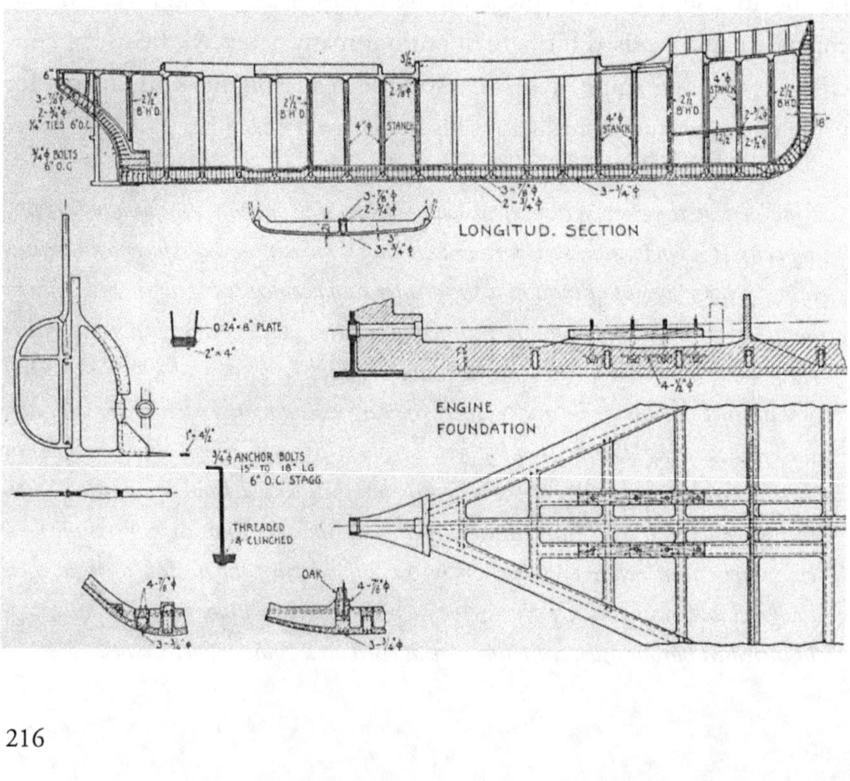

Longitudinal section of the Namsenfjord

U.S.S. Selma being launched at Mobile, Alabama, in June 1919

Sea Directorate and consultants on board. A strong gale was blowing over Oslofjord, also called Christiania fjord, on 31 August, when Namsenfjord carried its first cargo from Moss and delivered it in Skien. When the boat returned, it was ordered to be docked for inspection by the Sea Director with his assistant and "the concrete ship committee" on 10 September. Right after the inspection, Namsenfjord carried freight during several trips between Moss and Drammen. On 6 October, the official seaworthy-certificate was issued.

Fougner later built the boats Stier and Askeladd at the wharf in Moss.

Another concrete boat pioneer was Ingeniør Harald Alfsen. He developed his own launching method. The boats were launched into the water upside down and the principle of flotation ensured that they floated up in the right position. One of Alfsen's boats must have been taken care of in later years, namely the freighter XI, that was built at Porsgrunds Cementstøberi in 1917. In the years around 1920, several concrete boats were built in the counties of Øst- and Vestfold in Norway [52].

Then came the lightweight concrete boats (see also chapter 4.4); first was **Atlantis** with a capacity of 300 tons, built in 1918. Then **Selma** was built. Selma had a length of 132 meters and a wall thickness varying from 100 to 127 mm. The reinforcement cover was 16 mm. The aggregate for Selma was produced in a rotary kiln at the Atlas Cement Company in Hannibal MO, USA, and the boat was built at the governmental wharf in Mobile, Alabama [192].

Selma, which was launched in June 1919, had got her name from the town of Selma in Alabama [193]. Selma was the biggest ship that had been launched sideways. There were some worries about this, but 20 of the employees at the wharf chose to stay on board during the launching.

In July 1953, The Engineers Testing Laboratory Inc. in Houston, Texas, got the assignment to inspect Selma in the harbor of Galveston, after 34 years in the sea. The report dated 10 September 1953 stated that the hull was in good shape, both where the concrete had constantly been exposed to sea water and where it only had been exposed to salt and washing. The little corrosion that was found could have been there from the time of construction [193].

Building of concrete boats in USA stopped in 1918. Only the boats under construction were completed. As a result of this only 12 ships were built; deadweight in tons, in instances where this is known, is also provided:

Polia	(cargo)	2 460 tons
Atlantus	(cargo)	2 542 tons
Cape Fear	(cargo)	3 078 tons
Sapona	(cargo)	3 078 tons
Selma	(oil-tanker)	6 380 tons
Palo Alto	(oil-tanker)	6 380 tons
Peralta	(oil-tanker)	6 380 tons
Cuyamaca	(oil-tanker)	6 380 tons
San Pasqual	(oil-tanker)	6 380 tons
Moffit	(oil-tanker)	6 380 tons
Dinsmore	(oil-tanker)	6 380 tons

The cargo ships were delivered between October 1919 and January 1920. The oil tankers, with the exemption of Moffit and Dinsmore, were delivered between May and September 1920.

Ref. [193] is of the opinion that 14 boats were built. It is opined that due to the material shortage in USA during the Second World War, 104 concrete ships were built, with capacities varying from 3 200 tons to 140 250 tons. 24 of these were ocean going ships, while 80 were barges. In total, these concrete boats made up a fleet with a total capacity of 488 000

The 57 meters long Crete Joist

tons, at a cost of 167 million USD. As a comparison, this would have been equal to 46 Liberty ships.

During the Second World War, more than 100 boats were built in a series in Great Britain. This was part of the crisis plan because all the steel had to go to the making of other war materials. The so-called "Crete" boats were produced in several places in Great Britain, in segments that were 62 cm wide and 6 cm thick. Three of the "Crete" boats ended up in Norway. The most well known is "Crete Joist," which was built in 1927. This boat grounded in Fevåg during the Second World War. The Germans tried to blast it to utilize the reinforcement, but the concrete was too solid. Some years later, the harbor authorities in Trondheim tried to bring the boat to float. The plan was to tow it to Trondheim to use it as a wave breaker, but the boat was too very strongly grounded [52].

Some more early concrete boats:

M.S. Linnea was the first engine driven concrete boat built in Sweden. She was launched in December 1918. The builder was A.B. Betonverket of Malmø and the owner was A.B. Betonelt also from Malmø. The boat had a torpedo shaped hull and a length of 40 meters.

Of the other concrete boats built in Sweden, we mention the coastal sailor Yg, which was built in 1928. It participated in several regattas.

S.S. Armisice was the first seagoing, engine driven concrete boat built in England. The ship was built according to the Lloyds Register. On the virgin trip from Londonderry in February 1919, she was loaded with 1 000 tons of oatmeal in bulk. The boat was built by Ferro-Concrete Ship Construction Company in Barrow-in-Furness. The boat was built in wooden molds and concrete had a mixture proportion of 1 : 1.3 : 2.7.

M.S. R.P. Durham was a 2000 tons oil tanker. Durham and a sister ship, which was built in Aransas Pass in Texas for France and Canada Oil Transport Company, New York. Durham was launched sideways on 24 July 1920.

In a correction to his book in 1922, Fougner stated that the biggest concrete boat in Europe was no longer S.S. Bartels (see below), but Persivoranza that had just been launched in Lavagna in Italy. The boat was designed by Signore Galliquani and had a displacement of 3 000 tons.

Fougner also opined that by 1922, concrete boats had been built in China, Denmark, England, France, India, Italy, Norway, Spain, Sweden, Germany, and USA as well as in other countries.

A small boat story; the ship wreck of Håndverksbåten (S.S. Bartels) outside the island of Øland in the morning of 30 June 1994;

There are some variable opinions about the final destiny and the details around the compensation in the case. The case of the shipwreck went through several levels in the courts of justice. Examination of evidence happened in Sødra Roslags Tingratt (court) in Sweden on 24 April 1996. The sentence was pronounced in Stavanger Byrett (town court) on 17 October 1966. Juridical complains were made on 30 May1997, which were directed to the court on 30 May the same year. The case was with the Skien and Porsgrunn Town Court on 13 April 2000 and the complaint case at Agder County court in Skien during the period 27 March–04 April 2001.

We will try to stick to some simple facts:

- Nic. Fougner's book *Seagoing and other Concrete Ships* [191] published in 1922 discussed the boat. It was named **S.S. Bartels** and was **Europe's largest concrete boat** when it was launched in Denmark in June 1920. The boat had a displacement of 3 300 tons, of which 1 500 tons was the deadweight of the ship. The ship was perfected at the Norwegian Veritas. A lecture about the boat was given at the Danish Engineering Society on 18 February 1920.
- In his book, Fougner criticizes the boat and is of the opinion that the double bottom is less rational than a massive bottom.
- The boat was 72 meters long and 11.6 meters wide.

- After launching, the boat was put into coal traffic in England.
- Fougner writes in 1922 that the boat went on ground "some time ago" and he opines that the leakages could have been avoided with a thicker bottom.

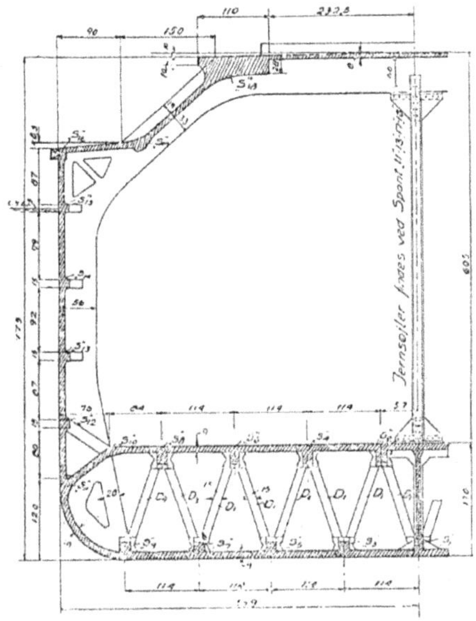

Fig. 57.—S.S. *Bartels*, Midship Section.

Illustration from Fougners book[191]

- In an article in *Byggmastaren* (Sweden) in 1982 [194a], there is a mention about the somewhat "variable" life of the boat. They claim that it was built in Germany in 1918, and that concrete was chosen because of the steel crisis. They opine that Bartels was first used as a cargo ship and that it among others sailed across the Panama canal. It is also claimed that the ship was used in troop transport during the First World War. After some time, the boat came to Karlskrona, where it was a living quarter for the Swedish navy. In the 1930s, it was bought by Ekersberg Warf to be used for storage.
- *Byggmestaren* further reveals that the boat went through extensive repairs in 1982 before it was used as a shelter for local craftsmen in the Hammarby harbor in Stockholm. A group of local investors had bought the boat, with the intention of using it for craftsmen shops, offices, and apartments.
- At some point in time, it was repaired by filling the bottom up to the base of the lowest deck with some 500 tons of concrete. After this, it tilted a bit toward the starboard side.
- In the court trial after the shipwreck, simple drawings in the scale of 1:100 were presented. The drawings were dated 1982 and 1984 and in addition there was an undated drawing probably from a later time.
- The boat was bought at an executors sale for the third time in 1993 initiated by a Norwegian investor group. They planned to tow the boat to Bergen and engaged a towing company from Langesund in Norway to tow SS Bartels' Handverksbåten.
- The towing operation with the boat M/S Vrangsund from Brevik started on 28 June 1994 at 1300 hour from the Hammarby harbor. Nearly immediately, the tow boat got into trouble because Bartels started moving in all wrong directions, variable from side to side. Only 20 minutes after the start, they were afraid that it would tear down a bridge (Vesterbroen). At this point in time also, there was contact between Bartels and the tow boat. The technical experts in the later court cases (I was one of them) did not agree on the effect of this small collision. The towing operation continued down the Sødertelje Channel.

- The towing operation started in the open sea at 06:20 on 29 June, and till 12:00 the same day the weather was fine and there was little wave activity. Then, close to 19:30, the wind had increased to 4 meters per second and the wave height had increased to about half a meter. After one and half hours, it was less windy, but the tow started to move from side to side. At 23:00, the speed of the wind increased and the tilt toward starboard also increased. At 23:30 there was a crisis. They were now 8 nautical miles east of the north tip of the island of Øland. The tilt had increased to 5–7 degrees. The speed had been reduced further, and via Stockholm's radio they contacted the Swedish Coast Guard and a possible rescue ship with pumping equipment. They got in contact with a rescue vessel at the west side of Øland, but this boat did not have enough capacity to help. On the way toward land, the towing wire hooked up because of the slow speed.
- At 3 o'clock they had reached the west side of Øland. They contacted the Coast Guard to get instructions about where to ground the ship. Due to the westerly wind, they wanted to ground it on the east side of the island, at a shallow depth, so that it could be sighted. The tow boat tried to reach Angjarnsudden and to get the tow inside the Knølen pole mark. They did not manage to reach this goal, and SS Bartels sank at 04:30 in the position N57 degrees 21.55 minutes and E 17 degrees 09.22 minutes, about 3 nautical miles from the destination.
- In a later investigation by divers, it was reported that SS Bartels' Handverksbåten was lying at 21 meters depth in the front and 14 meters in the aft – one nautical mile from the northernmost point of Øland, with the front pointing south.
- It is not easy to determine the exact reasons for the damages and the grounding. Some possible causes can be seen from the description provided above. A combination of exposed reinforcement and carbonation resulting in corrosion damages from several groundings, collisions etc, combined with thinner walls over a rather long period of time can either have been one of the primary causes or contributed to the final grounding.
- The tow company was acquitted of possible recklessness in the final court.

We do not know what happened to the wrecked ship. However, SS Bartels was very useful for many different purposes, for 74 years.

It is hard to find what exactly is the shortcoming as far as the construction of concrete ships is concerned or if the interest for concrete ships will return again. However, we have chosen to include an article from the Financial Times in 1979 [194b].

The article describes a tanker under planning for Tampimex Oil, a British company, and stated that the planning had started in 1970. The ship was going to be 950 ft (285 meters) long and take 126 000 m^3 of LNG. The development company was English–American–German and the ship should be classified in the Lloyds register. Dyckerhoff and Widmann seems to be

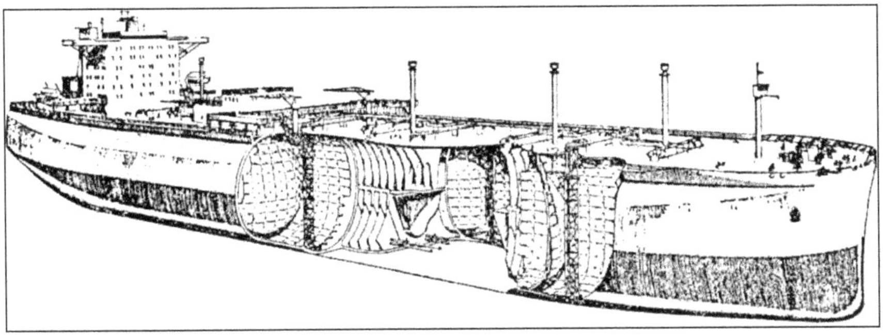

128 000 m^3 concrete LNG tanker

the German expertise that participated in the project and their expert, Dr. Ulrich Finsterwalder, had been involved in the building of concrete ships and barges for the German Government during the Second World War

Concrete Barges

In his book dated 1922, Fougner stated that he had built a number of barges in concrete, first in Manilla in the Philippines, and later in Norway. In 1916, he designed and built two barges with 100 DWT for AS Lægterkompaniet (the barge company), which was at that time the largest barge company in Christiania (now Oslo). He mentioned about the building of 60 and 200 tons barges. He further stated that in 1922, there were more than a dozen concrete barges in operation in Norway and Sweden.

He gave the following table with data about the barges (Fougner gave his data in feet and inches):

	250 tons	***100 tons***	***60 tons***
Length	24 m	19.2 m	14.1 m
Width (maximum)	6 m	5.55 m	4.5 m
Depth	3 m	2.25 m	1.65 m
Weight reinforcement (tons)	10	5	2
Weight concrete (tons)	122	49	25
Cost in 1916 (GBP)	1 750	450	260

Through the years, many barges have been built for a number of reasons in many different countries. As an example, we mention Philippines, where at the end of the 1950s, barges were built in prestressed concrete and later

"Mulberry harbor"– plan [195]

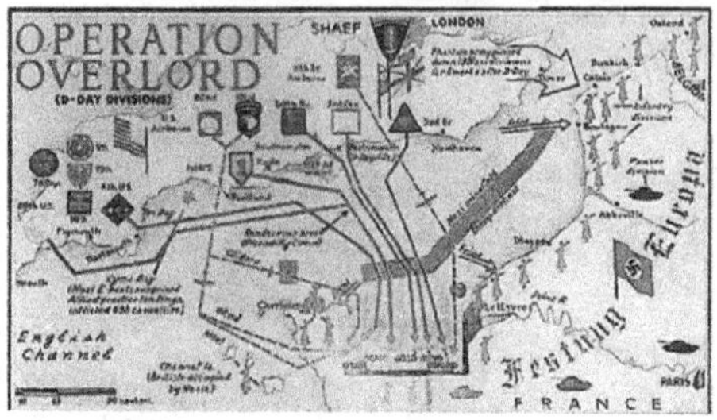

D-Dag – map

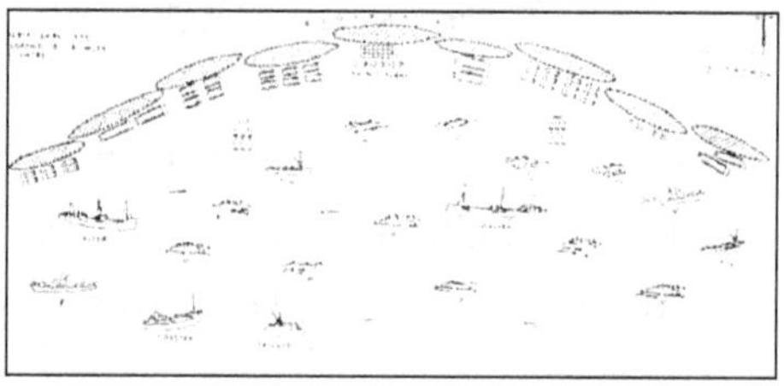

19 of the same type were built in the years 1964–1966.

Around the world, probably thousands of concrete barges have been built. Many of them have also been used for other purposes such as warehouses, living quarters, and floating harbors, or they might now be tourist attractions. Many old stranded barges are found along the English Channel.

Possibly the most special, historic, interesting, and possibly also the most well-known barge building was the building of the components of **"The Mulberry Harbour".**

In 1942, the English Prime Minister Winston Churchill initiated the planning of a floating harbor to secure the landing for the invasion in France during the Second World War. The concept was to make a breakwater in a half circle, with openings for sailing in and out, and a floating harbor on the inside. The breakwater consisted of "barges" or "caissons" in concrete. They were floated over the English Channel and later sunk to the bottom. The harbor inside consisted of four piers, built on floating concrete barges with the top structure in steel. The concrete barges were mainly standardized to 204 ft (61 meter) length, 60 ft (18 meters) height and 50 ft (15 meters) width [195].

A great number of concrete barges were built along the English coast and up along the rivers. On the D-Day, 6 June 1944, the armada of troops, boats, and barges sailed over the channel to Normandy in France. The harbor project for safe landing got the code name "Mulberry harbor."

The English navy needed all the steel plates they could get. That was why concrete got into the picture. The construction mainly took place during the winter of 1943, to get ready for the invasion in June 1944.

More information about the "Mulberry harbor" can among others be found on the internet home page of the consulting company "NCE" [196].

About the same time as the construction of the barges for "The Mulberry Harbor," a similar construction project, with nearly the same barge size, went on secretively on the other side of the globe. In August 1942, the casting of two oil barges started in Japan. They were 64 meters long and 10 meters wide and got the name "Takechi Maru." In February 1943, they were hauled out of the 200 meters long dock, while naval officers and concrete engineers waved their flags as a sign of their success [73].

Later, more barges were built. These boats or barges went to Sumatra with oil. The drawings for the barges have been lost, but the designer was Shojiro Takeshi (who died in 1951). Today, Takechi Maru 1 and 2 are used as breakwaters outside the fishing harbor in Kure near Hiroshima.

We would also like to mention a few barges built for special purposes.

- **The Arco Barge**

The barge, or rather, the gas terminal (LPG), was designed by ABAM Engineering and cast in a dock at the precast concrete factory of CTC (Concrete Technology Corp.) in Tacoma, Washington, USA. After launching, it sailed 16 000 km across the Pacific Ocean to the Java Sea in Indonesia in 100 days.

The barge had a storage capacity of 36 000 tons of propane, in addition to being the living quarter for a crew of 50. It also had its own heliport [197].

Casting of barrel shells for the ARCO BARGE, 1974

From the dock in 1974

"Artist view" of the barge in position

From the launching of the barge.

Owners were Atlantic Richfield and Pertamina.

Key elements in the design were three parallel barrel shells that span 14 meters in the direction of the short side and 61 meter in the direction of the long side [198]. The loading on the barrel shells were approximately 100 times more than what the companies involved had been used to for similar roof structures. The barge was match-cast (side edge against the side edge of the foregoing component) and later poststressed together in the dock. The displacement was 65 000 tons.

They used 25 000 tons of concrete for the barge, with an average compressive strength of 9 800 psi (70 MPa) and a standard deviation of 6% [197].

The Archo Barge or Ardjuna Sakti had a total length of 140.5 meters, width of 41.5 meters, and depth of 17.2 meters. After 20 years of service, thorough tests were conducted on the structure and the conclusion was that it had an unlimited life expectancy.

The terminal was used until 2006, when the reserves in the oil field were exhausted [197].

In February 1973, the shipping company Fearnley & Eger along with Elkem-Spigerverket A/S and Norcem A/S in Norway formed a committee to inves-

The LNG terminal in Italy

tigate the possibility of freighting oil from the fields in the North Sea in a concrete barge. They hired the Norwegian Veritas and the consulting company Dr. techn. Olav Olsen for the design. In the course of the work, they also contacted R.G. Morgan from the University of Bristol and Ben C. Gerwick from the University of California. They were also aware of a doctoral project from NTH, Trondheim, which was a 218 000 tons DWT concrete tanker.

The terminal, which was built in Algeciras in Spain has a total weight of 290 000 tons [201]

The committee gave their report on 5 March 1974 and presented a cost estimate for 3 concrete barges, each with a capacity of 100 000 tons and 2 tow boats, the total cost of which was estimated to be 250 million NOK [199]. A similar estimate for a barge in steel was estimated to be 400 million NOK. The dimensions of the barge hull, consisting of two horizontal cylindrical shells, was length 212 meters, width 48.8 meters and height 25.2 meters. The draft was about 16 meters.

The report from the Norwegian Veritas [200] reports that three alternative shapes for the hull have been investigated;

1. Two parallel tubes, which became the chosen solution.
2. A hull with a section in the shape of an unsymmetrical ellipse, with double bottom and ribs.
3. A section consisting of six cells, three above and three below.

All the alternatives had been analyzed according to IMCO with respect to oil spillage in the sea.

The biggest offshore oil gas terminal for LNG

The Norwegian journal, *Teknisk Ukeblad* in October 2009 [201], discusses a special "barge," cast in Spain, which is now anchored as a LNG terminal in Italy. The company Aker Solution has been responsible for the design and the management of the building. They have utilized their knowledge and experience from building offshore structures in the North Sea, and they were also the precursors of some of the structures mentioned above.

The construction period was one year more than anticipated. During the summer of 2008, the barge was launched from the dock in Spain and towed the 3000 km to its final destination, 40 km south west of Venice. The terminal has an area of 180×80 meters. The design is based on a life expectancy of 25 years.

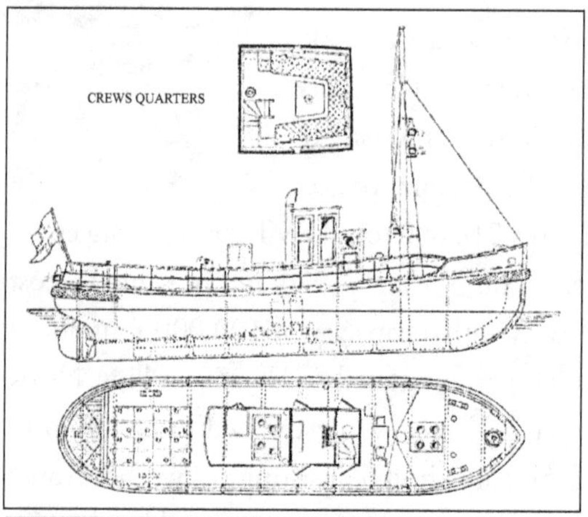

Fougners "Staal-Beton 19"

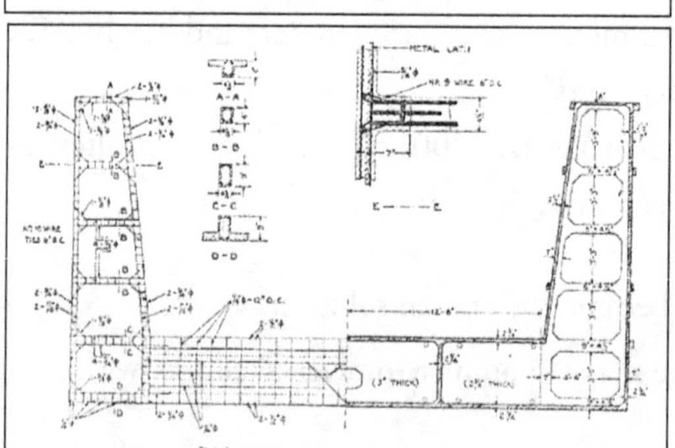

Fougners floating dock, 1917

•The first tow-boat in concrete

Fougner built the tow boat "Staal-Beton 19" in his own cost. The boat was launched on 17 October 1917. During the first winter, the boat was used as an icebreaker at the harbor in Moss and to break ice at Oslo-fjord. When Fougner wrote his book in 1922, he mentioned that the boat had functioned well for four seasons and that there had been no need for repair till then.

The first floating dock in concrete

Fougner built the first floating dock in 1917 [191]. The dock was built on the basis of a contract with the company Maritime A/S in Christiania (now Oslo), and on 10 May 1918, Fougner received a letter from the Managing Director of the company, A. Bryde, in which he expressed his satisfaction, though the dock had been frozen in the previous winter. The dock was stationed in Bestumkilen in Oslo and had the following main measurements:

Internal width:	7.5 meters
External width:	10.8 meters
Length:	24 meters
Total depth of side walls:	6 meters

Fougner mentioned that the first dock had a lifting capacity 90 tons and that he had also designed floating docks with capacities of 4 000 and 8 000 tons.

The first lighthouse ship in concrete

The first lighthouse ship was "ILJERNSFLUA." It was stationed in the sailing route to Christiania (now Oslo) in 1918. For the construction, 6 tons of reinforcement and 52 tons of concrete were used. The ship was 12.6 meters long, 5.4 meters wide, and 2.7 meters deep.

A concrete lighthouse ship was also initially installed in the Irish Sea, East Canada, and Botnic Bay [198].

Ferrocement boats

Inspired by the Italian Luigi Nervi's ideas on ferrocement, a number of ferrocement boats were built around the world in the 1960s and 1970s. One of the most well-known was the designer and engineer Peter Ellen's elegant regatta boat "Helsal".

"Helsal" racing sloop

"Helsal" was launched in March 1973, and participated in a number

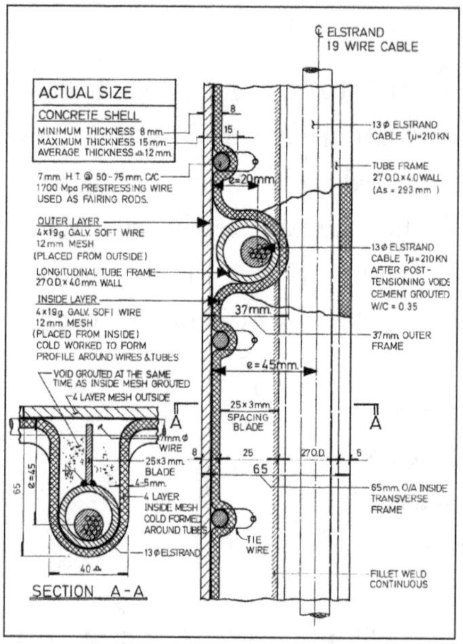

One of Peter Ellen's drawings from "Helsal" [202]

of regattas, among others in the famous Sydney–Hobart regatta, and it also won a number of prizes. From 1979, it went into charter traffic in the Philippines until it grounded on a coral reef 120 miles south of Manila in 1981. The damages on Helsal were not so severe that they could not have been repaired. Helsal had an IOR rating in 1973 of 20.7 meters and had a mast of 32 meters, sail area of 558 m^2 on the spinnaker, 235 m^2 on the genoa-sail and 125 m^2 on the main sail [202].

A rather special feature of Peter Ellen's ferrocement design was that the boat had poststress reinforcement. Ellen was also very proud of the fact that the boat, even after the grounding in 1981, showed no signs of corrosion damages. The mortar that was used had a water/cement ratio of 0.4 and contained a lot of admixtures. The mortar had a compressive strength of 50–80 MPa.

Peter Ellen's "Helsal" is in many ways a special case. As mentioned in the introduction, a number of ferrocement boats were built around the world in the 1970s. The main motives behind the building of these boats were often adventures, interest in boating, enthusiasm, and not least the need to use a cheap building material for a large private boat. We give a small, rather typical example, cited in *Concrete* dated May 1979 [203], "A six-year labour of love";

- Name: Devcon Phoenix
- Building site: Phoenix Wharf, Greenwich, England.
- Length: 57 ft (17.1 meters)
- Weight: 25 tons.

- Construction time: 6 years.(1973–1979)
- Reinforcement: 9000 ft (2 700 meters) galvanisert nett
- Mortar in the hull 2 : 1 – cement:sand – mortar, pressed into the net, together with 70 friends in a bit more than one day in the summer of 1976.
- Motive: to sail across the Atlantic Ocean and to charter out the boat in the Caribbean.

The canoes

Enthusiastic students from Berkeley at Lake Nicol

A small and curious item among all the different types of concrete boats is the concrete canoes. The 22nd annual "America's Cup of Civil Engineering," organized by the American Society of Civil Engineers, was conducted at lake Nicol in Tuscaloosa, Alabama, in 2009, with 22 universities from USA and Canada as participants, all in concrete canoes. The winner was from Berkeley in California, with a canoe that weighed 104 kg and was 6.1 meters long. This is a three-day competition, in which sprint properties as well as the perseverance and design of the canoe counts. ACI (American Concrete Institute) was as usual a sponsored the competition [204].

This chapter about concrete boats only touches upon the many floating structures that have been built in concrete. According to [205], in the years 1950–1982, 1130 floating structures in concrete were built and registered. The impressive structures such as floating tunnel elements and intermediate and permanent floating platforms for oil and gas production, floating bridges etc., that have been built in the last 30–40 years come under another category, apart from the examples mentioned earlier.

4.2 Ready mixed concrete

Ready mixed concrete, or rather factory mixed concrete, is today an obvious part of the picture in all industrial countries, and in many countries it makes up over half of all the concrete that is cast. The general manger of the Norwegian Ready Mixed Concrete Association, Hallvard Magerøy, has written a very interesting article about the beginning of ready mixed concrete in 1997, in which he refers to a lecture by the first president of ERMCO (The European Ready mixed Concrete Organization) Dr. Max Bonner [206]. We have quoted some parts of Magerøy's article.

The idea to make a mixture of aggregate, binder and water ready mixed for building in fresh useable consistency is a bit more than 100 years old. In 1872, a British engineer by the name of Deacon shall have said; "There is no doubt that it would be of great advantage to deliver concrete or mortar ready mixed in the mould at the building site".

Deacon never came as far as actually to see his idea in practice. It was the German, Robert Guthman from Berlin that first put Deacon's idea into practical use. He established the first known factory for delivery of ready mixed mortar already in 1875.

The first factory for production and marketing of ready mixed concrete came just before the turn of the century by a retired "Regierungsbaumeister" Jürgen-Heinrich Magens from Hamburg.

Magens was born 29 May 1857 in Strohdeich in Niedersachsen in Germany. He was educated as a civil engineer. In 1896 he established his own building company in Hamburg.

With support from Hamburg's building control authorities Magens started shortly after the establishment of his company, research on how to keep the concrete fresh over a longer period in time without sacrificing the quality of the concrete. The purpose of the research activity was to develop a concrete that could be transported over longer distances, and by doing so mix the concrete another place than on the narrow building sites in the metropolis.

Magens' first technology to keep the concrete fresh over longer time, was to cool the concrete down towards freezing, before it was transported out with horse and cart, or by railway. This idea comes from the fact that concrete do not harden at temperatures near freezing. For this idea, Magens got his first patent on the topic "Keiserliches Patent nr. DPA 146243, Klasse 80b" 10 January 1903.

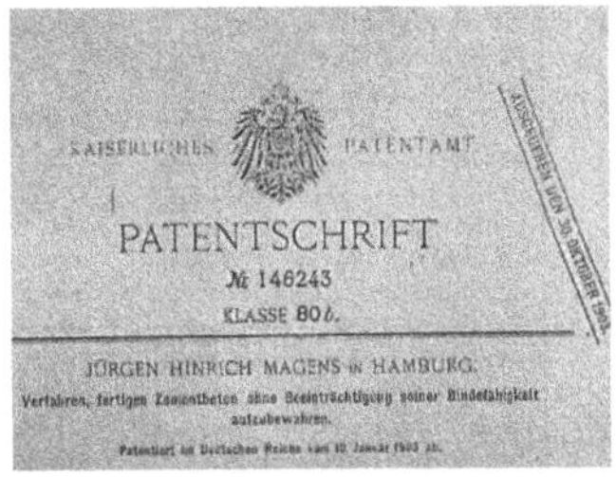
KAISERLICHES PATENTAMT
PATENTSCHRIFT
№ 146243
KLASSE 80*b*.
JÜRGEN HINRICH MAGENS IN HAMBURG.
Verfahren, fertigen Zementbeton ohne Beeinträchtigung seiner Bindefähigkeit aufzubewahren.

Keiserliches Patent Nr. 146243, Klasse 80b

This day, is by many regarded as the birthday of Ready Mixed Concrete, even if Magens must have produced this type of concrete earlier.

Shortly after Magens found out that the same effects could be achieved without reducing the temperature down to freezing. He achieved this by using chilled aggregate, for example by storing the aggregate cold or spraying of the aggregate, and then storing the mixed concrete cold. For this idea Jürgen-Heinrich Magens got his second patent—"Keiserliches Patent Nr. 163501, Klasse 80b," 4 September 1903.

KAISERLICHES PATENTAMT
PATENTSCHRIFT
— № 163501 —
KLASSE 80*b*.
JÜRGEN HINRICH MAGENS IN HAMBURG.
Verfahren, fertigen Zementbeton ohne Beeinträchtigung seiner Bindefähigkeit aufzubewahren.
Längste Dauer: 9. Januar 1918.

Keiserliches Patent Nr. 163501

In 1906, Magens got the idea that by prolonging the setting time of concrete the transport time would also be prolonged. This was done by vibrating the concrete during transport. This way he got his third patent;"Keiserliches Patent nr.DPA 192029, Klasse 80b" 6 January 1907.

KAISERLICHES PATENTAMT
PATENTSCHRIFT
— № 192029 —
KLASSE 80*b*. GRUPPE 3.
JÜRGEN HINRICH MAGENS IN HAMBURG.
Verfahren, fertigen Zementbeton ohne Beeinträchtigung seiner Bindefähigkeit aufzubewahren.
Patentiert im Deutschen Reiche vom 6. Januar 1907 ab.
Längste Dauer: 9. Januar 1918.

Keiserliches Patent Nr. 192029

Through his activity in the Hamburg's architects and engineering society, Magens had great confidence and reputation as a tradesman. His lectures and discussion contributions about factory-made concrete at the German Concrete day – Betontage in 1909 and 1910, also contributed strongly to convince the contractors about ready mixed concrete.

In 1912, the company Schilling Co. and other companies in USA, established ready mixed concrete plants.

In 1926, several ready mixed concrete factories were established in Europe, by the start of a plant in Denmark. Later England and Norway (Ludvig Lorentzen Ferdigbetong A/S) followed up in 1931, Sweden in 1932, and France and Switzerland in 1933…

Mortar transport in Helsinki, Finland, 191 [149]

A forerunner to the production of factory made concrete was factory production of mortar for masonry works and transport of the mortar to the building site. The picture below shows such mortar transport from Pargas Kalk's production factory at Sørnes in Helsinki in Finland in 1912.

Norwegian Factory Concrete Association (NFBF) was established at Hotel Bristol in Oslo on 10 June 1964. The initiative to establish the Association was taken by the Chief Cement Chemist Johs.Rutle in a meeting at Slemmestad in 1962.

A jubilee script at the 30th year anniversary in 1994 describes the beginning of the association [207].

> *Up to the 1960s, the concrete factories were very secretive about information about their production equipment, their mixture design, their control activities (or lack of control) etc. No external people were allowed into the plants. If someone wanted to study ready mixed concrete, they had to go abroad.*
>
> *In the beginning of the 60s, some people in the trade found this secretive attitude to be unreasonable. They started to argue for a more open attitude, and cooperation through a technical association. Johannes Rutle had great confidence among the producers, and was keen on the cooperation thought. In 1962, he arranged a meeting at Slemmestad. Most factories were invited to visit the cement factory and the Eternite factory, with a dinner afterwards. Present were amongst others Åge B. Hvam (Betong Compagniet), Sverre Markussen (Franzefoss), Harry Bråthen (Skedsmo) and Edvin Holm (Svelviksand). After a couple of glasses of red wine, Grill Fasting from Hamar, took the word, and concluded that when they could have such a good time together, it was about time that they established a common technical forum, - an association. One after the other of "the big boys": Åge B. Hvam, Sverre Markussen and Harry Bråthen, stood up and supported the thought. In the end Åge B. Hvam were elected Chairman of a committee to work further with the case.*

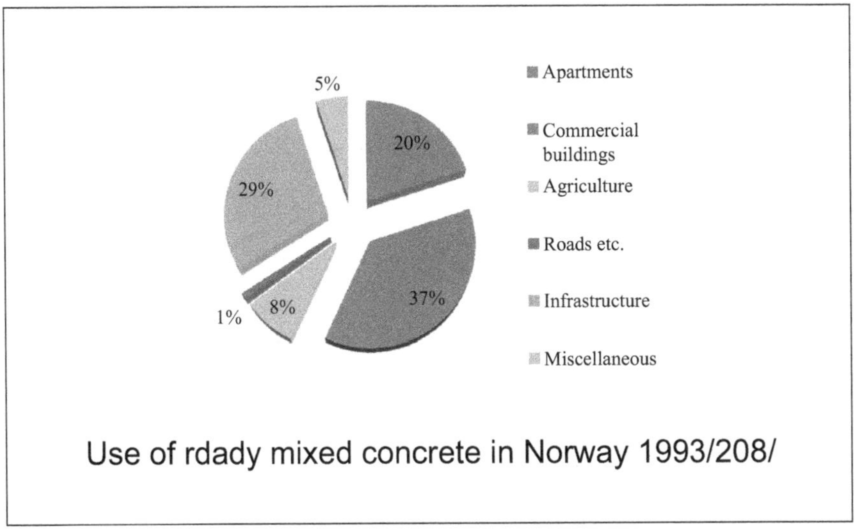

The use of ready mixed concrete in Norway in 1996[208]

At the first general body meeting of the NFBF (Norwegian Factory Concrete Association), a "control committee" was elected. Hans Bøen became the Chairman for the committee. He later became the first Manager of NFBF, and continued in this position from 1967 to 1987. Another member was Olfert Karsen, who later became the first director of the "Control Council for Concrete Products" in Norway.

In 1950, the total ready mixed concrete production in Norway had reached 120 000 m^3.

This was doubled to 240 000 m^3 in 1955, and in the period 1955–1970 doubled again and from then on doubled every fifth year [208]. The best year in Norway was 1997, during which 2.5 million m^3 of concrete was produced.

The use of ready mixed concrete has spread into many sectors of the building industry.

The general manager of FABEKO (Norwegian ready mixed concrete association), Hallvard Magerøy says in an article in 1997 [206] that the annual world production of ready mixed concrete was 700 million m^3 and that it was a trade sector that had more than 300 000 employees.

Sui Tongbo mentioned in 2006 [209] that more than 300 million m^3 of concrete is annually produced in China alone and that the volume is increasing. It is therefore reasonable to believe that the annual production of ready mixed concrete in the world has surpassed 1 m^3 per year.

An estimate made in the spring of 2009 states that we now produce more than 8 billion m^3 of concrete in the world.

4.3 Precast concrete

Vitruvius talks about precast blocks (see Chapter 1.5) as early as 25 BC.

In modern concrete history, we know that William Aspdin's house had precast parts though it was built in 1847 (Chapter 2.1).

Chapter 3.1.2 mentions about precast blocks with exposed aggregate in Cerrigdrudion Dam in 1910.

Precast concrete is a very wide market area, where the importance of the various products varies considerably from one country to the other.

Concrete blocks are the most important product in a number of countries, particularly in countries with a low level of concrete technology and where clay masonry is less in use because of lack of acceptable clay quality or because burning of clay was restricted due to the fact that emission of CO_2 is regulated.

Concrete pipes are globally very important, but the technology used for the production has wide variations. In some countries, the spinning of pipes is the main method used, and in Japan concrete pipes are even called "Humes-pipes", named after one of the spinning methods. Spun pipes have also been produced in the Nordic countries, but here the normal production method is by far vertically cast concrete pipes in a vibration machine.

Of the many methods, as a curious case, we mention an example of spun pipes observed during a visit to a small concrete pipe factory outside Dar es Salaam in Tanzania in 1993. In a small shed, there was a truck chassis. From the axel, a belt connected the rotating truck axle to a small aiding axle. On this, two other belts supported a pipe mold, one at each end of the mold. The mold was made of a certain number of 2×4 inch planks, forming as a periphery to the pipe. At the ends were stoppers made of wood which are of the height of the diameter of the pipes to be produced. Concrete in soup consistency was shoveled into the mold. Then the spinning system was rotated in the lowest gear. More concrete were shoveled in if necessary. Then the speed and the gear were increased. After some minutes in fourth gear, the pipes were ready. Normal dimensions were from 100 mm to 1500 mm, with lengths from 100 cm to 180 cm.

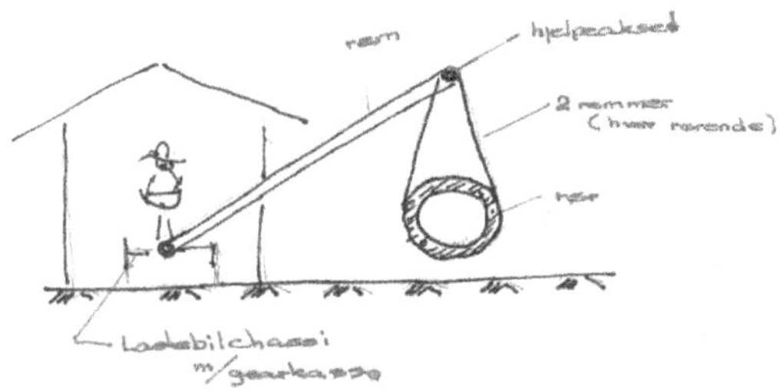

Sketch of the production of spun concrete pipes in Tanzania

At the storage area; finished spun piles in Tanzania

The example illustrates that advanced technology can also be adjusted to suit more simple or primitive local condition.

Paving products are another major product, which in particular has been produced by machines with a very high degree of automation in the larger industrialized countries.

In Norway, the industrialized production of precast concrete products started in 1895 by the establishment Christiania Monier & Cementvarefabrikk A/S in Oslo and Stavanger Cementvarefabrikk A/S in Stavanger. The Monier factory were producers of not only cement- based products but also of "gypsum-based" ceiling decorations that were common in apartments and commercial buildings hundred years ago. Another important product (in cement concrete) were window sills and beams.

There are, as mentioned earlier, a very wide range of precast concrete products, and the production volume might consume 10%–20% of the cement volume in some countries.

Of the more special products, not contributing so much to the total volume, we mention some products from Japan, which are shown below. Maeta Concrete Ltd. in Sakata in Japan, in 1993, combined advanced technology in mix design with old Japanese lacquer technology, and pro-

High tech concrete products

duced ashtrays, sushi plates, flower vases, and loudspeakers using high performance concrete.

Many producers are concerned with the high tonnage price of their products. It is easy to conclude that Maeta Concrete succeeded in that respect.

The single sector of precast concrete that probably represents the most important precast area in many countries, at least with respect to product size, is the production of structural building components in a factory, also called element production.

Significant production of **precast concrete components** started in the mid 1960s in countries like Norway.

Component production in other countries started long before this. In an article in the Swiss journal *CEMENTBULLETIN* in 1976, there is a reference to the English journal *Concrete* dated April 1972 [210].

The article briefs about William Henry Lascelles who tried to patent a precast concrete floor in 1875. The system was based on precast concrete plates on wooden beams.

Lascelles convinced one of his friends, the Architect R.N.Shaw, to use the system in a private house in Croydon in 1882. The ceiling plates were produced in sizes 2×3 ft (60 × 90 cm).

Ceiling plates in precast concrete with flower pattern [210]

Dabney [72] mentions about the English Ernest L. Ransome, who immigrated to California and started producing concrete blocks. After some time, he found out that reinforced concrete beams could compete with iron beams, which had to be transported from the east coast. His first project was the Artic Oil Works in San Francisco in 1888. The next project was at Stanford University in Palo Alto.

Two family houses in Croydon in South London built in 1882 using the Lascelles system. The other parts of the building also had precast concrete components, as for example window sills

Zollmann [92] discusses about a special incident. In 1939, due to material shortage, steel roofing had to be changed to something else. A natural choice was concrete. In Florida, they started to produce roof plates in precast concrete with 9 cm thickness, 60 cm width, and spans of 2.4 and 2.7 meters. Also, the production of channel components started about this time. The width here was also 2 ft (60 cm). Toward the end of the Second World War, such components were produced industrially in Birmingham in Alabama, Cleveland in Ohio, and in New Jersey in USA.

In an article in the *PCI Journal* in 1994 [211], Norman L. Scott traces the development of precast concrete components in USA. What he re-

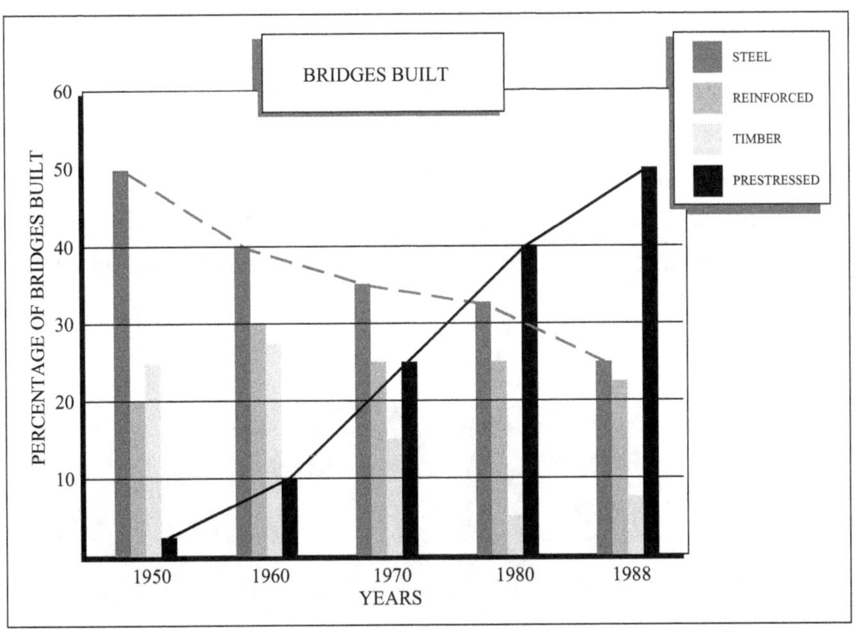

Development in the market share of various materials for bridges in USA [212]

The development of DT elements in USA from 1951 until today

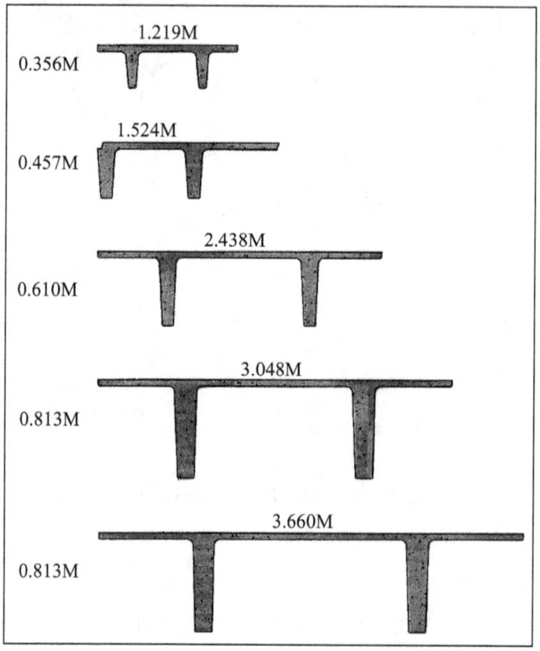

fers to as "architectural precast concrete" is what is regarded as reinforced (not prestressed) concrete components in Europe. He states that these components started to become common in the American building market in the 1920s, but they were not a common sight in facade structures until around 1960, about the same time as in the Nordic countries in Europe.

He also shows an interesting development in the market share of different materials used for bridges in USA.in the years 1950–1988.

The significant increase in the use of prestressed concrete for bridges was due to the development of standardized beams.

The single type of precast concrete component in USA, that has been most popular over time, is the Double T-component (In some countries called DT or TT components). The first one came as early as in 1951 and the size development has been interesting.

Edwards [212] claims that the first prestressed concrete DT elements were produced in Florida in the mid 1950s. This was a success and within

The largest American DT elements are too big for most European roads

5 years DT components were produced in 5 factories in Florida and in total in 20 factories in USA.

A similar development can be seen for hollow core planks. The early components of this type came into the market in the mid 1950s. The first machine in USA was imported from Canada. A typical American trend is the use of these components in facades, while in Europe the planks have mostly been used for floors and roofs.

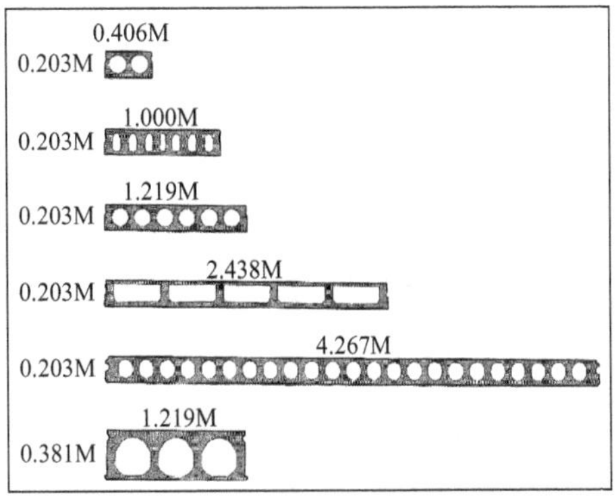

The development in precast concrete for hollow core components [212]

The Dodd Extruder

The first hollow core machine was designed by David Todd in 1958. In the 1960s, machines such as Spancrete, Span Deck, and Spirol came into existence. However, it must be noted that nonprestressed, reinforced hollow core floor planks were produced before this, using tube mandrills that were withdrawn from the mold after casting.

The advertisement below reveals a bit about the work that the pioneers did to get acceptance for the new technology that they initiated.

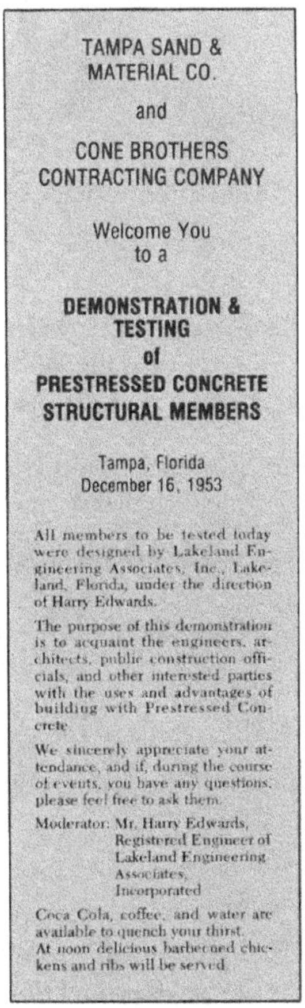

Typical advetisement from the concrete environment in Florida, anno 1953

An article in *BD-orientering*, information from the government's Housing Directorate in 1959 [213], explains about the early precast concrete apartment buildings in Norway.

The article discusses the methods that the developer A/L Ungdommens Selvbyggerlag and their contracting company A/S Ungdomsbygg had used from 1953 to 1959. The purpose was in the end to use a system with 100% precast concrete in the structure.

The early precast concrete apartment buildings in Oslo [213]

They mention about four steps in the development, through the following projects:

- Årvoll Vest (1953–1954) – 224 apartments in four-story-blocks. Precast exterior walls, stairs, landings, and balcony plates, while the floors and load bearing walls were cast in situ.
- Tonsen (1954 – 1955) – 260 apartments in three and four story blocks. Here they started to use precast floors (Non-prestressed hollow cores).
- Bøler (1956 – 1957) – 600 apartments in four story blocks.
- Plogveien (1958 – 1959) – 452 apartments in three and four story blocks. All structures above the first floor were precast concrete.

Of the details of the technology that was introduced half a century ago, we mention the following:

- No fixed module system, but trials to use a module of 2.4 meters.
- Maximum component weight of 2.8 tons.
- Erection with tower cranes with a capacity of 35 tons.
- Average production sequence: 1 apartment per day.

Ungdomsbygg's factory at Skedsmo (25 km north east of Oslo), which was established in 1957, later became an independent company "A/S Betongelement." The factory had in 1959 a production area of 1000 m^3.

The author of the article, Erling Bakke, shares his thoughts about the plans he has for the future.

As the prefabrication includes more and more of the various parts of the building structure, it seems clear that it is the component production itself that becomes the key question to such a building method. On sight there are two main questions that are important for the development of the precasting technology.

1. *First of all, we have the economical side of the production: the possibility to create low-cost products. Here we must accept that we still are at a very primitive stage of production development. The production today in the element factories (this accounts, not only for our own factory), are to a high degree based on hand-work. A further mechanization of the working processes is therefore a very important task. On the other side, we also have to focus on making more material-saving designs. In this connection, we also have the challenge of standardization. It is the large series that makes cheap production and creates the possibilities for further mechanization.*

2. *An important matter for good results and an economical precast concrete building is a continuous production based on long term contracts. The financing situation we have in the building industry today does not give the right working conditions for the precast industry.*

Many things have happened over the last 50–60 years, but possibly representatives from the precast industry might have even used these plans even today.

Like many other European countries, the Norwegian precast concrete industry had got a good foothold in the building market in the 1960s; they mastered large projects and even exported structures to Sweden. B.Brynildsen & Sønner A/S in Moss was a "leading lady" on the export side. In 1968, they delivered all the components of the structures to the Swedish Co-Op national warehouse outside Stockholm. The project consisted of a main building of 504×252 meters plus two side buildings. The total ground area of the buildings was equal to the size of 12 football fields, a project comparable to larger projects even today. At that time, this was the largest precast building in northern Europe. The drawings of 1968 reveal that there are about:

- 450 main columns of 50×50 cm of different lengths in 117 variations.
- 252 I-shaped main beams with 18 meters length in 21 variations.
- 504 saddle shaped secondary beams with 18 meters length in 12 variations.
- 17 masonry-carrying facade beams.
- 23 special secondary beams.
- 37 rectangular beams.
- 23 T beams.
- 132 DT elements with widths of 240 cm.
- And 147 other components of various shapes.

All the 1585 concrete components in 279 variations were sent by rail, across more than 500 km from Moss in Norway to Stockholm in Sweden.

The structure was designed by KFAI (The Cooperative organization's architect and engineering bureau). Here is a small story from the component planning.

The main building at KF's warehouse at Bro outside Stockholm, Sweden, during construction in 1968

KFAI had on their drawings shown a number of threaded casing in the bottom flange of the 504 secondary beams. The purpose of the casings was to fix partition walls etc. During the planning of the component production, the precaster found that by putting in a smaller number of additional casings, the number of beam variations could be reduced from 70 to 12. KFAI would not pay for these extra casings, but the casings were still put in, as this was rational decision for the production and the logistics. About a year after the erection of the beams, a call was made from KFAI to Norway, asking if it was possible to drill in some extra casings in the beams, without disturbing the capacity,because they wanted to do some modifications inside the building.

The drawings were studied, and with great satisfaction they could call back to Sweden and say that they did not have to drill,the casings were already there.

On 1 February 1972, the Norwegian precast concrete producers association founded their own technical office, (with the author as the first manager) with their office at the Norwegian Concrete Technology Institute at Blindern in Oslo. At this time, Norway had 35 precast concrete factories, out of which 15 produced prestressed concrete components. While these factories produced 75 000 tons of precast concrete components in 1964, this had grown to 300 000 tons and 145 million NOK in 1971. Of the 35, 31 producers had joined the association.

It was a unique group of people that was the pioneers in the Norwegian precast concrete industry, as also in many other countries. They were not only competitors, and very strong competitors too at times, but saw the advantage of "creating a bigger cake", through active support and work in the branch organization.

Even if Norway, in 1972, had experienced a fantastic growth period as far as precast concrete elements are concerned, among the Nordic countries Norway was in the last position.

Denmark – produced 1.1 million tons precast concrete components even in 1967, and had in 1971 increased this to 1.7 million tons. The introduction of module systems and the prioritization of houses built on an industrial scale, combined with the tradition of preferring heavy building materials in a much higher degree than Norway and Sweden, were probably the main reasons for the growth. Danish politicians and authori-

Precast concrete element statistics – Norway 1972[214] Production volume of precast components in tons. Sales in million NOK (the flat part of the curve is due to changes in statistical methods)

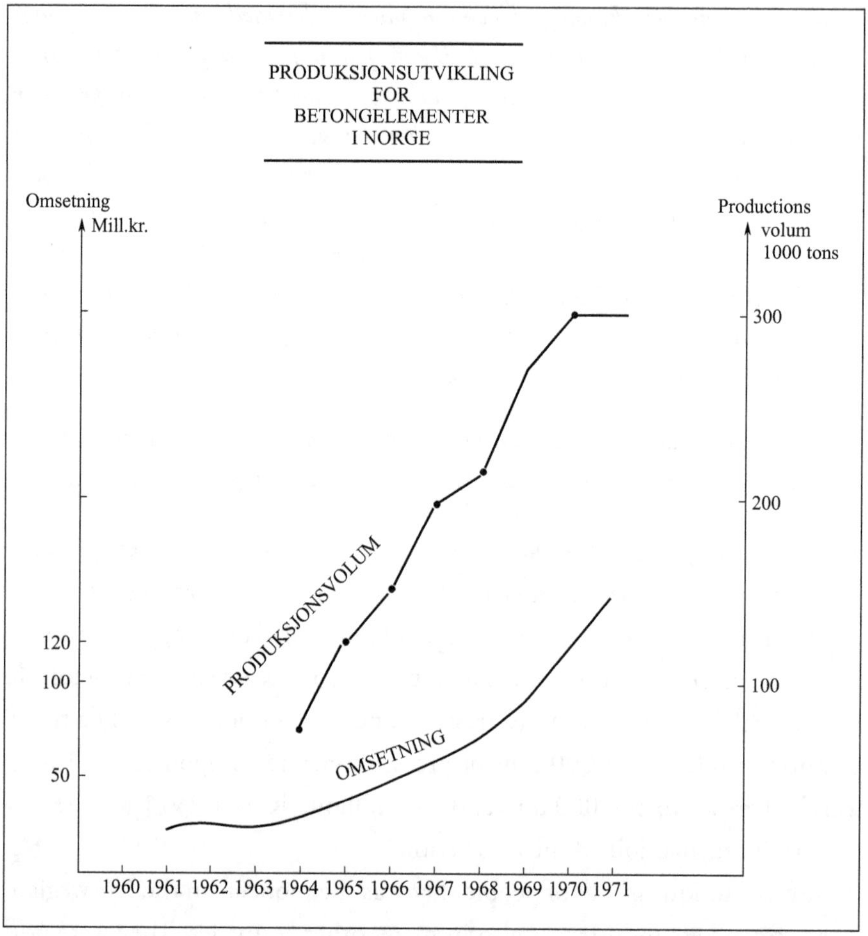

ties had strengthened the development by prioritizing industrial building. Half of the volume of precast concrete components was for apartment buildings. Danish companies like Larsen & Nielsen and Jespersen also had a very high reputation in the export market, both in direct export, and licensing. Several Norwegian companies produced systems based on a license from Larsen & Nielsen.

Already in 1968, 80% of the Danish apartment buildings were in precast concrete, amounting to 2.4 million m^2 [215].

Sweden, in 1970, produced precast concrete components for a value of about 400 million SEK, equal to about 1 to 1.3 million tons. Swedish companies were extremely good in a number of areas, particularly in prestressed concrete for industrial buildings and similar structures and had got an international reputation for this because of important engineers like Carl Gustav Bernander and Mogens Lorentzen.

Testing of trapezoidal roof components, 1949

Finland was on a fantastic development drive, not least due to an impressive common branch effort and the launching and development of their BES system and was in the beginning the leading among the Nordic countries in precast concrete technology and related machinery.

Later, Finland developed their precast industry to be one of the leading countries in the world in industrialized precast concrete production. The first industrialized precast concrete production in Finland started in 1941 with lightweight concrete elements and the first production of prestressed elements, based on Danish technology, started in 1950 [216].

Through the years, a number of development initiatives have character-

"Open" 32.4 meters long "sadle-beams," Lujabetoni Oy, 1967

Precast water towers of capacity 6 000 m3 – 1977

ized the Finish element industry. Here are a few examples that are less known internationally [217];

Great Britain, in 1971, produced 18 million tons of precast concrete components in many (300) relatively small production plants and a few giants. The largest companies had worldwide activity. The largest producer represented a production value in 1971 of 13 million GBP.

Germany was lagging a bit behind, but had still about 20 production plants with sales of more than 10 million DM. In total, there were over 500 plants that produced precast concrete components of one sort or the other. Many of these had the same owner or belonged to the same chain [214].

In France, a considerable amount of important system development had taken place, and precast buildings based on French systems were produced both elsewhere in Europe and in USA.

In Norway, industrial buildings were the most important market segment. The market share figures for 1971 shows the following distribution for the deliveries [214]:

Housing 12%
Industrial buildings 72%
Schools, hospitals 10%
Commercial buildings 5%
Agriculture 2%

An investigation conducted in 1972 [218] (the first relatively complete concrete element statistic in Norway) showed the following key figures:

- 31 plants
- 1175 employers
- 75 205 m^2 of production area
- 14 prestressed concrete producers with 7 249 meters of prestressed benches
- Consumption of 8950 tons of reinforcing bars and 1 894 tons of prestressed wire.

John G. Richardson, an English precast concrete enthusiast, book and article writer, illustrator, and well-known lecturer, visited Norwegian and Swedish precast concrete plants in 1973 and gave them a lot of credits in his article in the English journal *Precast Concrete* [219].

He gives a detailed description of the various plants he visited, and it looks like three things in particular had greatly impressed him.

- The importance shown toward health and safety.
- Female workers working side by side with men.
- When the mechanization in these plants is compared with that in English plants, the latter looked like "foot walkers".

After some random engagements with precast concrete in various parts of the world for the last 40 years, from time to time I question myself about where is it best to learn more about concrete. The answer is simple – I do not know.

- Sweden has for years been ahead of most countries with respect to precast concrete stairways and was early in the forefront with respect to industrial production of prestressed components.
- Denmark was the world leader with respect to precast concrete in housing in the 1970s.
- After their BES (Concrete Element System) work in the 70s, Finland established a leading position in mechanization.
- In the 70s, you could find some of the highest quality in "façade-components" in England.
- Belgium can probably show you more impressive work in precast facade-components than any other country, and has shown impressive structures with respect to heavy components for bridges etc.
- France has been the world leader with respect to precast concrete pipes and similar components.
- Italy can show impressive precast light roof structures, and in the last decenniums there has been an impressive development in the machinery sector.
- USA is the world leader in heavy structures, for example with respect

to heavy DT-components, and the use of precast concrete in parking garages.

- Japan has been leading in the use of precast spun components for piles and masts.
- Australia can produce exposed aggregate precast concrete that none can match.

The list could have been longer. The point is that precast concrete is produced all over the world and local traditions and circumstantial developments make the technological development manifold.

4.4 Lightweight concrete

Stephen J. Hayde, father of the modern lightweight concrete [192]

Lightweight concrete is actually an old type of material. We know that the Romans reduced the density of concrete by using lightweight aggregate in a number of structures. Most known is probably the Pantheon Temple in Rome (Chapter 1.5). The volcanic aggregate, riddled with pores, was sorted by weight and the weight of the concrete reduced as the height of the dome increased.

In Chapter 1.4, we have also referred to Mexican structures with light aggregate dated to 850 to 1100 years BC.

The use of lightweight concrete really picked up speed only in the beginning of the 1900s when lightweight aggregate with relatively constant properties were produced. Bremner and Ries [192] have written an interesting article about the person, whom they think was the father of lightweight concrete – Stephen J. Hayde.

Hayde was born in Keokuk, IO, USA in 1861 to Irish immigrants from Tipperary. He worked with his father in Kansas City, MO, before he for some years moved to San Francisco. After coming back again to Kansas City, he became famous as a producer of quality clay bricks and as a contractor. Brick production in the beginning of the 1900s was primitive with a major variation in quality. The bricks were, after burning, sorted into various qualities by hand. The lowest paid bricks, and often a loss for the producer, were the ones that had been placed too close to the heat burnt too fast. They might expand up to 1/3 of the original clay volume. When special types of clay are used, Hayde observed that the burnt products might get a porous inner core, with a hard ceramic shell. Hayde was inventive and started to use some of the waste from the brick production as concrete aggregate, after having crushed it to an acceptable size. In a letter to a nephew in October 1914, he had written about his invention, and wrote, among others, that the material was so light that it floated on water. Hayde was so certain that he had found something clever that he engaged a patent attorney in January 1914. On 18 February 1918, U.S. Patent No.1 255 878 was issued in the name of Hayde. In the patent, he mentions the various clay materials used and that they have to be heated up to 1220℃ for two hours.

In 1917, Hayde found out that the best way to produce this aggregate was to use a rotary kiln. The timing of Hayde's invention was perfect. It was the time of war and material shortage. The first seagoing concrete boat, Namsenfjord, was built in Norway in 1917, and in USA they were looking for materials with a density lower than normal aggregate to be used in concrete boat production, (Chapter 4.4).

Hayde contacted the director of the U.S. Emergency Fleet Corporation in Washington in a letter dated 18 February 1918. This resulted in a development project, in which they managed to produce concrete with a density of 1 697 and a compressive strength of 28 MPa using haydite.

A government employee, Carl Boynton, tried to copy Hayde's technology in the very interesting and expanding shipping market, and took out patents in his own name. However, he later lost the court case against Hayde, and the appeal court in the District of Columbia, on 1 June 1928, stated that Boynton's knowledge was based on Hayde's work. Unfortunately, the decision came too late for Hayde to feel happy about it. He died 16 days earlier

Chase Park Plaza Hotel, St. Louis, Missouri

The double towers Marina City, Chicago, – 60 floors, 176 meter high, 1959–1964.

during a train journey between Montreal and Kansas City. Hayde's patents were in force up to 1947. Unfortunately, due to various financial circumstances, his family never got the opportunity to benefit from them [192].

At ACI's 15th Conference in Atlantic City in 1919, the chief engineer of Turner Construction Co. in New York presented a paper on lightweight concrete. He stated that this type of concrete was highly advantageous for high rise buildings, with considerable savings in the reinforcement.

The first full scale factory for the production of lightweight aggregate was started in Kansas City, Missouri, in 1920, under the name – Haydite Company [193].

But, it took time before the industry had any full scale project to show. The first came in 1922 and it was the Westport High School in Kansas City.

The first sky scraper with lightweight concrete was really an addition. In 1928–1929, Southwestern Bell Telephone Company wanted to build additional floors on their 14-story- office building in Kansas City. Analyses showed that they could add another eight floors by strengthening the foundations. Then lightweight concrete was evaluated instead of ordinary concrete. The result was that another 14 floors were added to the building, so that they in total they got a 28 floor building [193].

The development continued.

In 1929, Chase Park Plaza Hotel in St. Louis was built with 28 floors using lightweight aggregate concrete in both beams and slabs and for fire protection.

Of the many projects and test programs with lightweight concrete, probably the so-called "Operation Cue tests at Yucca Flats" was the most curious. In Yucca Flats in Nevada, houses were built using many building materials – masonry, timber, aluminum, lightweight concrete, and ordinary concrete. All the buildings were tested in an atomic bomb test in 1955. Only four buildings passed the test. Two houses with lightweight concrete blocks and two with precast lightweight concrete [193].

Of the more famous lightweight concrete buildings we also find Marina City in Chicago. The double tower is a landmark near the river near the "the circle" in Chicago.

In September 1967, Ivar Sørli wrote a kind of a state-of-the-art article about *Armerte konstruksjoner med lett tilslag (Reinforced structures with lightweight aggregate)* [220]. Sørli stated that USA had been "the leading lady" in the area and argued that this was mainly based on two reasons – they have more high buildings and therefore the need for lower material weight and they often lack good ordinary aggregate.

Sørli also mentioned important developments in Germany, with the following examples.

• The building of barges in lightweight concrete during the Second World War

The Dyckerhoff Bridge near Wiesbaden, span 96 meters. A gift to the city at Dyckerhoff's 100 years jubilee [220]

Stolmasundet Bridge

• Prefabricated garages with concrete quality B 300/1600

• The recently opened Dyckerhoff Bridge with a lightweight concrete of quality 300/1600 in the middle part of the bridge.

Sørli was of the opinion that Norway had been in the forefront with respect to standardization of lightweight concrete during the issue of *"Beregningsregler for armert betong med Leca som lett tilslag" (design rules for reinforced concrete with Leca as lightweight aggregate)* in November 1963.

The world record for the longest bridge span with lightweight concrete is held by a Norwegian bridge, the Stolmasundert Bridge, between the islands of Stolmen and Seløy in the community of Austervoll in Hordaland county.

The bridge was opened in 1998, with a length of 467 meters and a main span of 301 meters, where 184 meters in the middle is made of high strength lightweight concrete [119].

Earlier the same year, the Raftsundet Bridge was opened. This bridge is part of the mainland connections to the islands of Lofoten and crosses the main coastal shipping route to northern Norway. The large span of the bridge and the special windy climate posed a lot of challenges to the designers and the consulting engineers at Aas-Jakobsen A/S. This bridge has a total length of 711 meters and a main span of 298 meters. In the middle, high strength lightweight concrete of quality LC 60 has been used to save weight and to make a longer main span possible. This bridge was also a world record when it was opened [118].

Lightweight concrete has been used in all parts of the world. In March 1968, a skyscraper in Sydney in Australia, which was at that time the highest building in the world with lightweight concrete, was opened [221]. The building is a circular tower with a radius of 40 meters and a height of 180 meters.

The building was designed and built by Australian concrete technicians, but the idea for the design came from the Italian Engineer and Architect Pier

Raftsundet Bridge

Luigi Nervi's idea for the tower construction

Luigi Nervi. The gigantic building has 46 floors, with a concrete quality that is similar to what at that time was called a compressive strength of 250kg/cm^2. This quality of concrete was used for stories above the eighth floor.

The building has 20 exterior columns connected by beams. The exterior precast concrete components have exposed concrete aggregate (white quartz for the columns and above the windows and black below the windows). The precast components are cast together through the in situ cast floors.

4.5 Admixtures

An international dictionary defines admixtures as a material used for modifying the properties of concrete. This a very broad definition and can be useful in a historic evaluation.

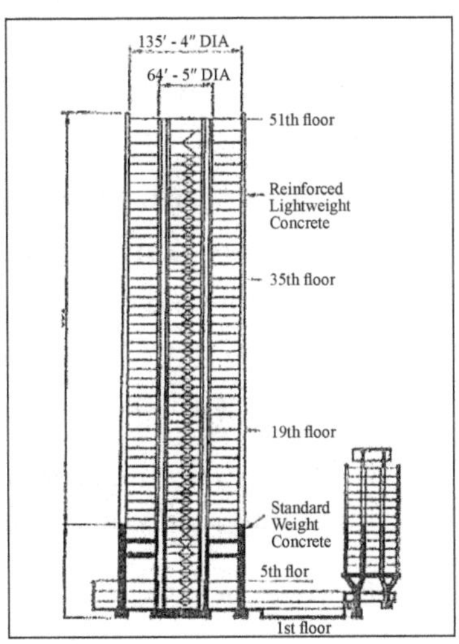

A section through the Australian skyscraper

Aïtcin [8], with reference to Martinet, states that the Egyptians used admixtures of egg, milk, wine, beer, and animal blood in their anhydrite-based concrete. In an article in 1996 [222], Bryant Mather traces the origin of air entraining admixtures as well as the use of fibers in concrete.

He claims that the Romans seem to have known the advantage of the use of air entraining admixtures in concrete. Markus Vitruvius Pollio (see Chapter 1.3) had written about a mortar recipe, in which he adds lard, milk, or blood, in the first century. These substances will definitely produce air pores. We assume that this was done to plasticize the concrete or to make it more frost resistant. Mather says that he believes that when Moore in 1940 and Wuerpel in 1946 referred to the use of special air entraining admixtures to increase the frost resistance, they were the first to come up with the idea of using special air entraining admixtures for this purpose.

However, the use of fat as a grinding aid in cement production has been known from the 1800s and such cements also had improved frost resistant properties.

Briant Mather refers to George Hoff when he points out that straw fibers were used by the American Indians, long before the Europeans came [81]; he also refers to the use of straw in sundried bricks by the Egyptians and the of the fibers of bamboo in the Great Wall by the Chinese.

Harry Porter wrote a report in 1910, in which he mentions a 8-fold increase in the tensile strength of the concrete by the addition of cut nails to the concrete. The test was done in connection with the bridge work in New York City.

In an article by McKibbin in *Concrete* in 1998 [223] which is actually about an analysis of ancient mortars, he mentions about even more exotic admixtures in the old concrete recipes.

- Straw and hair were often used to reduce cracking.
- Ox blood, cow dung, and beeswax were used to regulate workability, setting time, and water absorption.
- Urine and beer to get air pores.

It is reasonable to believe that it is only with difficulty that we can analyze ancient concrete with modern standardized concrete testing methods.

Idorn, referring to Chinese Miao Ji-chang et al. [64], states that rice flour was used in the great Chinese wall to get air pores to improve the workability of the mortar. The same statement has also been given by a number of other authors.

In a memo, Sui Tongbo [16], states that old Chinese concrete technology often also included the use of organic substances in the mixtures, for example, lime combined with rice flour or lime and heavy oil. In addition, blood and other organic substances were used to produce Sanhetu mortar (See Chapter 1.2 and Chapter 5). Sticky juice from rice or starch was definitely used in the mortars for the great Chinese wall. In the period between the Lower and the Southern Dynasties, this type of material was also used to build the Shaolin temple and several towers during the Song and Ming Dynasties. Sticky rice in the mortar is also used in the Nanjing Wall (described in the book *Song Hui Yao).* The famous book *Tian Gong Kai Wu (Heavenly creations)* discusses about the use of mortar with admixture of sticky rice and ox blood during the construction of a stone bridge over Lu Gou Qiao in Beijing during the Qing Dynasty. The result was an artificial stone as hard as granite. The stone was so hard that it could not be chipped with iron chisel and they had to use explosives to demolish it.

The Chinese wall is 8 851.8 km long. If we had reduced the height of

The Great Wall of China

the wall to 5 meters and the width to 1 meter, the volume of the wall is comparable to a wall of the size mentioned here that can cover the distance around the globe. Several walls were built earlier, but the wall we refer to here, was started by the first Emperor who ruled the whole of China – Shi Huangdi about 200 years BC.(The first Emperor Dynasty was the Qing Dynasty, from 221 to 206 BC., with their capital in Xi-an [17]). The Emperor ordered the building of a wall from east to west and under the direction of General Meng Tian. 300 000 farmers, soldiers, and criminals were put into work. Tens of thousands died during the 10 years of the construction of the wall [224].

The wall is also called the Ming Wall because the construction took place during the reign of several dynasties and was finished during the Ming Dynasty. The wall goes from the Liaoning province in the east to Jiayuguan in the Gansu province in the west and passes through 10 of the provinces or autonomous regions in China. Of the nearly 9000 kilometers, 6259.6 km is a human made wall; 359.7 km are ravines, and 2232.5 km natural shelter [225].

An investigation conducted [226] refers to a wide variability in the use of admixtures in different countries.

Low consumption (10%–40% of the concrete volume):

- Finland
- France
- Italy
- Netherland
- United Kingdom

Medium consumption (40%–75% of the concrete volume):

- Australia
- Belgium
- Denmark
- Sweden
- USA
- USSR
- West Germany

High consumption (75%–95% of the concrete volume):

- Japan
- Norway

More than one-third to two thirds of this consumption were air entraining agents, plasticizers, or superplasticizers.

Air entraining agents were introduced in the 1930s to improve frost resistance and workability.

A number of stories regarding the origin of air entraining agents exist. Most of them mention about the observation of improved frost resistance of some concretes and the investigation that followed as well as the subsequent detection of oil spillage. Of the stories I have heard, one of them mentions about an oil spillage in an old cement factory, another about diesel that by accident was dropped into a concrete tub during concreting for dams, and yet another about oil spillage in a mixing station. Which one of these is correct is hard to find out. All the stories, however, are believable, and it might even be that the idea popped up about the same time at several places. Under any circumstances, we can only regard this as the beginning of modern use of air entraining admixtures. The origin was several thousand years earlier.

Plasticizers, especially the ones based on lignosulphonates, got their breakthrough in the beginning of the 1970s. The development work, however, started long before this.

In two interesting articles, the authors James W. Adams [227] and Pat McKeogh [228], discuss the beginning of the use of plasticizers. It started in Rothschild in Wisconsin in USA. The prehistory is that seven investors, based on a right given to them by the state of Wisconsin in 1893, built a dam over the Wisconsin River and in 1906 established Rothschild Water Power Company.

In 1909, they found out that there were not enough houses around to consume the electricity they produced and on 2 February 1909, the Marathon Paper Mill Company was established. After a while, the river became very polluted by the waste from the paper production. The red color of the waste gave it the name "Red Liquor". In addition to the fact that the waste smelled bad, it also caused the death of the fish. Up through the years, many had asked if this waste could be used for something. The breakthrough came in June 1927, when they were contacted by Guy Howard from a consulting company in Seattle, Washington. Howard had found a method that in an economical way could extract lignin from the waste. The research work started in 1927; they produced the first lignin in a pilot plant in 1932 and the full production in a lignin factory was achieved in 1937. This factory was, when it started, the first commercial lignin factory in the world. The use of lignin as a plasticizer in concrete is only one of its many uses. In a patent in 1937, Edward Scripter refers to an addition of 0.15% of lignin in a concrete, which would lead to a possible 12% water reduction and a corresponding increase in strength.

The Norwegian company Borregaard bought the technology from Rothschild in 1967 and the company is now called American Can Company. The Rothschild factory was later, on 9 January 1991, bought by Borregaard Industries and is today part of the company Ligno Tech USA.

However, Borregaard was, when they started their lignin production in Norway, not the first producer. Toten Cellulose had already been in the market for some years. Toten was the single supplier to the admixture company Betokem. When Geir and Odd Tjugum broke away from Betokem in 1976 and started their own company, Rescon, the owner of Betokem, Johs. Rutle (see Chapter 2.4 and 4.2), tried to block the deliveries from Toten to Rescon. Borregaard had at the beginning a production capacity of 15 000 tons of lignins (all types of end-products) per year. They delivered their first 5 tons of the product Borresperse Na to Rescon in bags in 1977. In 1980, Borregaard increased their lignin production. They made a contract with Betokem to supply 40% of their requirement. Rutle knew at this time that Toten was on its way out of the market, and accepted this. Toten Cellulose disappeared from the market in the early 1980s and Borregaard then became the only national supplier of lignins to the Norwegian market [229].

The lignosulphonate production in the world is today about 1 million tons of dry material and Borregaard produces more than half of this. About half of the production goes for concrete production.

In later years, polycarboxylatene (PCE) has taken over more and more of the traditional lignosulphonate market as plasticizers for concrete production.

Regarding the early use of plasticizing admixtures, we cite this story from the "jubilee-book" *Fra Farris til Troll (From Farris to Troll)* [49]. The story comes from an old veteran in the Norcem concrete factory in Greenland, Sven Skogsrød:

> *"When I started in the ready mixed industry 43 years ago, 1952, many things were different ··· We cooked our plasticizing admixtures our self based on resin and caustic soda. The blend was doused with a water- scoop. This definitely had plasticizing effects, but I also observed a deep etching hole in the concrete floor where the bucket was".*

In 1958, Kenichi Hattori in Japan started the work that led to the development of **superplasticizers**. He reveals that he found the advantages of his research in concrete in 1960, and that the product Mighty was introduced into the market by KAO Soap Co. in Japan in 1964 [101]. This also gave the real start to the development of the type of concrete that later has been named High Strength Concrete or High Performance Concrete.

Hattori's product (sulfonated naphthalene) came into the Norwegian market in 1976, first under the name NB76 and then Mighty 150 and Mighty 100 (the powder alternative) and later under other different names. Already in the beginning of the 1970s, other superplastizisers had been introduced into the Norwegian concrete market, which had been used in limited quantities; among others Melment (sulfonated melamine) from SKW in Germany and Lomar D from USA (sulfonated naphthalene).

The development did not stop with this. In the 1990s came the polymers; at the turn of century, the polycarboxylates came to be known for high water reduction in concrete production.

Polymer concrete might be recognized as a special case when we talk about admixtures of concrete. Authors writing about polymers at least use the same stories about the historic origin of polymers as those for admixtures in general [230, 231].

We have previously mentioned that when the Chinese built the Great Wall they used gluten-rich rice paste together with lime in their mortars. It has also been claimed that some of the admixtures that the Romans used were monomers.

The development of "Bakelite" by Dr. D. Baker at the end of the 1800s marks the beginning of the modern development of polymers [230].

Small tests with polymer concrete started in the 1940s and 1950s. The first account is not really about concrete, but about mortar [231, 232].

The German company Bayer experimented with monomer additions to mortars in the beginning of the 1950s and the breakthrough came in 1957.

From 1966, tests with financial aid from the government started in USA.

ACI's committee 548 – Polymers in Concrete was established in 1971.

ACI had their first technical session about polymer concrete at a conference in 1972 in Hollywood, Florida.

The first international conference conducted by ICPIC (International Congresses on Polymers in Concrete) was held in May 1975 in London.

4.6 Silica fume[233]

The use of silica fume in concrete is an important Norwegian contribution to the history of the development of concrete. Norway and Norwegian concrete experts have contributed considerably as far as the use of silica fume to make concrete with high durability and strength is concerned.

An important reason behind the development in Norway is of course the important position that the Norwegian ferrosilicium industry holds in the global scene. Silica fume is the gas fume that is obtained during the production of ferrosilicum and silicium. After the Second World War, environmental consciousness increased and the industry started to try to prevent this fume gas from polluting the atmosphere. The Norwegian industry was in general cooperating through the organization Fesil, in which Elkem was the "leading lady".

Norwegian plants that were in this sector when the concrete adventure

The lake Mjøsa Bridge from Biri to Moelv, built in 1983–1985, was the first major bridge project in the world that used silica fume in the concrete [235]

with silica fume started include Elkem with their plants in Kristiansand (Fiskaa) and Salten, Hafslund in Sarpsborg, Tinnfoss at Notodden, Bremanger Smelteverk, Ila Lilleby with their plants at Holla and in Trondheim, Meråker Smelteverk, Borregard with their plant in Orkanger and also the plant at Finnsnes.

The first silica fume was produced at Elkem Fiskaa plant in Kristiansand in 1950 [234]. In the beginning, they only produced a few tons per year, until they learned how to scale up the process.

The first industrial filter came in 1966, which was also at Fiskaa. The industrial breakthrough in the cleaning technology came in the early 1970s [234].

The early tests with the use of silica fume in increasing the durability and strength of concrete had supposedly led to a patent application by James William Sharp, Permanent Cement Company in Oakland, USA during the Second World War; the patent was regarding the blending of 3%–5% silica fume in Portland cement to get a more plasticized concrete [234].

In the book *Condensed Silica Fume in Concrete* published in 1987, Malhotra et.al. [236] have in the preface to the book traced the history behind the use of silica fume. We cite a few lines from the preface: ··· *The first investigations on condensed silica fume began in the late 1940s and since then Norway has made significant advances on various aspects on silica fume ··· Contributions made by other countries were relatively minor, at least in the period before the 1970s ···*

V. Mohan Malhotra has been one of the most important contributors in making the silica fume technology known to the world. In addition to his own research and publications, he has played a very crucial role in conducting international conferences on the topic. He arranged the first "CANMET/ACI International Conference on fly ash, silica fume, slag and other pozzolans" in Montebello in Canada in 1983. The seventh and the last so far was held in Warsaw in Poland in 2007.

Some of the following documentation, references are not given, originate from an internal company memo called *Silica fume in cement and concrete—a resume of some of the development up until today,* written by the author, and dated 17 September 1979. Unfortunately, some of the underlying documentation has been lost through the years. The memo is a 12 pages log of a major part of the history up to 1979. Most of the contents in the notes up to March 1972 are repeated below. From then on, only the more important happenings in the development have been repeated.

Here is summary of the history:

1945–1952: Important test conducted in USA with pozzolans with a high content of SiO_2 [237].

1948–1951: Professor Markestad and Dosent Bernhard began their tests at NTH in Trondheim.

1952: Dosent Bernhard wrote an article in Betongen I dag (Concrete Today) about tests with silica fume in concrete [238].

1968: Professor Markestad presented a paper at the Nordic Concrete Congress in Gothenburg, Sweden about silica fume [239].

03 September1969: Practical tests with silica fume in concrete started at the concrete laboratory at Dalen (Norcem's cement factory, which was also the laboratory for the experiments with concrete at Norcem) Key persons in this work were Knut Bratlie and Paul Olstad. The work had been initiated by a letter from Elkem dated 15 July 1969.

It is important to remark that both with these first tests and in the tests done by Bernhard and Markestad earlier, they started with very high dosages of silica fume in the concrete. The reason for this was of coarse that this was normal dosages from other known pozzolans at the time. They thought that these dosages were necessary to observe the expected effects. This, together with the fact that superplasticizers were not known when they started, and that they even did not know the advantages of combinations with plasticizers, made the blending in the first tests very different from the practical use we have today. Both Bratlie and Olstad therefore became very important in the transformation from the first tests to more practical dosages of silica fume, and the combinations with admixtures. Bratlie had considerable experience from concrete production, and was very important when the results from the laboratory should be tried out in practical use. Olstad had considerable experience from material testing, and also became a key person in the later development of the admixture product CORROCEM that had silica fume as one of its main ingredients.

01 October 1970: Letter from Professor Markestad to Elkem A/S, the reply was dated 13 May 1971, with a request for 10 000 NOK for tests on "*anvendelse av mikro-sil til betong (the use of mikro-sil in concrete).*"

12 October 1971: An assignment numbered 40834 and titled "Investigation of Elkem-silica in concrete" was submitted to NTHM (Norwegian Technical University, Material testing department) by Elkem A/S and Norcem A/S".

16 December 1971: Report from SINTEF regarding the use of silica fume. The reports hints at the use of silica fume in special mortars in England.

21 December 1971: Report to Norcem from STI (The State's Technological Institute, Oslo), and letters dated 16 December 1971 and 05 January 1972 from STI to Norcem regarding the use of silica fume in mortars.

1971: NGT (Norwegian Geotechnical Institute, Oslo) publication, *Betong i Alunskifer (Concrete in Alun shist)* Nr. 86 Published the results of tests conducted in 1947–1963. The results showed impressive improvement in the durability as a result of the addition of silica fume to the concrete.

Medio 1971: The manager of the service office of the Norwegian Ready Mixed Concrete Association, Hans Bøen, contacted Elkem to discuss the use of silica fume in concrete.

Autumn 1971: The use of silica fume obtained at Elkem in Kristiansand was beginning to be used at Norcem's ready mixed concrete plant at Nenset in Greenland, under the management of Knut Bratlie.

09 August 1971: Japanese patent application regarding the use of silica fume in concrete to increase water tightness.

24 January 1972: Report from Knut Bratlie: *The use of Elkem Silica in ready mixed concrete.* The report was based on his own investigations at Norcem as well as tests at NTH in Trondheim.

12 March 1972: A memo from Knut Bratlie to Norcem's Concrete Division, where he points out advantages for the division, and claims that if the division uses silica fume at all their plants, and in addition the ready mixed concrete plants in Oslo uses silica fume in their concrete, this will mean a loss in cement sales for the Norcem Cement division equal to 7 million NOK.

24 May1972**:** A special study *Investigation of Elkem Silica as admixture to Standard Portland cement for the use in masonry and plaster mortars* was conducted at the Institute of House building technology, NTH, by A.Sandborg.

Primo May: General Manager Holbye at the Norwegian Concrete Technology Institute, Oslo, categorically stated at the Annual assembly of the Norwegian Ready Mixed Concrete Association that silica fume, because of its special properties, cannot go through the silos of the concrete plants!!

June 1972: Norcem established a separate project group for silica fume, with Director Per Lorange in charge of the project and Knut Bratlie as project manager. The cooperation with Elkem continued until 1975.

Grain silo in Buvika, Middle Norway – the first slip forming with silica fume in the concrete, 1979 [235]

Grain silo in Buvika, Middle Norway – the first slip forming with silica fume in the concrete, 1979 [235].

16 January 1973: Silica fumes were first used in the production of concrete products. There was a report from Torgils Opheim regarding the use of silica fume at Norcems's plant in Larvik.

18 January 1973: Silica fumes were first used in precast concrete structural components. Knut Bratlie sent a report regarding the use of silica fume together with limestone powder in the concrete at Norcem's DINA-plant at Hønefoss.

07 February 1973: A trial was conducted with the use of "frakkulert" (compacted) silica fume at the concrete plant at Nenset.

Medio February 1973: The Elkem Fiskaa plant in Kristiansand became the first plant that obtained an excess of collected fume.

01 October 1973: Managing Director of Elkem, Lorch, sent a reminder to the Managing Director of Norcem, Heiberg, to speed up the negotiation regarding silica fume.

20 November 1974: Knut Bratlie and Kåre Johansen of Norcem submitted a report regarding *various types of filler effects in concrete,* among others, the difference between compacted and noncompacted silica fume.

15 December 1974: Report from Tore Ekholt from the laboratory in Dalen reported about tests conducted during the period 14 August to 15 December regarding different types of silica fume from Fiskaa.

14 January 1975: Paul Olstad delivered a lecture at a smoke gas symposium in Kristiansand.

31 January 1975: Kåre Johansen reported about the tests with silica fume in a cement screw at Norcem plant in Fredrikstad.

12 February 1975: Per Jahren, Vice President of Norcem's Concrete division, presented a memo to the Vice president of the Cement division Per Ivar Åbyholm which stated that all the agreed tests had been completed and that further decisions had to be taken by the Cement division. The Concrete division wanted to use silica fume at all their 21 plants.

04 April 1975: Chief chemist Willam Loe at Dalen reported about grinding of silica fume pellets in cement.

09 June 1975: Negotiations between Norcem and Fesil took place, wherein Norcem quoted a too low price estimate for getting the rights for distributing silica fume all over Norway.

12 November 1975: New negotiations between Fesil and Norcem took place. Fesil was discontented with the Norcem offer and indicated that a price of 85–100 NOK per ton must be the minimum.

December 1975: A lecture on *Possible uses of silica fume* was delivered by Hallgren from Airea, Sweden, in Houston, Texas.

01 January 1976; IPFECO (European ferrocilicium producers) announced an idea-competition in Europe regarding the use of silica fume.

27 February 1976: Birger Warris, Cementa, Sweden, in a telephonic conversation with friends in Norcem stated that the Swedish ready-mixed company AB Ferdigbetong had purchased all the available fume production from the Vargöens plant in Trollhättan, Sweden.

10 June 1976: The article t*he smoke, the problem of the ferrosilicium producers – a waste that can be useful* written by Schei, Elkem, was published in the Norwegian journal *Kjemi (Chemistry).*

26 August 1976: During the board meeting in NFBF (Norwegian ready mixed concrete association), Hægstad from Trondheim Mørtelverk referred to his own positive testing of the use of silica fume in concrete and suggested that NFBF should sign an agreement with Fesil regarding the supply to the industry. Trondheim Mørtelverk has already made an agreement with the company Ila Lilleby.

22 September 1976: NFBF sent information to its members regarding the cement saving properties of silica fume.

21 January1977**:** Fesil offered to cooperate with NFBF. An analysis about the requirement and distribution in the industry was initiated.

24 January 1976; Fesil asked Norcem to stop further negotiations.

30 January 1976: Norcem's Chief chemist William Loe prepared a memo about the advantages and disadvantages of using silica fume for presentation at the NFBF member meeting in Røros.

08 March 1976. The Control Council for Concrete Products entered the scene. In a letter to the concrete industry, they asked for comments regarding the use of silica fume in concrete. They got answers from NTH, dated 17 March 77, NFBF, dated 02 May 77, and Norcem, dated 24 May 77.

15 April 1976: The "Control council" called for a meeting with the industry regarding the use of silica fume in concrete.

October 1976: First paper about silica fume *Betong og SiO_2- støv (Concrete and SiO_2*-fume) *Adding methods– consequences* was presented at the "Norwegian Concrete Day" by William Loe.

12 December 1976: Members of NFBF provided information regarding the use of silica fume and official rules.

1978: NBR (Norwegian Building Standard Organization) decided to make modifications in Norwegian Standards NS3420 and NS3474 so that an 8% addition of silica fume can be used in concrete without special permission.

November 1978: Fesil issued their brochure "FESIL-SILICA".

We would end the rather detailed summary of the various happenings here. This was a time of heated discussions and great differences in opinion about this new product. Unfortunately, it also led to the use of silica fume, which we later realized was disadvantageous. Tests in Trondheim (NTH and FCB) had shown that most effective use of silica fume to save cement, was in low and medium strength concrete. You do not need much knowledge in economics to see that the biggest savings for the concrete producers was to use silica fume in large volumes at just these strengths. The specifications at the time were mainly based on compressive strength. The result was the use of silica fume with water/cement ratios up to and above 1.0. Obviously, the concrete later had serious problems with shrinkage and carbonization. Considerable corrosion damages resulted; quite a few balconies had to be repaired or taken down at the end of the 1970s and the beginning of the 1980s. Important and highly qualified technologists at the Technical University in Trondheim were so impressed with the strength results that they even denied that the water/cement ratio had any effect on the concrete durability. On the other hand, the experienced concrete technologists from the Technical University were so frightened by the cracking they observed that they proposed to forbid any use of silica fume. I even received a memo from an important chemist who was convinced that it was the silica fume that was causing all the problems and that it was dangerous to use it. As a curious case we wish to mention that one of the researchers at FCB (the research institute for cement and concrete in Trondheim), in a paper, claimed that it was dangerous to use silica fume in combination with a plasticizers. Today, we know that the opposite is the case.

There is no doubt that many mistakes were committed when silica fume was first used in concrete, but investigations have shown that at the beginning of the 1980s, we got into a stable period during which Norwegian concrete technology and the Norwegian ferrosilicium industry flourished and also received international acknowledgement for their development. A short summary of the further developments;

1980:

* 6 May: First lecture on silica fume, *the use of silica fume in concrete,* was given by chief chemist William Loe and Per Jahren in a meeting of the Norwegian Concrete Society.

- May: A Nordic seminar on the use of silica fume in concrete was conducted in Ålborg, Denmark.
- Elkem established Elkem Materials.

At the first international conference on silica fume in Canada in 1983, I referred to a literature survey from July 1982. This survey had registered a total of 132 publications about silica fume in concrete, including internal Norcem reports. 107 of these were Norwegian and less than 20% of the reports were in English [240].

The development started in Norway, but by the end of the 1970s the interest very quickly spread and became international. An ACI publication dated June 1996, among others, names the following reports of this first period [241]:

- In 1981, Oberholster and Westra published a report on the use of silica fume to control alkali–silica reactions.
- The first North American report came from Buck and Burkes in 1981.
- Malhotra and Carette from CANMET in Canada published the results of a research in 1983.
- Professor Aitcin from Sherbrooke University in Canada published his results in 1983.
- Norcem in USA with John Wolsiefer (later Norchem) published results in 1984.

In February 1986, Sellevold and Nilsen published the *World review*, in which they reviewed reports from various sectors regarding the use of silica fume. The report has 161 references from 19 different countries.

An Internet search in Google on 22 September 2009 on "silica fume" showed 55 200 hits.

Silica fume is often used as one of the most important components

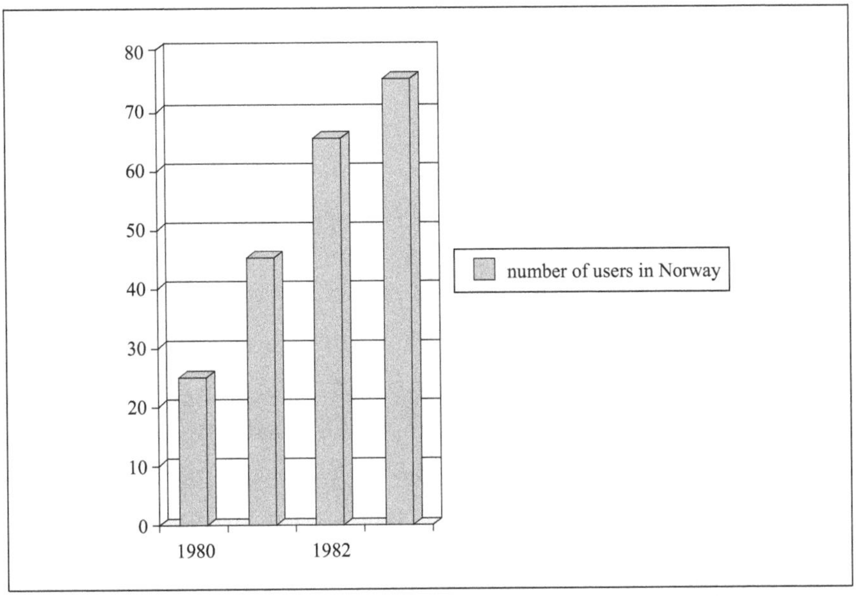

From 1980 to 1984, the number of users of silica fume nearly tripled in Norway, and in the beginning of the 1980s silica fume was a fully accepted commercial product. In 1983, the Norwegian consumption of silica fume had reached 65–70 000 tons [235]

when high strength concrete is produced. While evaluating the development of interest in silica fume, it might be interesting to look at the parallel development in the interest for high strength concrete. The illustration below was given by Steinar Helland, Norway, with reference to A. Ipatti from Finland in 1995 [190].

In the beginning of the 1980s, it started to be relatively normal to add silica fume to the concrete in Norway. In 2005, I made an investigation funded by Elkem Materials A/S, in which 21 random objects were studied,

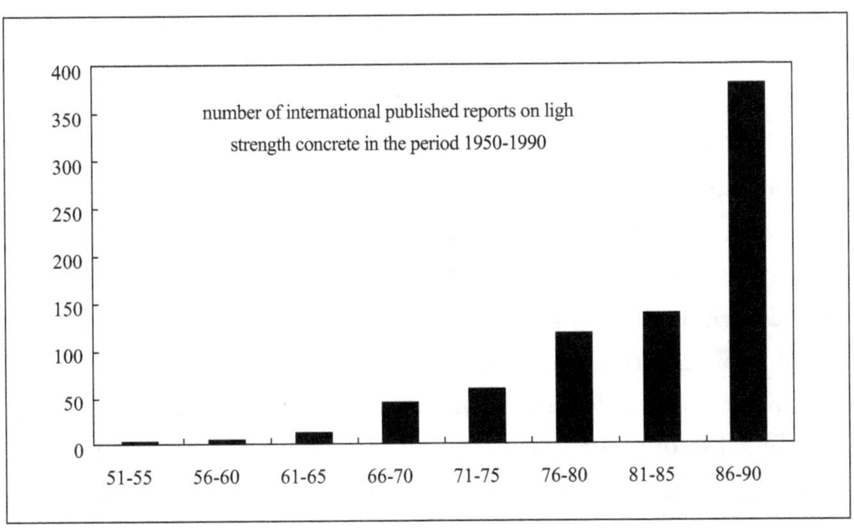

The accumulated number of international reports on high strength concrete in the period 1950–1990

The main build of the Central Hospital of the county of Buskerud in Drammen, Norway, built in 1979–1982 [235]

all of them more than 20 years old. The idea of the investigation was to evaluate the status of the objects in relationship to the purpose for which they were built [235].

The conclusion of the investigation was clear; 19 of the 21 investigated objects had a material quality/condition, after 20 years, that was *in excess of what could be expected for concrete in general.* All the objects functioned according to the intensions and all the objects had over time increased their concrete strength.

The building of the Key Towers was started in October 1988 and it was finished in January 1992. The towers were designed by Cesar Pelli and have 57 floors and a height of 289 meters. The building has a floor area of 144 000 m² [242].

In a report in 2007, Dåstøl and Fidjestøl [234] claim that we today use about 400 000 tons of silica fume in concrete, corresponding to a volume of more than 10 million m^3 of concrete.

A milestone in the use of concrete in normal skyscrapers was the Key Towers in Cleveland, Ohio, U.S., where silica fume was used together with Portland cement and granulated blast furnace slag. The water/cement-ratio was 0.24 and the compressive strength after 56 days was 105 MPa [101].

4.7 The technology spreaders

To write a chapter about the many individuals who have contributed to spread concrete technology is of course impossible if you want to be strictly objective and at the same time be reasonably moderate in the volume of the presentation. This chapter must therefore be regarded as a subjective contribution, mentioning persons who have had particular importance in the spreading of the concrete technology and who have in a special way contributed to our own understanding of the subject. All the mentioned technologists, however, have that in common, and they have had a very wide geographic field for their activity. They are well known and their publications are used across the world.

Arthur R. Andersson (1910–1995)

Arthur's father Eivind emigrated from Slapøy in Nordland in Norway. It has been told that he landed in USA with only one Norwegian kroner in his pocket.

Arthur finished High School in 1928. The economical crisis that came in 1929 gave him a good reason to start studying. He was educated as civil engineer at the University of Washington in 1934. After that he moved to Massachusetts Institute of Technology (MIT), where he got his Master's degree in 1935 and Doctoral Degree in 1938 [146].

Arthur's father was a contractor, so he grew up with concrete. His father's wish was that he should join his brother Tom and start a contracting company at home in Tacoma and name it "Anderson Brothers – General Contractors," which ultimately they did. Arthur stated that in the beginning they hand mixed the concrete, one spade of cement, three spades of sand and five spades of gravel was the standard mix. Slump measurements and cylinder testing were required and the compressive strength was about 2000 psi (13.8 MPa). In 1927, they got a steam-driven concrete mixer. That was some progress.

Arthur's first publication was about underwater concreting, which was published in the *ACI-Journal* in January 1937. After he had finished MIT, Arthur went to Germany and in 1938–1939 he worked for Bauer & Scharte & Max Klone, designing steel bridges. After his visit to Germany, he was engaged by MIT and started to work on a project that had many important consequences. He worked under Professors A.V. DeForest and A.C. Ruge for developing "strain gages" for stress measurements in materials. As a result, he was involved in the designing of boats and submarines in the U.S. navy.

With his knowledge about stress measuring and recordings as the base, he started his own consulting practice in 1946 in Springdale, Connecticut. He developed his own system, (Anderson model 301 Strain meter) and a special system for bridges (Anderson 302 Bridge Balancer). Consequently, Arthur got the job of doing the instrumentation for the first prestressed

Arthur and Tom Anderson with the concrete chair "Møbius" in 1956

Precast concrete carports made of prestressed concrete barrel shell (1953)

Production of long, prestressed, cylindrical, hollow piles in Tacoma

concrete bridge in the U.S, designed by the Belgian Professor Gustave Magnel (See Chapter 3.1.3).

Arthur's brother Tom was a civil engineer who had most of the time worked locally in Tacoma, Washington. In 1948, their father managed to convince his sons to realize the dream about a family company. Both of them had substantial experience in the field of concrete and Arthur had learned a good deal about prestressed concrete. Before they started, they decided to go Europe, to learn as much as possible there. In October 1950 they travelled around in France, Switzerland, Belgium, Sweden, and England, visiting precast concrete plants, building sites and laboratories. In France, they among others met Freysinnet, in Belgium they visited Magnel, and in Sweden they visited Strängebetong.

In July 1951, Arthur and Tom Anderson started their precast concrete operation in Tacoma that later became Concrete Technology Corporation (CTC).

The list of assignments, developments, and patents etc. is very long. Here are some special examples.

The name Anderson can also be found on their own prestressing systems, vibrator systems, and more.

Arthur was not the only one behind all the inventions and developments, but he was the visionary, the one that brought all the ideas to Tacoma and became famous for his ideas nationally and internationally as a result of his work in PCI (Precast Concrete Institute), ACI (American Concrete Institute), and FIP (Federation Internationale de Precontrainte).

In 1959, Bob Mast came to Tacoma and ABAM (Anderson, Birkeland, Anderson, Mast) was established as an independent design and consult-

The first double curved beams for The Disney Monorail in Orlando in Florida was produced in the plant in Tacoma and transported by rail from the northwestern corner of USA to the southeastern corner in Florida. The curvature of the beams, both vertical and horizontal, differed from one beam to the other. The casting mold was a masterpiece in practical engineering work [243].

ing company. Over time, many people contributed ideas and expertise for more complex jobs and inventions.

As an example, the *ABAM*-book [197] states that when they designed and built the monorail at Disney World in Florida, the design team comprised Bob Mast, Charlie Dolen, Erik Prestegaard, Ole Krokstrand, Terry Nettles, and Bob Richardson.

ABAM later became Berger/ABAM. This conglomerate is today present in more than 90 countries.

The above is just a small example of the development work that took place in Tacoma. (See also a bit about the famous ARCO-Barge in Chapter 4.1).

As an author and a speaker in the USA and a number of other countries, Arthur was very persisting and enthusiastic. I also had the pleasure of being with him and Charles Knight (at that time President of PCI) for a lecture trip, when ATC (Anderson Technology Corp.) was established in Tokyo in 1981, together with Japanese partners.

Arthur was President of ACI (American Concrete Institute) in 1966 and of PCI (Precast Concrete Institute) in 1970–1971.

He received a number of awards for his work, among others, PCI Medal of Honor, the FIP Medal and the T.Y. Lin Award. He became honorable member of ACI, PCI, and JCI (Japan Concrete Institute) and in 1976 he was voted into the National Academy of Engineering in USA. In particular, in the 1960s, he gave a number of lectures in several of the countries in Europe (e.g. FIP meeting in Paris in 1966), Japan in 1967, and the Soviet Union in 1968.

Arthur Anderson delivered an excellent lecture, *Advances in Concrete Construction* [91], on the Norwegian Concrete Day at Grand Hotel in Oslo in 1975. In the lecture, he first gave a historic resume of the developments in the field of prestressed concrete, before he gave history of his own activity. The highlight of the day was probably the dinner afterwards, when people after the main course, through away their jackets and told concrete stories. Then Arthur was really happy. This was his way.

Arthur was always full of stories. The following story was narrated in the *ABAM-book* [197] and they say that Arthur used it to convince a developer of a convention centre in Oklahoma that they had to design the roof for snow loads.

> *Two friends were talking. One of them wondered why his friend was not afraid of flying. Never, answered the friend. I always bring my own bomb. You, see, I am a mathematician. The probability for being on a plane with a bomb is 1 to 100 000. The probability for that there will be two bombs on the plane is less than 1 to 1 billion. So I always bring my own bomb.*

The same story was probably used in other contexts too, among others, on the lecturing trip to Japan in 1981.

One of the key people in ABAM was Bob Mast. Bob was President of PCI in 1995, and in 2001 he received PCI's highest award, the PCI's Medal of Honor [244]. Bob narrated the following story about Arthur [198] in 1956;

> *The Dillingham Company in Hawaii should build a large shopping centre. Art was engaged to change the design from in-situ concrete to precast concrete. When the President of Dillingham asked Art what kind of payment he wanted, Art answered; What about half of the savings. The President replied; that will be too much. We are going to save 1 million dollars. That was a lot of money in 1956.*

Arthur also got a professional fishing license and spent many weekends on board his boat "Peskador" in the sea outside the mouth of the Columbia River hunting for salmon. The fish were smoked in a separate smoking tent at the precast plant in Tacoma.

ACI has made a special award in his name, the Arthur Anderson Award. The first receiver of the prize was Adam Neville.

Adam Matthew Neville (born on 05 February1923)

Adam Neville has been an institution by himself in the field of concrete for more than half a century and had worked as lecturer, professor, consultant, and author in a number of countries and continents. It is with great difficulty that we try to summarize what he is mostly known for in a few pages.

To spread concrete technology, the book *Properties of Concrete* has been extremely important. The book has been sold more than half a million copies. It was first published in 1963, later in 1973 and 1981, and the fourth edition in 1996. The book has been translated into thirteen languages. The fourth edition has been translated into Spanish, Portuguese, French, Polish, Rumanian, and Japanese. This book with more than 800 pages is a heavy volume and possesses in depth knowledge of concrete, with its clear and understandable language and perfect illustrations it is easy read and an important reference for many technologists in the field of concrete. The Japanese even call it "The Neville Bible." Probably it is the most sold concrete book in the world.

Neville was born in 1923 and married his wife Dr. Mary Hallam Cousins in 1952. He had his basic Civil engineering education at Queen Mary College, London University, and his Doctoral degree from London University and Leeds University.

After being engaged as a lecturer in Southampton in 1950–1951, New Zealand 1951–5194, Manchester 1955–1960, and Nigeria 1960–1962, he continued his career as Professor and Dean at several well-known Universities:

- University of Calgary, Canada (1963–1968)

- University of Leeds, England (1968–1978)
- University of Dundee, Skottland (1978–1987)

For those that are interested in a more complete, detailed, and correct summary of his many titles, awards etc, we refer them to a website such as "WHO's WHO."

As a spreader of technology, he had also received awards, not only from England, but also from Scotland, Canada, the U.S, Brazil, Belgium, and Singapore.

He has also contributed as a consultant for building projects in UK, Albania, Australia, Chile, UAE, Hong Kong, Jordan, Netherland, New Zealand, Pakistan, Saudi Arabia, Spain, and USA. This work included a wide field within concrete technology.

In addition to having published more than 200 scientific and/or technical papers, he has to his credit, partly in cooperation with others, ten books, of which six are about concrete technology, two about prestressed concrete, one about design analysis and one about statistical methods for engineers.

I had the pleasure of sitting next to Adam Neville during a dinner at a conference on concrete in Seville in Spain in October 2010. I had told him that I was working on a manuscript on concrete history and asked him for some documentation. With his ironic smile, he among others told the following story;

In the beginning of the 1900s, an American concrete expert came to Norway, with the purpose to convince Norwegian people to build concrete houses. He tried to convince a Norwegian architect to build concrete houses, but in Norway all private houses were in timber. The architect said that he could not quite understand the concrete argument, and said; First you tell me to build a wall in timber. Then I shall build another timber wall outside this. Then I am going to fill the space between them with concrete. I do not quite see how this can be cheaper.

A few words also have to be mentioned about the journal author Adam Neville. Many people in the field of concrete have with interest followed his

Professor Adam Neville from a picture in the journal a "Concrete" May 1975 together with Mrs. Molly Laister and Mrs. Margaret Dixon, great grand niece of Joseph Aspdin (the Portland Cement inventor). Neville was at this time both chairman of the Concrete Society and manager of the civil engineering department at Leeds University. The occasion was that the first statue in Portland cement concrete, originally owned by the Aspdin family, which had been in the garden of Mrs. Laister, was via Neville handed over to Leeds University.

Picture of the statuette of a young angel, which was on the front page of the proceedings of a conference on concrete in Seville, Spain, in October 2009. (See later). The statuette must, according to Barfoot [245], have been made by Joseph Aspdin's frustrated son James. The statuette is meant to be of the prophet Samuel as a child and was designed by J. Aspdin about 1850.

many comments in *Concrete International,* in which he in an exceptionally knowledgeable and fascinating way gives *his opinion* about important issues in concrete. Together with the analysis and the comments, there was very often a story or an anecdote, which has beencome like a signature for the author.

Adam Neville, from Concrete International 2002

We cite a few lines from some of his articles *In My Judgment* in *Concrete International.*

- April 1999: "What Everyone Who is 'in' Concrete Should know about Concrete" [246]. (about ten years after this article was written, it has become "good Latin," , to claim that concrete shall have the right quality, not maximum quality).

- Neville, among others in his article says "*we should not try for the very best concrete but rather make concrete that is good for the specific purpose for which it is intended*"
  Another citation;
  "it is a fact that designers know less, and not more, about concrete than they did 20 years ago"
  About the quality of the raw materials;
  "But you can have excellent ingredients and cook a bad meal"
  And finally a citation about high performance concrete;
  "In my view, the division between high-performance concrete and ordinary concrete will soon disappear, and we shall be providing custom-made concrete or tailor-made concrete for most of our clients"

- September 1999: "How Useful is the Water-Cement-Ratio?" [247]
  The article published in September 1999 was followed by a number of letters from the readers, to which Neville replied in January 2000. Here he felt it necessary to point out that what he had expressed his own opinion in the article and not the compromises from a committee and added *it has long ago been said that a camel is horse specified by a committee.*
  In the article and the later replies, he pointed out the importance of the water/cement law and also its limitations. He ends a letter with the following sentence;
  "In my opinion, there is no single 'correct' view of concrete, just as there is no single correct way to mix a martini. I like them both, concrete as well as martinis."

- January 2000: "There is More to Concrete than Cement" [248]
  We quote from the conclusion of the unusually short article; concrete is not the same as cement, nor is it a "diluted" cement. Concrete is a material with three components: cement paste, aggregate and the interface between the two. Aggregate deserves more attention than it is often given···. These operations may be low-tech but good concrete is not low-tech material.

- December 2000: "Water and Concrete: A Love-Hate relationship" [249].
  Reading the article is compulsory for all concrete technologists, but here we only include the last sentence,
  My wife of 48 years' standing, who has contributed a great deal to my "concrete thinking," would not allow me to liken this love-hate relationship to marriage. Let me just say that there is no doubt that, for the foreseeable

future, we shall need not only water but also concrete, the two will have to live together."

- November 2001: "Core Tests: Easy to Perform, Not easy to Interpret" [250]
  The article gives a thorough analysis of the use of core tests on finished concrete structures. The main message is the importance of having a clear opinion about how realistic the methods we use are before starting to test and before discussing the results. To underline the uncertainties, Neville gives a citation from the golf-ace Jack Niklaus who is supposed to have said: *People tell me I am lucky. The funny thing is that the harder I practice the luckier I get.*

- November 2002: "Autogenous Healing – A Concrete Miracle?" [251]
  Neville does not doubt the ability of concrete to heal by itself – the ability, through a chemical reaction over a long time, to close smaller cracks. The article is about the problem of generalizing this and the uncertainties that are found in the evaluations. He hints that if cracks smaller than 0.2 to 0.3 mm close over time, then that is a safe limit. He ends with the following sentence:
  So, autogenous healing of concrete is not a miracle, but it is an interesting and, under some circumstances, useful self-healing process; we can thus look to concrete to heal itself in the words of St.Luke in Chapter 3: "Physician, heal thyself".

Mohan W. Malhotra

Mohan W. Malhotra was born in India on 30 October 1931 and got his B.Sc.degree from an Indian university in 1951. He received his education as a civil engineer at the University of Western Australia in Perth in 1956.

Malhotra came to Canada in 1962 and was employed by CANMET (Canada Centre for Mineral and Energy Technology) in Ottawa in 1962.

As a researcher and a developer at CANMET, he has set his footprints in the field. However, it is as a spreader of technology that he has been most known around the world.

Malhotra retired from CANMET in 1999 from the position of Program Manager for Advanced Technology Program at the International Centre for Sustainable Cement and Concrete, but he did not retire from the work that he had been doing in the field of concrete.

Mohan Malhotra pictured in Nanjing in China in May 2007, together with the author, in connection with a lecture series at Beijing, Nanjing, and Lhasa arranged by Malhotra with the China Building Material Society to promote the sustainable development of cement and concrete technology.

As a researcher he has presented a number of important papers and publications.

Of the more well known publications we mention:

High-Performance, High-Volume Fly Ash Concrete, which was written together with P.K. Mehta. He is also the coauthor of books on silica fume and superplastizicers.

In addition, he is responsible for the development of several test methods for concrete.

The Conferences

Mohan W. Malhotra has been in charge of the arrangement of a number of conference series and seminars in many countries and continents. He has also been the main editor of the conference proceedings. Here are some of the most important (publication number from ACI).

International Conference on Fly Ash, Silica Fume, Slag and Natural Pozzolans;

1. Montebello Canada 1983 (ACI SP 79)
2. Madrid, Spain 1986 (ACI SP 91)
3. Trondheim, Norway 1989 (ACI SP 114)(the highest number of delegates – over 400)
4. Istanbul, Tyrkey 1992 (ACI SP 132)
5. Milwaukee, USA 1995 (ACI SP 153)
6. Bangkok, Thailand 1998 (ACI SP 178)
7. Chennai (Madras), India 2001 (ACI SP 199)
8. Las Vegas, USA 2004 (ACI SP 221)
9. Warsawa, Poland 2007 (ACI SP 242)

International Conference on Superplasticizers and Other Chemical Admixtures:

1. Ottawa, Canada 1978 (ACI SP 62)
2. Ottawa, Canada 1981 (ACI SP 68)
3. Ottawa, Canada 1989 (ACI SP 119)
4. Montreal, Canada 1994 (ACI SP 148)
5. Rome, Italy 1997 (ACI SP 173)
6. Nice, France 2000 (ACI SP 195)
7. Berlin, Germany 2003 (ACI SP 217)
8. Sorrento, Italy 2006 (ACI SP 2399
9. Seville, Spain 2009 (ACI SP 262)

International Conference on Advances in Concrete Technology:

1. Athens, Greece 1992 (published by CANMET)
2. Las Vegas, USA 1995 (ACI SP 154)
3. Auckland, New Zealand 1997 (ACI SP 171)
4. Tokushima, Japan 1999 (ACI SP 179)
5. Singapore 2001 (ACI SP 200)
6. Bucharest, Romania 2003 (no ACI- publ.)
7. Las Vegas, USA 2004 (ACI SP 222)
8. Montreal, Canada 2006 (ACI SP 235)
9. Warsaw, Poland 2007 (ACI SP 243)
10. Seville, Spain 2009 (ACI SP 261)

International Conference on Durability:

1. Atlanta, USA 1987 (ACI SP 100)
2. Montreal, Canada 1991 (ACI SP 126)
3. Nice, France 1994 (ACI SP 145)
4. Sydney, Australia 1997 (ACI SP 170)
5. Barcelona, Spain 2000 (ACI SP 192)
6. Thessaloniki, Greece 2003 (ACI 212)
7. Montreal, Canada 2006 (ACI SP 234)

International Conference on Marine Environment:

1. St. Andrews, Canada 1980 (ACI SP 65)
2. St. Andrews, Canada 1988 (ACI SP 109)
3. St. Andrews, Canada 1996 (ACI SP 163)

International Conference on Non Destructive Testing:

Ottawa, Canada 1984 (ACI SP 182)

International Conference on High Performance Concrete and Materials:

1. Florianopolis, Brazil 1996
2. Granado, Brazil 1999 (ACI SP 186)
3. Receife, Brazil 2002 (ACI SP 207)
4. Olinda, Brazil 2005(ACI SP 229) (the conference was dedicated to V.M.Malhotra)
5. Manaus, Brazil 2007 (ACI SP 253)

V. Mohan Malhotra is an honorable member of ACI (American Concrete Institute), Concrete Society – UK, Institute of Concrete Technology – UK, ASTM Committee C09, Brazilian Concrete Institute, Canadian Institute of Civil Engineering and The Engineering Institute in Canada.

The list of awards is long, among others from a number of American and Canadian institutions such as ACI, ASTM etc., and from countries like Brazil and India. ACI has also made a special award in his name.

V. Mohan Malhotra was the first to receive the Environmental Award from the Norwegian Concrete Society during the "International Conference on Sustainability in the Cement and Concrete Industry" at Lillehammer in September 2007.

In the last few decades he has been more and more focused on the environmental challenges we have and sustainable development and climate change in particular.

He has been responsible for a number of conferences on the sustainability challenge in the cement and concrete industry.

- International Symposium in Ottawa, Canada 1998.

- International Conference in San Francisco, USA, in 2001 (ACI SP 202). The conference was supposed to have been conducted from 16 September, but was cancelled because of the terrorist attack on 11 September, but the papers of the conference were published.
- International Symposium in Toronto, Canada 2005.

In addition to the conferences, Malhotra has also been the key person in arranging "smaller" seminars and workshops on the topic of sustainability, for example:

- 2002: Barcelona, Spania and Lyon, France
- 2003: Milano, Italy
- 2005: Hydrabad, India

Together with Sui Tongbo and his team from the China Building Materials Academy (CBMA) he was also co-designer of the "International Workshops" conducted in various provinces in China;

- August 2007: Beijing, Nanjing, Lhasa
- October 2008: Beijing, Shenyang, Jinan
- May 2009: Beijing, Xi-an, Mianyang.

After Sui Tongbo worked as director general of Sinoma Research Institute, he continued leading the team to cooperate with Mohan Malhotra in May 2013 for the 4th serial 3-city international workshop in China to promote the advancement of concrete technology for sustainable development. Next international workshop on concrete sustainability is in planning to be held in China in January 2016.

2007 International Workshop on Cement & Concrete Technology for Sustainable Development, Beijing-Nanjing-Lhasa. Chairmen of the workshop: Mohan Malhotra (right) and Tongbo Sui (left) at CBMA.

- May 2013: Beijing, Linyi, Xi'an
- January 2016: Beijing, Ji'nan, Chengdu

The list speaks for itself. It is hardly possible to find another person who better suits the label "global technology spreader".

V.Mohan Malhotra opens a "Sustainability-workshop" at Beijing, China, in May 2009.

P. Kumar Mehta

Mehta received his education as chemical engineer from Dehli in India. He continued his education and got his Master's Degree in Ceramic Engineering from North Carolina State University and Doctor of Engineering in Materials Science and Engineering from the University of California at Berkeley.

He is now Professor Emeritus in Civil and Environmental Engineering at the University of California at Berkeley. When he retired in 1993, after 30 years as lecturer and researcher in concrete technology, he was awarded the Berkeley Citation – the highest award that the university could give for his work over many years.

Many positive things can be written about Kumar Mehta as he is a very likable and positive human being in general; the following pages discuss him as a technology spreader.

He is the author and coauthor of more than 250 scientific papers and four books.

Probably, his most well known work is the popular university textbook *CONCRETE – Microstructure, Properties, and Materials.* This book has been translated into Chinese, Japanese, Spanish, Portugese, and Persian. Together with his coauthor he in 2006 published the last and modernized version of the book.

P.K.Mehta has through the years received a number of awards;

- ACI's Wason Medal for Material Research
- ACI/CANMET's award for his work, knowledge, and understanding of physiochemical factors that influence the properties of concrete in marine environments.
- ACI/CANMET's award for research with supplementary materials.
- Honorary member of American Concrete Institute.
- At the ACI fall convention in 2001, he was invited to give the Tuthill Memory Lecture about *Greening of the Concrete Industry.*

P.K.Mehta in the middle before the "birds nest" at the Olympic village in Beijing during a lecture trip regarding sustainable development in the cement and concrete industry in October 2008. To the left of Kumar are Mohan Malhotra and Theodor Bremner from Canada; to his right, the author and Koji Sakai from Japan.

In 2006, he received the Lifetime Achievement Award from Coal Combustion Products Partnership for his "outstanding research for decades on the use of fly ash in structural concrete."

Mehta has been a pioneer in a number of areas, but internationally in the last few decades, he is probably most known for his work with substituting materials in concrete and his interest in the environment and for a sustainable development.

Mehta also has 9 patents in cement and concrete technology.

5 The cement industry (in the Nordic countries)

5.1 General

5.1.1 Norway

The following is a translation from Gartmann's book, published during the centennial celebrations of the cement company Norcem [6].

> *In November 1888, the company Christiania Cement Aktiebolag was registered to "work with production and sales of Cement, lime and stone from the properties of the farms of Slemmestad in the Røyken community." Behind the new company were one German and two Swedish business men that built the first factory at Slemmestad during 1889 and got started a modest production, some cement, but mainly bricks. This happened in 1890.*
>
> *The activity did not go well, so the share capital of the company had to be increased, and they had to let Norwegians into the company. The money came from the individuals that later should take over all of it, and to establish A/S Christiania Portland Cementfabrik at the same place in 1892. The refinancing took place in 1891, and from then on this was a Norwegian cement company.*

In the beginning of the 1900s, about 100 000 tons of cement were produced in Norway, which grew into 200 000 tons in 1920. Toward the end of the 1960s, this had increased to 2.7 million tons.

In the beginning, cement was distributed in barrels. The weight of one barrel was fixed at 425 "skålpund" or 180 kg gross and 171 kg net weight. In some plants, hemp sacks were used in addition to the wooden barrels, but with the First World War there was a shortage of hemp, so the barrels were used till the 1920s. At Slemmestad, the barrels were used until 1930 [6]. The use of paper bags started in 1927. In the beginning, the content of one bag was equal to one third of a barrel and was 56.6 kg net weight. This was rather soon modified to 50 kg so that 20 bags became equal to one ton. The bulk transport started in the 1960s.

The idea of cooperation among the Norwegian cement plants came into existence in 1927, when the active president of the Dalen cement plant, Alfred Holter, convinced his competitor at Slemmestad to join him in establishing the Norwegian cement association. The purpose of the association was to expand the market through information, development work, and promotion. In the beginning, the association had 23 employers.

The merger of the three Norwegian cement companies took place in 1968.

In the spring of 1972, the old men in the cement industry decided that they needed younger power. They employed a 32 year old enthusiastic and highly qualified economist as their new president. The "young man," Gerhard Heiberg, came into the picture in August 1973. With him started a new period of expansion in the industry with more focus on international activity.

In 1986, Norcem in Norway and Cementa in Sweden merged and established the company Scancem International to coordinate their international interests.

Norcem became, through merger, a part of the Aker-conglomerate in 1987.

In 1995, Aker merged their cement company with Swedish Euroc, in which Cementa was a part, to form Scancem.

Scancem became part of the multinational company Heidelberg Cement in 1999.

5.1.2 Denmark

The Danes were the first cement producers among the Nordic countries The plant was in Ringsted in 1968 and the second came in Rødvig two years later. In 1874, the Cimbria factory in Mariager fjord was established. In the Aalborg area, three plants were established, among which Aalborg Portland Cementfabrik was the largest. Two plants were also started at Nørresund. In the period from 1916 to 1918, the Danish plants together produced 3.5 million barrels of cement. As a comparison, at the same time, the production in Norway was 1 million barrels at Slemmestad, 900 000 barrels at Dalen, and 300 000 barrels in Kjøpsvik.

The Aalborg plant soon became a leader in the Danish cement industry and was the "leading lady" among the Nordic cement producers and to some extent also at a global level.

5.1.3 Sweden

The history of the beginning of cement production in Sweden is interesting and also an example of how coincidences might be important for the results and the later history. Much of the information has been taken from the centennial commemoration book, "Cement i 100 år," written by Alf Åberg and published in 1972 [60].

A key person in the development was Otto Fahnehjelm. He was very fascinated by the possibilities of Portland cement and, as a newly educated civil engineer, he went to the Swedish island of Gotland in 1869 to investigate. First he investigated the possibilities of establishing a cement factory at Klintehamn, a bit south of Visby. The result was disappointing and so he continued the investigations in Visby. This was promising, something that later was confirmed by the Swedish Geological Investigations. Fahnehjelm then went to England and France to study cement production. Fahnehjelm's investigations were continued by the Engineer A.W.Lundberg who was the area architect in Visby. He also got in touch with English industrialists, who could have been interested in a cement factory at Gotland.

At the time of Fahnehjelm and Lundberg's investigation in Gotland, some work was being done in southern Sweden, in Skåne. Burning of lime on an industrial scale had been going on in Limhamn since 1622. Production of clay brick from local clay had been going on at Lomma since 1682. The transport conditions between the two places were good. The industrialist from Malmø, Frans Henrik Kockum, had become the owner of the brick factory at Lomma in 1854. He was related to the owners of the Limhamn estate, Peter and Ludwig Kockum. Otto Torell was professor in zoology and geology at Lund University and had given lectures that Kockum had found very interesting. In the beginning of 1870, Kockum contacted Otto Fahnehjelm who had just returned from his studies on cement abroad. Fahnehjelm's evaluation of the possibilities of cement production in Skåne was very positive.

- The raw materials were good.
- Timber for the barrels was cheaper in Sweden than in England.
- Coal was not very expensive, and there was the possibility of using peat as energy.
- The possibility of exporting cement to Russia, which was then importing cement from England, was good.

Lundberg, with whom Fahnehjelm had worked with at Gotland, showed up in Skåne after his study trip to cement factories in England, France, and Germany. Kockum persuaded him to stay in Skåne.

Skånska Cementaktiebolaget was established in 1871 and they produced their first barrel of cement the next year. The factory was in Lomma, not far from Malmø, and the production was based on limestone from Limhamn and clay from Lomma. The factory was in operation until 1905. The more modern Limhamn factory was after many discussions ready in 1890. Many personalities were involved in the process along with the German cement expert, Professor Wilhelm Michaelis. His evaluation of the availability of raw material is said to have been the most important criteria in the final decision about the establishment. Lundberg became the first general manager of the factory, with a salary of 3 000 riksdaler + 2% of the profit. Fahnehjelm became technical manager, with a salary of 2 500 riksdaler + 5 øre per barrel of the cement sales.

The U.S. steamer Franklyn is loading cement in Limhamn in 1906

Cementa's trademark, the red cross, became known in a number of export markets

The cement factory at Visby in Gotland was established in 1883.

Later, a number of factories were established. Sweden had altogether 14 cement factories. When the two cement companies Cementa and Gullhøgen merged in the 1970s, about hundred years after their establishment, Sweden had 7 plants in operation, with all together 24 kilns and a total capacity of 5.4 million tons of cement. The Swedish cement consumption was 3.7 million tons. The rest were exported. Soon after, it went steeply downwards, which led to the reduction in the number of plants to 3, with a total of 4 kilns and a production of 2 million tons [6].

The Swedish cement activity was export oriented from the beginning. In the 1890s, 60% of the production was exported, mainly to the Baltic countries and Russia. In 1906, Cementa had important deliveries to San Francisco in connection with the rebuilding after the big earthquake. All together 6 400 tons was shipped in a single load. The factory in Limhamn could not manage this big order alone, so they had to get help from the factory in Klagshamn to be able to meet the requirement. It has been claimed that this was the biggest single shipment of cement at that time in history.

The best year with respect to the percentage of export was 1893 when 60% of the production was exported. In the second half of the 1890s, the export part of the production was between 28 and 45%. After the turn of the century and up to the Second World War, the figures vary from 1% to 37%, with the highest percentage (more than 30%) in 1900, 1911–1913, 1922–1923, and 1925–1928.

The building of the Pargas plant in September 1914 [149]. The factory building was 105 meters long.

5.1.4 Finland

In 1869, Herman Brummer established the first Portland cement factory in Finland. However, they functioned on a very limited scale in the 1870s with only one foreman and 10 workers. The production stopped from 1887 to 1891 and in 1894 the whole factory was closed down.

It was not until 1914 that the Finnish cement industry started its full fledged production by establishing a cement plant in Pargas or Paraisten Kalkkivuori Oy; the plant was internationally known as Pargas. The company, which was not far from the town of Åbo, was actually founded in Sørnes in Helsinki in 1912.

The next cement company in Finland was Lohja, the plant of which was established about 50 km east of Helsinki.

The Lohja factory produced its first cement in January 1919, somewhat delayed according to the plans. One reason for this was that part of the production equipment was highjacked in Germany in May 1917. The second kiln in Lohja started to function in the spring of 1924.

The Pargas company also established cement plants in Vilmanstrand and Kolari in the north of Finland.

The Lohja factory has now been closed down.

5.1.5 Cooperation among European cement producers

The European cement producers have been cooperating through CEMBUREAU (The Cement Statistical Bureau). Today, Cembureau has 28 member countries whereas one is associated, Croatia is one. In 2008, the Cembureau member countries produced 310 million tons, 11% of the world's total production of cement, to a sales value of 22 billon Euro, with 72 000 employees [6]. Cembureau's headquarter is in Brussels.

The initiative for the establishment of Cembureau came from the president of the Dalen factory in Norway, Alfred Holter. The export activity carried out by the European cement producers got into trouble in the 1920s and 1930s, with decreasing profit and major differences of opinion among many members. Holter travelled between the different member countries, trying to find some solution, without much luck in the beginning. In the fall of 1928, representatives from eight European countries met for a conference in London, where they formed a committee to look at possible solutions. The result was that the cement industries in Belgium, Denmark, England, France, Germany, Norway, and Sweden decided to establish a common office in den Haag in Netherland – International Cement Bureau.

The purpose of the office was to try to find a solution through which the producers were guaranteed a certain amount of sales in their home markets in Europe. This work led to the development of European cooperation.

In November 1936, Director Malcolm Stewart of Associated Portland Cement Manufacturers Ltd. in England invited the representatives from the European countries to a meeting in London. A working committee was established and in February 1937 an agreement to form a new organization – Intercement – was signed.

Alfred Holter was appointed as the managing director of the new organization and the office was located in Paris. The Intercement agreement was basically about the members getting certain export quota. In addition, the members should not export to the other members' home market without permission. The agreement led to a certain stabilization of the prices, but there were considerable cooperation problems. In particular, Belgium and Germany were dissatisfied. This led to the disbandment of the organization at the outbreak of the Second World War in 1939. Denmark, Norway, and Sweden, however, had good cooperation and still continued their cooperation.

After the Second World War, it again Holter who took the initiative to begin a European cooperation movement. On the basis of this, Cementa in Sweden was invited to a meeting in Stockholm in October 1946. The negotiations during this meeting led to the formation of CEMBUREAU. The first office of the organization was located in Brevik (The Dalen plant office) in Norway, with Holter as the president.

When Holters died in 1948, the office was moved to ble Malmø in Sweden and Director Axel Iveroth of Cementa became the president of Cembureau.

Before the Cembureau office was established in Brussels, it was functioning for sometime in Paris.

5.2 The Norwegian cement plants

In the beginning of the chapter, we had a glimpse of the turbulent start of the Norwegian cement industry. The Slemmestad factory was reorganized again in 1892. The company dominated the Norwegian cement market single handedly, until the establishment of three new cement companies; in Lier near Drammen (Ce-No), Dalen in Brevik, and Kjøpsvik in Nordland in the years 1916 to 1918.

5.2.1 Slemmestad

In 1888, Slemmestad was an agricultural area. A German–Swedish company saw the possibilities of establishing a cement factory when they saw the local lime deposit and bought up agricultural lands with the aim of starting cement production. Limestone had been exported from this area for a very long time.

The Slemmestad plant began production in 1890. The first rotary kiln at Slemmestad started operating in 1909. The first two rotary kilns were 31 and 34 meters long. As a comparison, the last rotary kiln installed at Slemmestad was 163 meters long. Later, the cement kilns became shorter.

The precalcination technology has reduced the length of the kilns to half, without reducing the capacity.

Like the other cement plants, Slemmestad became a corner stone, with interesting working possibilities and important initiatives in the development of the local society.

Around the turn of the last century, 250 men were employed in the mining, the production, and the packing. This number grew to a maximum of 584 in 1920, after which it was reduced again. In 1940, there were 293 employees in the company [6]. After 1945, the number increased again and in the 1970s, it grew to over 500. At the 75 years anniversary in 1967, the factory employed 532 persons and had a capacity of 1.1 mill tons of cement.

All cement factories seem to have their own interesting history. We refer to Gartmann [6] for a more fascinating and detailed history of the Norwegian plants. However, to give a little insight into the early history of the Norwegian plants, we among others quote this little extract from the functional regulations at Slemmestad.

It is forbidden to talk during working. A functionary that smokes cigars, enjoys alcohol of any kind, visits billiard saloons and political meetings adds to the possibility that one might doubt his honor, serenity, uprightness and honesty. Simple attire is necessary. The personnel must not use bright shiny colors, and should use proper stockings. Galoshes and coats must not be worn in the office, because the personnel have got an oven for their disposal. Only in bad weather are scarfs and hats allowed. In addition, the personnel are recommended in the winter to bring with them 2 kilos of coal per day per person.

Part of a painting of the cement plant at Slemmestad, painted by the artist Jacob Weidemann in 1955.

Awe for God, cleanliness, and punctuality are bases for good business.

The personnel only have to be present on weekdays between 6.00 in the morning, and 6.00 in the afternoon. Sundays are used for church going. Every morning there is a prayer meeting at the office.

It is expected that everyone contributes with overtime work, if the business demands this. The functionary with the highest age is responsible for the cleaning of the office. All teaching boys and juniors report to him 40 minutes before the prayer, and will be at his disposal also after office hours.

It is allowed to enjoy food between 11:30 – 12:00, but the work must not be stopped for this reason.

Every member of the personnel, have to take care of their own health. In case of illness, payment is stopped. We therefore recommend strongly that everyone save a considerable amount of their salary for possible cases of illness, so that they will not be of any burden to the community, when they no longer can work, or his strength is reduced.

As a curious case, we also mention that all the workers at Slemmestad got a delivery of one block of ice each week. This continued up until the 1930s, when they instead could get a loan to buy a refrigerator [6].

The author Rudolf Nilsen wrote the following about Slemmestad in *Norges Kommunistblad (Norwegian Communist Paper)* dated 1 May 1925 [252].

A constant smoke somewhere out along the fjord, – that is Oslo's imagination about Slemmestad. In addition we know that the four chimneys that are spewing out smoke day and night, summer and winter towards the sky, belongs to a cement factory. But otherwise do people in Oslo know very little about Slemmestad, because it does not belong to the beautiful places along the fjord where you like to spend your Sunday. Smoke frightens, and we have enough of it in the city. That is why Slemmestad has become a small town by itself, an industrial town, a miniature, about like the small medieval town around its knight castle.

The Slemmestad factory was in operation for 95 years. The clinker mill stopped in 1989.

5.2.2 Ce-No

The cement factory Ce-No in Lier started its trial production in 1917. The Ce-No factory did not function for a long time and was closed after about 10 years of troubles. One important reason was probably that the plant began functioning with vertical shaft kilns, while the rotary kiln technology was on its way in.

The establishment and the production was strongly criticized by the press, not least by the competitors, who, scared of possible competition from the company, did not put an end to their negative views about the future of company. The ice in the Drammen fjord during winter was a definite handicap and it was openly mentioned. The newspapers in Drammen, however, were full of hopes for the new company in Lier; it was supposed to be of great importance as there was a shortage of cement in the market, but the plant had problems with the machinery. The planned railway took longer time than expected and there were plenty of functional conflicts to write about too.

The full name of the company was Cementfabrik Norge CE-NO, Portland Cement A/S. The company was planned in 1914 and registered in May 1915 with Christiania (Oslo) as the location and the trial production started in 1917. When the plant finally started to produce, the annual production was only 50 000 barrels, one quarter of what was planned. The factory was closed down in 1927, after ten years of troublesome production.

5.2.3 Dalen

Dalen Portland Cementfabrik (DPC) A/S was built on the shores of the Eidanger fjord outside the town of Brevik. The factory was established in 1916 and the machinery was ordered shortly afterwards.

Dalen's early history is turbulent, fascinating, and interesting. For those who show special interest in this topic, we recommend the anniversary book of 1969 written by Kåre Fasting and Fridtjof Gartmann [107].

The history is dominated by strong and enthusiastic personalities. From the beginning, the first and foremost personality associated with company is the Director Alfred Holter. He was born in Christiania in 1884. Holter took his exam from Christiania Cathedral School in 1902 and was in the next year a lieutenant in the army. In 1903, he started his university education at the technical high school in Hannover and took his exam as

Dalenfjæra, where it started in 1916

Dalen at the 50th year anniversary in 1968

marine and machine engineer in 1908, after having served the military even during his study period at home in Norway. Alfred Holter worked for a short period at Christiania Portland Cementfabrik at Slemmestad, before he was employed by Norsk Hydro. First he was working at the main office in Oslo, but in the fall of 1911 he came to Rjukan Saltpeterfabrikker as department engineer. This factory was then buying burnt lime from estate owner G.R. Wankel's lime kilns at Dalen.

When Holter started to plan his cement factory at Dalen, he had a lot of help both financially and practically from persons that he had known during his work in Norsk Hydro. A "preliminary plan for the establishment of a cement factory at the property Dalen near Brevik," was signed by Alfred Holter on 8 May 1915. From the beginning, thoughts about export were always there; export of timber from the Baltic Sea to England, coal from England to Dalen, and cement to Petrograd were definitely one of the freight wise competitive advantages. In a contract dated 28 January 1916, the name Aktieselskapet Dalen Portland Cementfabrik was used for the first time. The contract, among others, stated that Holter was committed to take the position of managing director for at least two years after the company had been established. The share capital of 3.2 million Norwegian kroner was fully sanctioned and the general meeting to establish the company was held in Brevik on 7 August 1916.

The machinery for the plant was ordered from Krupp in Germany and the company Høyer Ellefsen was the main contractor for the building. Budget deficits and other problems at the beginning were formidable and

Alfred Holter

"Grey Bear" was chosen as the DPC symbol and logo

the economic position was also very problematic.

The first cement in Dalen is registered to have been produced on 28 June 1919. Many typical start-up problems led to the situation that it took five years to get the production up to the planned level.

In 1919, Holter travelled to Argentina. Even if the export to Argentina was not particularly large or profitable from the beginning, this was an important part of the history of Dalen and Holter's further reputation. Over time, Alfred Holter played an important role not only in Norwegian, but also in Nordic and international cement export and cooperation. The Norwegian cooperation started already toward the end of 1918. An agreement on the principal sharing of the market was signed, with 52% to CPC (Slemmestad), 30% to DPC (Dalen), and 18% to Ce-No. Holter was not too worried about the new modern plant at Dalen getting only 30% of the Norwegian market because he focused on export and succeeded at it. Already in the five years between 1923 and 1927, DPC exported 2.3 million barrels or 391 000 tons of a total production of 544 000 tons. (The production at Slemmestad was about the same, but the export only half of that of Dalen in the same period).

Holter was also a pioneer in establishing Norsk Portland Cementkontor A/S (the Norwegian Portland Cement Office) in 1922 and an enthusiast in other areas of development; Dalen was the first plant to use paper bags and, on the initiative from Holter in the road-sector, the so-called Holter-slab was developed. This patented road slab was exported to places as far as Egypt and Brazil. That Holter, who had experienced tough competition in the cement industry and at the international arena, in 1926 managed to get the principal approval of a body of cement producers across the world, shows how reputed he managed to get internationally. The final agreement did not come to fruition until 1937, when the International Cement Export Conference (Intercement) was held, with an office in Paris and Alfred Holter as president.

A small story would speak of Holter's ability to adjust to the circumstances. This is during the outbreak of the Second World War in 1940. Holter had travelled to Copenhagen on Saturday 6 April, with plans to travel further to attend meetings in Paris on the 9th. During these days Germany invaded Denmark.

He could not sit in the trap that the Germans had laid for the soldiers at Copenhagen. He got hold of a rowing boat, used a bed sheet as sail, and got across to Sweden. From Gothenburg, he got to Brussels and Paris. From there he fled further to the USA.

Holter was a member of the Norwegian government's "Relief Commission" in the USA in 1940–1941 and director of the department of provision in London in 1942. Afterward he returned to USA, and until the end of the war he was the managing director of the Atlantic Coast Fisheries Co. in New York.

On 15 July 1948, Alfred Holter launched his large motor boat "Havskåren." An explosion in the engine caused fire and Holter was severely burned. He died two days later in the hospital in Brevik.

Another great personality from Dalen was Per. M. Backe. He was born in 1914 and had been involved in the reorganization of the Norwegian Airline Company and the formation of Scandinavian Airlines, SAS, and had been the director of both companies. Before he came to Dalen in 1959, he had also worked in the Norwegian Industry Association as consultant, with office in Paris. Backe was the one, who, after 40 years of trials, managed to get a ten years agreement among the three Norwegian cement companies. Under Backe's management, the Dalen plant was modernized

Development in the cement supply at Dalen (in 1000 tons)

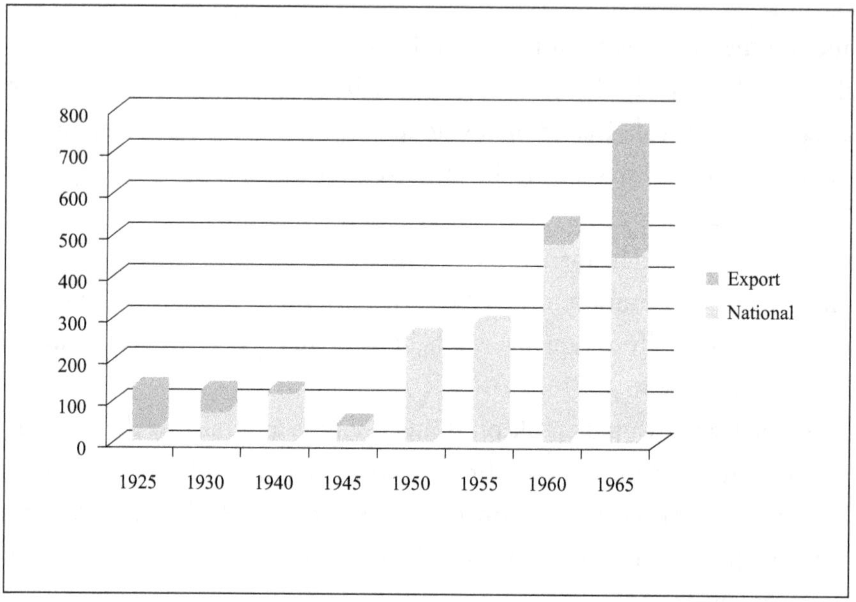

and extended. The export picked up speed again after the war, and the cooperation among the cement companies led to the merger and the establishment of Norcem in 1968.

5.2.4 Kjøpsvik

In the book "Sement i Norge 100 år (Cement in Norway 100 years)" [6], Gartmann writes thus in the introduction;

> *it was probably not God's opinion that there should be a cement factory in Kjøpsvik. But the good case had many enthusiasts and fighters among them, that did not take this so serious. The factory had to be bolted to a peninsula without road and quay, without electricity and houses, and without labor force that was trained to other purposes than fishing. The place was outside any routes in communication, and the mail only came twice a week. If area-planners had had anything to say in 1918, Tysfjord had probably not got any cement factory.*

Possibly, most cement factories are proud of their location and fine local environment. In connection with the 60th year anniversary in 1978, a book about the memories from the place was published [252]. The book is not a document with lot of historical data, but it is a very enjoyable book to read, with interesting stories about the local community. As in the previously mentioned book by Gartman [6], this book also has stories about the very colorful priest, chairman of the shooting club and school manager Kolbjørn Varmann, and his importance to the locality. Varmann later became Norwegian minister of communication and county president in Norway's northernmost county, Finmark. During the dinner at the 60th anniversary in 1978, we were seated at tables of four. I was lucky to sit next to, at that time very old, Varmann, with very long white hair and beard. I cite here a shortened version of one of the many stories he had narrated.

> *One day the parson got a visit from one of his parishioners from the bottom of the fjord, from Hellemobotn. The man had travelled far to report that a new child had been born, and that it was time for baptizing. With a visitor from so far away, it was reasonable to invite him into the living room for coffee and small cakes. In the introductory conversation, the parson asked what name the child should have. Seftref, the man answered. Varmann thought that the name was a bit strange because it was normal and in particular for the people in Hellemobotn to use names from earlier generations. No, the man answered, this time he would like to use a name from "the book." Varmann was even more surprised, because he thought that he knew "the book." The official Bible was taken to the table, and the man was asked to show where he had found the name. The man opened the book, found the right page, and pointed. It was correct; there was the name, the abbreviation for "Sunday after three-fold".*

After the modernization in the beginning of the 1990s, it also became an efficient factory. But it was a long way from the ideas and plans that the sheriff had when he travelled to Oslo in 1917. The company Nordland Portland Cementfabrik A/S was established in October 1918. The original intention was to establish the factory as a northern Norwegian venture, but other interests came in, not only Norwegian but also from the Danish machine supplier F.L.Smidth & Co. That the factory was located in Kjøpsvik was not the result of any area planning. At the start, the place had no road connection, a simple and bad phone connection, a population of about 100 people, and mail delivery two times a week. The history of Kjøpsvik is also the story about an

The cement factory in Kjøpsvik is not only the northernmost in the world, but probably also has the most spectacular location

area with very low industrial activity. In addition to the cement factory, a completely new industrial society was also built, where the borderline between the tasks of the company and the society were not as different as in other places.

Cheap electricity, good raw materials, and a fantastic courage on the part of the pioneers were not enough and, in particular, the first 15 to 20 years of the life of the factory was characterized by constant deficits and refinancing.

During a refinancing process in 1923–1924, the share capital came down to 10%. The factory was probably saved by an enthusiast at their biggest creditor, F.L.Smidth, in Denmark. Danish Paul Larsen showed his enthusiasm for Kjøpvik through a letter to his management in the autumn of 1925; "In the whole cement industry there is no match to what have been achieved at the Nordland-factory, where an activity established during a costly high market activity, in a district comparable to 1/3 of the production of the factory, has had the ability to produce with a profit."

The profit he was referring to was a figure of 11 500 kroner, booked in a year when the factory did not deprecate anything. In reality, the factory did not make any profit, but Paul Larsen would not desert his north Norwegian friends. For him, this adventure was one of his own [6].

The modernization in the beginning of the 1990s changed the production to dry production and Kjøpsvik became an international competitive plant.

5.3 The cement museum at Slemmestad

The cement museum at Slemmestad is worth a visit for those who are interested in the local history, cement history, concrete, or geology. The museum is located on the seaward side of the cultural house, "the bagging factory" at the center of Slemmestad. As the name indicates, the museum has been functioning in the old bagging plant from 1920. Paper bags were produced there until 1977.

The culture house "the bagging factory" is located in the middle of Slemmestad

The library and the cement museum at Slemmestad, with the artist Arne Durban's statue of "The cement worker" to the left of the entrance

At the back, or more correctly, on the sea side, is the library of Slemmestad with the cement museum inside. The library is normally open from 11 o'clock, with somewhat variable closing time for the different days of the week. Check the opening hours at www.royken.folkebibl.no.

As part of the museum is also a geology center built up in cooperation with the paleontological museum in Oslo. The place is also used as a school for young students in the local area and the Oslo district.

The cement museum was established by the five former employees who established the foundation "Cementmuseets venner (Friends of the cement museum)." A key person in this work was Cato L. Christoffersen, f.1927, a former personnel manager who was employed by the company on 01 March 1972. The foundation was supported by Norcem, the national cultural fund, and the local community. They opened the museum in 1991. Today the local community is responsible for the running of the museum.

Barrel labels

One of the many things that can be found in the museum is old barrel labels. The cement barrels from Slemmestad were exported to many countries and each country had its own labels. Slemmestad's logo was the red lion. In the 1920s, the company

Cato Christoffersen in front of the drawing of Slemmestad as it looked at the end of the 1800s

received protests because they had used the crowned lion, which was also part of the Norwegian national code of arms. Then they had to change the logo. In the picture above, you get glimpses of some of the lion labels over the door to the library.

The museum also shows the principles of cement production, samples, production equipment, tools, office furniture, books, and protocols from more than 100 years of activity.

A fascinating proof of the development is the table with data about the various rotary kilns that have been in use. The first one started production in 1909 and had a length of 34 meters. The diameter was 2.1 meters and the capacity 2 tons per hour.

The last kiln that was put into production was Kiln Nr. 9 in 1967. This had a length of 163 meters, diameter of 6.0 meters, and capacity of 57 tons per hour.

The working rules of 1894 are curious (see the chapter about Slemmestad above) and they might seem strange today. It is, however, interesting to note that even at that time the rules had to be approved by the authorities. The rules are enclosed within glass and frame at Slemmestad, with approval according to the law of 27 June 1892. They were approved by the Royal Norwegian Government's Department of the Interior on 11 December 1894.

6 History of Cement and Concrete in China

Portland cement, the most commonly used construction materials, has been increasingly contributing to social and economic progress of human beings since the birth of the first Portland cement patent in 1824 by Joseph Aspdin. Importance is that the demand for Portland cement in the future will continue to increase by the year 2050 according to the WWF-A blueprint for a climate friendly cement industry [253]. There are no other binders that can compete with Portland cement considering the wide availability of raw materials, the large versatility of the performance and functionality as well as the reasonable economic viability of production and application. Nevertheless, modern society has been more commonly recognized the fact that the huge volume consumption of cement also leads to tremendous consumption of energy and natural non-renewable fossil minerals such as limestone. The CO_2 emission from global cement industry, accounts for 5% of anthropic CO_2 emission and 8% of industrial CO_2 emission [253,254].

Fortunately the whole world has recognized the double effects of cement and concrete and is taking tangible actions in meeting the challenge and addressing the issue of sustainability. This is particularly important for China as the world's fastest developing country producing and using 60% of world cement and concrete output [255]. China, as one of the ancient civilized countries in the world together with Egypt, India, Babylon and Greece has also a brilliant history in developing and using cementitious materials. It is thus very important to have a clear picture how the building materials evolved during the past thousands of years and what can be learnt from history so that wiser roadmap can be made towards the future sustainability.

This chapter is a review of the route of cements/binders in China covering two parts using the invention of Portland cement as boundary: Pre-Portland Cement Period and Portland Cement Period.

6.1 Pre-Portland Cement Period

Evolution of cement/binders since ancient time in China from Baihuimian, Yellow Mud, to lime, Sanhetu, and lime-organic composite binders has been briefly reviewed in Chapter 2.2 and Chapter 5.5.

In view of history of human beings, five elements (metal, wood, water, fire and soil) have been the most important substances, which was called Congenital Five Materials in ancient China[256]. The essence inherited throughout the Chinese history of culture is Harmony. This reflects a philosophy of using natural materials for buildings being harmonious with the environment and human beings.

6.1.1 Natural earth materials

6.1.1.1 Baihuimian

The earliest binder used in China in the record is natural materials-"Baihuimian" that received this name by archeological researchers due to its appearance as white colored powders. This can be dated back to as far as 5000–3000BC in Yangshao culture period of Neolithic times when cave dwelling buildings were popular in using natural building materials such as soil, wood and stone. Miao Jisheng et al [257] made a systematic investigation on the construction relics of around 5000–6000 years ago in China. They combined chemical analysis with phase identification and revealed that the composition of Baihuimian is the same as natural materials–Jiangshi, mainly containing calcium carbonate with some amount of quartz and plagioclase, etc. Comparison study through microscopic observation with modern lime mortar also proved that calcium carbonate phase in Baihuimian is natural mineral with large and perfect crystal morphology, which is different from the secondary fine calcium carbonate agglomerates produced by the carbonation of calcium hydroxide from the hydration of lime mortar.

6.1.1.2 Earth Materials

"Yellow Mud" is another natural binders used in 16th century BC in China for adobe wall and yellow mud reinforced with chopped straw popularized during 403–221BC. The yellow mud is mainly loess comprising of clay, fine sand and limestone particles with high plasticity and good cohesion. It after properly rammed forms integrated and dense structure, which was popularized in ancient China as wall and platform of palaces. One of the earliest structures as such was found to be built during the period of Shangtang (1766–1166BC) with the total area of the earth platform as large as 10,000 square meters[258]. It should be noticed that tamping is the key process for earth densification. Later this was further improved by timber molding combined with tamping for foundation, platform and adobe wall. Qin Dynasty (221–207BC) was regarded to use perfect such earth work. The figure below illustrates some tools and molds for earth works tamping and a schematic process of timber molding wall laying[259]. *Ying Zao Fa Shi* published in Song Dynasty (960–1279) made a clear description of the procedure in making these structures[260]. *Tian Gong Kai Wu* (Heavenly Creations, Ming Dynasty) also detailed the process of using earth materials to make

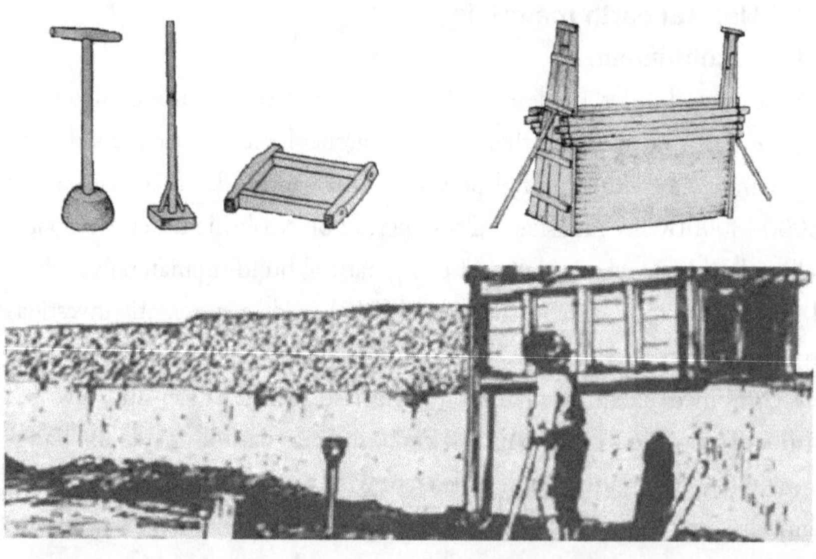

Schematic drawings of the adobe wall construction and the pestles and molds used for the earth work[259]

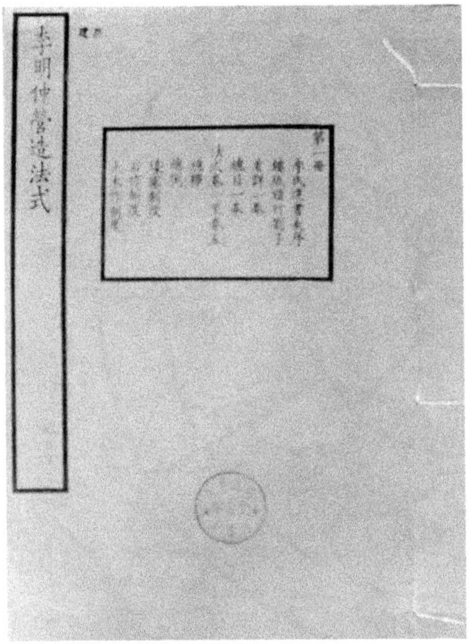

Picture of cover page of the copy of the book Ying Zao Fa Shi from library of Tsinghua University. Ying Zao Fa Shi, authored by Li Mingzhong is an architectural book of regulations on assessing the project and counting the materials, which was issued for enforcement to the whole country in 1103, Northern Song Dynasty. [260]

Traditional molding earth brick technique, illustrated from Tian Gong Kai Wu, authored by Song Yingxing (1587–1663) and published in 1637, Ming Dynasty [261] as Technology Encyclopedia of ancient China. Later it was spread worldwide and translated into many languages. The English edition was as Chinese Technology in the Seventeenth Century: T'ien-kung K'ai-wu, Yingxing Song, tr. E-Tu Zen Sun, Shiou-Chuan Sun, Dover Publications ISBN 0-486-29593-1

brick, wall, tiles etc. [261]. Zhang Huyuan *et al.* summarized in 2008 (Illustration shown in nexl page) the earth construction technology dealing with 4 types of earth materials in ancient China[262]. This technology can also be found in use nowadays in some rural areas in China and some African and South American countries.

Technology has been found to be improved consistently by using the combination of mud as flesh and stone and chopped straw as bone and muscle respectively, furthermore organic substances like rice flour, tung oil, animal blood, brown sugar, etc. were also included as organic admixtures to

Traditional earth construction technique [262]

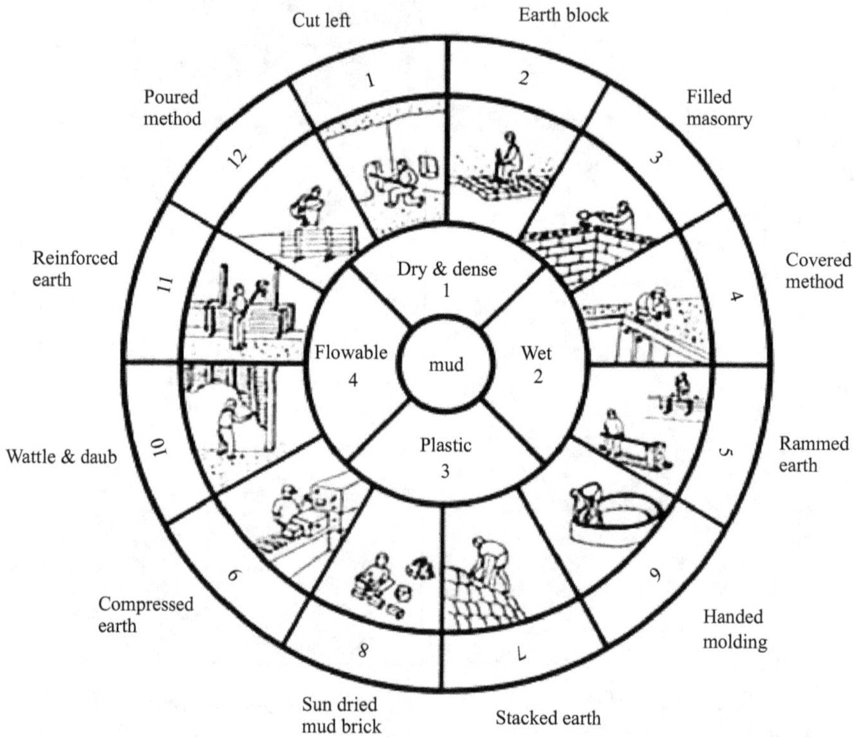

make a better earth masonry work. These can be considered as the enlightenment of modern concrete technology. Lime was also combined with the above-mentioned recipe to further improve the strength and durability. This will be explained later in this chapter.

In the cultural relics of Yangshao period in Dahe village, Zhengzhou, Henan province, people extended the technology by using straw mixed with mud as basis of floor and wall and painted the exterior with clay mixed with fine sand. Then fire was made inside the building and finally produced a pottery based integrated floor and wall inside that are hard and red like modern fired brick.

6.1.2 Artificial binders

6.1.2.1 Lime

Lime is another commonly used binding materials and the most important one in ancient China for different construction purposes, such as strengthening the foundations, adhering brick masonry as well as working as finishing materials for walls and ground. Qin Great Wall, one of the eight world famous wonders started to build from 214BC and involved 300,000 manpower, is a best manifestation of using lime. The construction continued in the following dynasties until Ming Dynasty. In the meantime, the development of lime as cementing materials advanced to a higher level, and many parts of the Great Wall are found with lime today.

The earliest documentary record in China for using burnt lime is *Zuo Zhuan*, China's earliest historical annals in history and a Confucian historical biography and classic in Spring and Autumn Period narrating the history during 722BC through 481BC. Burnt lime from clam shell was used in the tomb of the emperor Song Wengong (took office from 611BC through 588BC) as binding materials as well as damp proofing purpose. The book of *Tian Gong Kai Wu (Heavenly Creations)* in Ming Dynasty makes a detailed record on the method of lime manufacture, and the steam generated by reaction of slaking lime was also described [261]. Relationship between burning process of lime and its properties had been described in another document of Qing Dynasty (*Ying Zao Fa Shi*) [260]. These records demonstrate that China has accumulated abundant knowledge of preparation and use of lime since the Ming and Qing dynasty.

It is worthy of a focus on the immortal masterpiece on technology of ancient China *Tian Gong Kai Wu,* by Song Yingxing in late Ming Dynasty, that has been regarded as the first comprehensive work on agriculture and handicraft industry in the world and Encyclopedia of Technology by European scholars. The book is divided into 18 chapters covering the growing of grains, clothing materials, dyes, the preparation of grains, salt, sugars, ceramics, casting, boats and carts, hammer-forging, calcinations of stone, vegetable oils and fats, paper, the metals, weapons, vermilion and ink, yeasts, and pearls and gems. The Chapter 11, Calcination of Stone for Lime Making, detailed the procedure for making lime from limestone and/or shell with quality control requirement. Different applications of lime were also explained in this chapter. Plaster by mixing lime powders with water and in some cases further with paper fiber can be used for masonry wall. Modification of lime with tung oil and woven silk can be used to seal the boat for water-proofing. Mixing lime with river sand and losses in a ratio of 1∶3 and further admixed with rice flour and Carambola juice can produce a hard and durable cementing materials called *Sanhetu (Tabia)* which can be used

Chiseling shell from sea rock (left); Burning shell into lime with coal cake, from Tian Gong Kai Wu [261].

Transformation of lime burning kiln, earth kilns (left) popular decades ago later were replaced by modern lime kiln (right) [263].

for sealing tomb and building water tank. This will be discussed later. The illustrations inside the book help better understand the production processes [261]. As comparison evolution of lime kilns is also shown in above figures [263].

Recent archaeological studies through Radio C14 dating show that lime had been burnt and used since five thousand years ago, Evidence as remarkable smoothness and hardness of walls and grounds plastered with lime had been found in various cultural relics in the Longshan period in China's Yellow River Basin (about 2800–2300 BC years). The appropriate property and wide availability of raw materials make lime a traditional yet ever-new binder like Portland cement for the future buildings.

Another worth-mentioning figure in Chinese history related to lime is YU Qian (1398–1457), a national hero and statesman in Ming Dynasty praised lime as a metaphor in his poem *In Praise of Lime* to express his loyalty to the country, determination when facing difficulty and willingness of sacrifice for the country. We can also get a brief understanding of the origin and excavation of limestone as raw materials for lime making and the process of burning. The poem goes as:

In Praise of Lime

YU Qian (1398–1457)

Excavated from the mountains after thousands of times of hammering and chiseling,
Burning with fire seems very normal things.
Not afraid of being smashed to pieces,
As long as this can leave innocence in the world.

6.1.2.2 Sanhetu and Organic-Inorganic Composite Binders

Triple components consisting of lime, clay and fine sand were used in the 5th century in China Southern and Northern Dynasties era. More durable binders with superior strength and high water resistance after tamped, as Lime-Organic Cementing Materials were produced and used at least in that period for varieties of purposes such as making water tight wall, ship, water tank, floor, roof, tomb, bridge, tamped dam, military facilities and etc. In addition, natural polymer as tung oil impregnated brick (called "Golden Brick") was found very strong and dense in underground tomb buildings of Ming Dynasty, which can be regarded as the pioneer of modern polymer modified concrete.

Tian Gong Kai Wu [261] defined such cementitious materials as *Sanhetu (Tabia)* and the mixture design was given by mixing lime with river sand and losses in a ratio of 1:3 and further admixed with rice flour and Carambola juice. The resultant materials were so hard and durable that many documents in China analogized them "as hard as stone"[264], "as hard as ground stone"[265], and etc.

In Ming Dynasty, another type of *Sanhetu* was composed by lime, ceramic powder and crushed stone. *Sanhetu* was not only formed by lime, clay and fine sand, but also was made by lime, slag and sand in Qing Dynasty. At the same time, the book *How to build palace-style stone bridge* explained the process to make *Sanhetu* by blending 40% lime and 60% loess with fine sand. It is the same as concrete that we know today, which was prepared by lime and loess or other pozzolanic additions as cementing materials blending with fine sand, gravel or slag as filling materials. Evidence of C-S-H gel was found by Miao et al [257] in the *Sanhetu* buildings from Qing Dynasty.

It was a very interesting historical coincidence that Roman three-component mortar was employed in European continent, whilst the *Sanhetu* was used in the ancient China as we can see that many similarities exist between *Sanhetu* and Roman mortar.

Archeological investigation and scientific experiments reveal that sticky rice starch and calcium carbonate co-existed in buildings remained from Ming and Qing Dynasty in many provinces of China [266]. Simulation study and comparison test proved that this is a long bio-mineralization process through the penetration of atmospheric CO2 and the following carbonation reaction [267, 268]. Remarkable improvement was found for lime admixed with 3% of sticky rice glue. The compressive strength is increased by 30 times and surface hardness by 2.5 times. The water permeability resistance is also improved.

It was evidenced that starch derivatives play important role in this process by modifying the formation of calcium carbonate-the crystalline phase, the crystalline grain size and the morphology of the crystalline. The resultant carbonate by combining hydrated lime with glucan after carbonation is mainly very fine crystalline grain aragonite instead of normal calcite [269].

SEM observation of biomimetic synthesis of lime-sticky rice glue mixture revealed a very dense microstructure which better explained the excellent

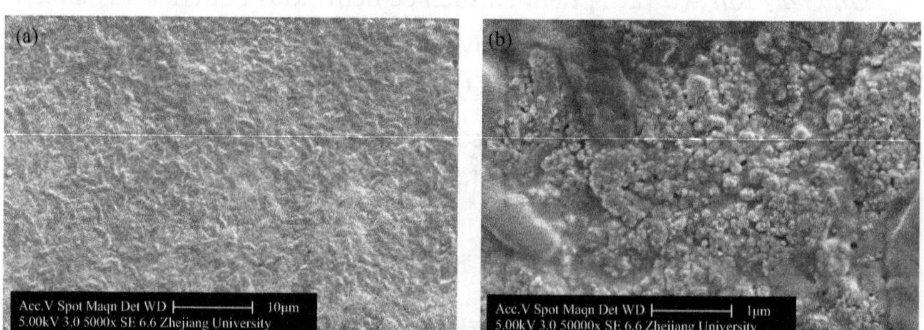

Scanning Electronic Microscopy of hardened mortar mixture of 0.01M calcium hydroxide-3% sticky rice glue in a ratio of 1 : 1. (a) ×5000 ; (b) ×50000 [270]

mechanical performance and durability of the ancient lime-organic cementing materials [270].

Leaning Pagoda with an inclination angle of 30°, located in Rongchang county, Chongqing, China, is standing like a miracle. The secret is the use of rice glue modified lime as binders for each story [271]

A good demonstration of the application of rice glue modified lime mortar is shown as the figure-Leaning Tower of China located in Rongchang county, Chongqing [271]. It is a 9.5m high stone Buddhist pagoda built 300 years ago in Qing Dynasty. This 11-stroey octagonal spire, though experienced earth quake, flood, and strong wind throughout the past 300 years and got an inclination angle of 30° , is still standing there like a miracle.

Local cultural relics expert revealed after careful investigation and analysis that one of the three main reasons is the use of rice glue modified lime as binders for each story. The tradition is to boil the rice into sticky porridge and further mix it into glue, and then blend the lime and sand with rice glue. The strong binders can be thus made with equivalently excellent performance to modern concrete.

It should be noticed that the invention and wide application of organic admixture-modified inorganic binders has made a milestone contribution in ancient China. The strengthening mechanism by bio-mineralization process and synergetic effect of lime-rice glue combination is of great significance in today with green construction materials and sustainable building as the main theme of the sector. No wonder Dr. Joseph Needham (1900–1995), a British scientist, historian and sinologist in his Science and Civilisation in China series highly appreciate the progress of the science and technology in China that exceeded the western countries during the era of the 3rd century through 15th century [272].

Unfortunately after the invention and wide use of organic-inorganic composite binders not much progress has been made later in China on the science and technology especially in the period of late Qing Dynasty. While in western countries dynamic improvement had been witnessed, which resulted in the invention of Portland cement on the basis of Roman mortar and concrete. A lot of fascinating stories about the invention of Portland cement have been introduced in the previous chapters.

6.2 Portland Cement Period

Discussing the production and use of Portland cement and concrete in China, three periods are generally agreed with the timeline of 1949 and 1978 as boundary. Before 1949, China cement industry mainly relies on imported facilities for industrial production and not much documents depict the use of concrete for that period. The founding of People's Republic of China in October 1949 brings the new dawn for the development of China cement and concrete industry. China has been stepped in a way of self-development. Thanks to the start of implementing the economic reform and opening to the outside world policy since the year 1978 that leads to a remarkable progress of the China cement and concrete industry.

Development of Portland cement in China have been well documented in *History of the Development of China Cement* by Dr. Wang Yanmou in 2005 covering a time span from the invention of Portland cement up to 2004 [273]. Zhou Zuitian and Han Changkai in 2011 through 2012 made an in-depth review of the development of China cement before 1952 in *History of China Cement* series [274].

6.2.1 Phase of Importing Cement Technology and Equipment

The use of Portland cement in China started with religious building from imported cement from UK, so called "Ying Ni-Ash from UK" or "Yang Hui-Imported Ash".

The 1st Portland cement plant built in China is Green Island Cement Factory in Macau set up by an Englishman as main shareholder in 1886, later Qing Dynasty. The process used imported shaft kiln. Green Island Cement Factory established another plant in Hong Kong in 1887. The annual capacity of Green Island Cement Factory was 120,000 t in 1908. The unit in Macau was closed down in 1936 while the company in Hong Kong is still active up to date.

The 2nd cement factory, Tangshan Cement Factory actually the 1st Chinese-owned cement plant in China started built in 1889 and was put into

Green Island Cement Factory, the 1st cement plant in China. Pictures from South China Morning Post on March 27, 1987 in commemorating the100-year anniversary of the establishment of the plant (Left: Green Island Cement, Macau, 1886, Right: Green Island Cement, Hong Kong, 1887)

production in 1891 in Tangshan, Hebei province. It adopted shaft kilns with imported equipment and unfortunately was closed down in 1893 due to the small capacity and poor quality. Based on the History of China Cement [274], the plant was restored under the order of government Qing Dynasty in 1900 producing high quality cement "Dual Lions" and received best quality prize in St. Louis, USA in 1904 and Italy in 1905.

During this period many stories happened. As depicted in Ref. [274], China lost its control over Kaiping Mining Co. including Tangshan cement factory from 1900–1905. This is something related to the Herbert Clark Hoover, who once worked in China from 1899 to 1901 and later became the 31st President of USA, & a UK company, an international lawsuit. In August 1906 cement works was returned to China. A new name of Tang-

Shaft kilns of Tangshan Cement Factory. Courtesy of Mr. ZHOU Zuitian & HAN Changkai [274]

*Chee Hsin Cement Company Ltd., 1st rotary kiln in China. Left: Trademark of Chee Hsin cement, with Horse, Sea, Lucky Cloud, Flame, Taichi Right: HQ office building of Chee Hsin Cement in Tianjin and demolished in 2005 * Courtesy of Mr. ZHOU Zuitian & HAN Changkai [274]*

shan Cement factory was given finally as "Chee Hsin Cement Co. Ltd. " which has been considered the foundation of Chinese cement industry.

With the following new investment in Chee Hsin, new rotary kiln and ball mill were imported from F L Smidth creating a new era of Chinese cement with the 1st rotary kiln (ϕ2.1×30m, 40 kt/year) in 1908, and the cement output took 90% of China cement market by an output of 270 kt/year in 1921.

Later Cement was intensively manufactured and used in China.

- The 2nd Chinese cement plant: Guangdong Shimintu Factory (85 t/d, i.e. 85 tons/day) 1909;
- The 3rd Chinese cement plant: Hubei Cement Plant (200 t/d) in 1909 using dry rotary kiln, later was controlled by Chee Hsin Cement in 1913, and then became Huaxin Cement, and now a joint venture with Holcium;
- Then cement plants were built in Shanghai, Dalian, Tsingdao, Ji'nan, Jurong, Taiyuan, Chongqing, Nanjing, ……

During the progress of early China cement industry, so many things happened. The most touching history revealed that a tragedy took place in Nanjing during the Japanese invasion to Nanjing in 1937-A history of "Jiangnan Cement" version Schindler's List was recorded [273, 274].

Jiangnan cement plant was built in Nanjing in 1935 (ϕ3×46m wet process, 715t/d) and a new plant (760t/d) set up in November 1937 with F L Smidth's cement equipment & German electrical device. At the same time a tragedy happened in Nanjing by Japanese invasion. A hero, a friend of China, a friend in need, a friend indeed, Mr. Bernhard Arp Sindberg, a Dane who helped save up to 30 000 Chinese together with a German Dr.

Guangdong Shimintu Factory, 1909. "Shimintu" is transliterating from "Cement" in English. Pictures reproduced from China Charm, No. 3, 2015 [275]. Front entrance of Guangdong Shimintu Factory (upper left) in early 1920s later became Mr. Sun Yat-sen's Generalissimo Mansion (upper right). West Section of Shimintu Factory (lower left) and stone tablet motto of Shimintu Factory "Qiu Shi" in Chinese means "Truth Seeking" inlaid in south side of the third floor of the North Building of the factory (lower right).

Karl Gunther, Acting Chief of Jiangnan cement plant during the massacre that lasted for 107 days [273,274,276].

Based on the Brief Introductions to Members of the International Committee for Nanjing Safety Zone [277], *"Bernhard Arp Sindberg was Representative of the property rights of Denmark F. L. Smith Company in Jiangnan Cement Plant. From April to May of 1938, he was forced by the Japanese authorities to leave Nanjing and immigrated to the U.S. by way of Shanghai and Denmark. After he left Nanjing, 11 villagers in the name of refugees from Qixia District co-signed a silk banner of thanks and asked Karl Geunther to pass it on to Sindberg through the Consulate General of the Kingdom of Denmark in Shanghai. The banner bore the words of "Brave Man of Justice'. In June 1938, he was again given inscriptions of 'Friend of China' in Geneva written by Zhu Xuefan, head of the Chinese Labor Delegation."*

Mr. Bernhard Arp Sindberg (left)-A Dane, born in Aarhus, Denmark on February 19, 1911 and died in California, USA, in 1983 helped save up to 30,000 Chinese together with a German Dr. Karl Gunther with his wife (right) during the Nanjing massacre. Mr. Sindberg was 26 in 1937 working in Jiangnan Cement Factory. [277]

Mr. Sindberg's heroic feat was re-discovered after the diaries of John Rabe were published in 1997. An undated file photo of a young Sindberg, provided by Edith Koerner Guenther to China Youth Daily for its first publication on December 9, 2009 [278]. Edith is wife of Karl Guenther who was a colleague of Sindberg at the International Committee for Nanjing Safety Zone initiated by John Rabe in Nanjing [277].

"The hall records the darkest history of the city, and although history cannot be changed, lessons can be learned to face the future···" Danish Queen Margret II made the remarks during her visit to Nanjing Massacre Memorial Hall, April 27, 2014.

"We are so proud of Sindberg for his courage during the wartime to save people's lives." Said Mariann Stenvig, Mr. Sindberg's niece in Nanjing, April 28, 2014 holding picture before the building Mr. Sindberg once lived in 1937.

More information can be found in "He saved 20,000 Chinese in Nanjing Massacre" (Xinhua http://www.chinadaily.com.cn/china/2011-02/21/content_12049645.htm) and "Danish man saved thousands from Nanjing war atrocities" (2014-04-28 09:11 China Daily http://www.ecns.cn/2014/04-28/111391.shtml) as well as "Sufferings of China Cement Industry during War Period" in Chapter 7 of *History of the Development of China Cement* [273].

6.2.2 Phase of China Self-Made Cement Technology and Equipment

China experiences a transition from importation to self-made and exportation in cement making, from the shaft kiln first and then to different types of rotary kiln.

6.2.2.1 First China self-made shaft kiln

1st China self-made shaft kiln (6000 t/year) was erected in Kunming Cement Plant, December 1940 as a milestone in China cement history. Jiangxi Cement Plant, the 1st vertical shaft kiln (VSK) plant using 100% Chinese equipment was built in February 1943. Later VSK became popular in China and dominated Chinese cement for many years until the year 2002. This can be evidenced by the continuous growth of VSK cement in China from 1953 through 2002.

1st shaft kiln made in China $\phi 2.5m \times 10m$, Kunming Cement Plant (Courtesy of Dr. Wang Yanmou [273])

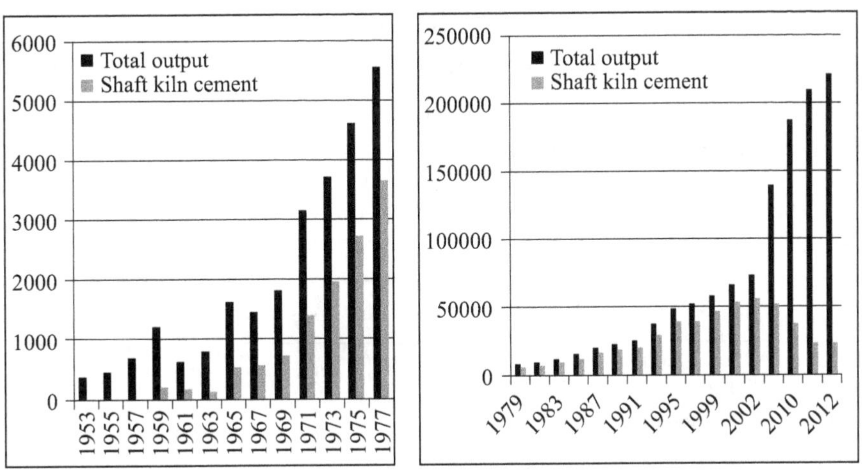

Y-axis: annual cement output ($\times 10^4$ t), X-axis: year

Change of vertical shaft kiln cement in China by year. Largest VSK cement output happened in 2002 with 555mil.t, and highest percentage of VSK cement is 83% in 1989.

6.2.2.2 First China self-made rotary kiln

The Chinese cement industry since 1949 has progressed tremendously.

Generally it can be divided into three phases.

1st phase: 1949–1975 1st hollow dry process kiln (ϕ2.4×60m) in Xinjiang in 1958; 1st wet process line (500 kt/year) established in Xiangxiang, 1967; 1st Lepol kiln (470 kt/year) in 1970, 1st dry process kiln with shaft preheater (ϕ4×60m, 300 kt/year) started design;

2nd Phase: 1976–1997 1st new suspension preheating and pre-calcining (NSP) cement line (ϕ2.4×40m, 400t/d, oil burning) established in June 1976, Shiling, Siping & 1st coal burning NSP line put into operation (ϕ3.7×52m, 1250t/d) in Benxi, March 1978;

3rd Phase: 1998–now 1st China made NSP line (4000t/d), then coming with 5000t/d, 8000t/d, 10000t/d and 12000t/d.

Two significant characteristics are worth mentioning from the year 2000. The first one is the quick replacement of VSK with NSP process from more than 80% of VSK cement to a sharp downturn of less than 10% within just 12 years. Thanks to the firm policy for restructuring the cement industry, through which leads to remarkable progress of China cement industry in every aspect. The other significant progress is that China started exporting cement manufacturing equipment from 1990s and has become one of the leading countries not only in exporting cement equipment but in conducting EPC service worldwide as well. The progress of China cement industry can be demonstrated in the table.

Comparison of some parameters of Chinese cement industry with data from CSI member companies

Item	Unit	1990	2008	2009	2009China
Gross CO_2 emission/t clinker	kg/t	916	860	853	~850
Gross CO_2 emission/t cement	kg/t	764	665	653	~556
Kiln fuel efficiency	MJ/t	4267	3658	3586	3426
Electricity consumption	kW·h/t	114	110	107	85.6
Use of alternative fuel incl. biomass	%	2	11	12	~0
Clinker in cement	%	83.0	76.1	75.6	65.4

Source: CSI 2011, average data from 46 companies with over 900 production facilities worldwide, 2009 China project outcome, calculated by author on basis of data from China Cement Association.

6.2.3 Cement and Concrete as Construction Materials

Not much document has been found for the use of Portland cement and concrete before 1949. The majority of this part therefore shows a picture of cement and concrete as materials starting from 1949, the funding of the People's Republic of China. The significant progress of cement and concrete sector took place mainly after 1978, the beginning of China implementing the policy of the economic reform and opening to the outside world.

6.2.3.1 Cement

Cement output in China is 1.5 million tons in 1950 and now it is 1600 times larger with 2.476 billion tons. The Chinese cement standard was issued for the first time in 1953 and has been renovated for 4 times since then. The current Chinese cement standard (GB175-2007) is equivalent to the existing EN-197 standards for both cement products and testing methods. Six types of general purpose Portland cements with varying strength grade have been classified in the standard based on the type and amount of blending materials (mineral admixtures). Standardization of mineral admixtures such as slag and fly ash began in 1986 and later becomes a series of standards as technical guide and support for using in cement and concrete.

The most dynamic increase of China cement production and consumption took place since the implementing of economic reform and opening to outside world in 1978. With global economic downturn and China's policy on industry upgrading and transformation, however, the over-heated China cement industry is at present encountering a big challenge and also another

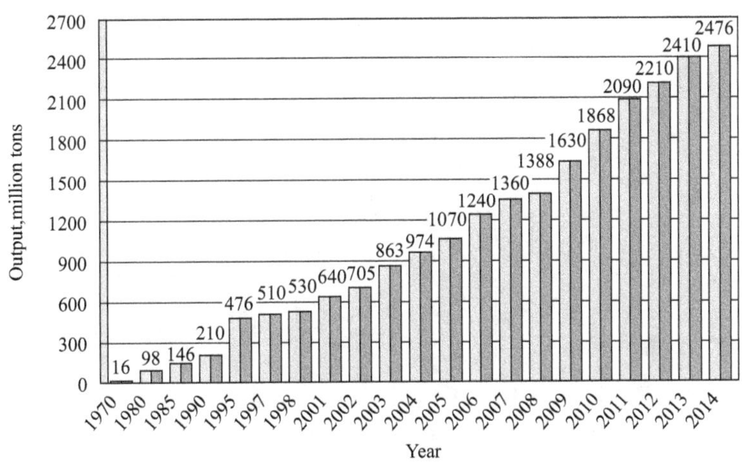

Annual output of cement, China

opportunity towards sustainable development.

In 2014 China produced 2.48 billion tons of Portland cement, accounting for 60% of world total amount. Yet when we take a close look of the steady decrease of the clinker factor of Chinese cement, the clear picture for 2014 with clinker factor of 57% is that China cements contain averagely about 38% of blending materials, mainly industrial wastes or by-products if 5% of gypsum is generally added. The direct CO_2 emission intensity of Chinese cement for 2014 can thus be estimated on basis of WBCSD CSI CO_2 and Energy Protocol as follows, assuming that the clinker direct CO_2 emission intensity is 0.86kg/kg of clinker:

$$\text{Direct } CO_2 \text{ emission intensity of Chinese cement} \approx 0.86 \times 0.57 = 0.49[\text{t }(CO_2)/\text{t(cement)}]$$

Thus

$$\text{The total direct } CO_2 \text{ emission of Chinese cement in 2014} \approx 0.49 \times 2.48 = 1.22[\text{billion t }(CO_2)]$$

Research and development of different types of special cements at the same time have been very active with more than 20 types in commercial production. The most interesting types of new cements developed and used in China are sulfoaluminate cements (SAC) developed in 1980s and belite-rich Portland cement (i.e. high belite cement, HBC) developed in 1996. Both cements feature lower energy consumption and CO_2 emission during clinkering process and unique performance compared with normal Portland cement [279–281].

About 50% of SAC have been used for prefabricated panels and the re-

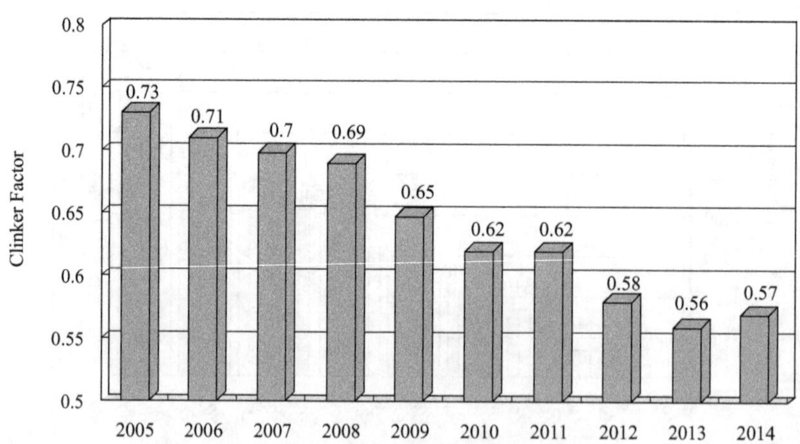

Clinker factor of Chinese cements (Calculated by author with original data from Chinese Cement Association, CCA)

maining 50% are used for various purposes such as dry mortar, fast setting and quick hardening cements, expansive agents, etc. HBC, due to its lower hydration heat and higher later age strength and durability, has been increasingly used for dam construction since its first successful application in the 3rd phase of Three Gorges Dam in 2004–2005 [281,282]. The use of HBC as low heat Portland cement in massive concrete already surpassed that of moderate heat Portland cement in volume in 2014 and is expected to further expand in the future.

6.2.3.2 Concrete

Concrete production and use in China was not well documented before 1978 though a number of applications of cement and concrete indeed supported a lot the economic development during this period. Most of the progress in concrete industry happed after 1978 in particular after 1987. A brief introduction of the evolution of concrete industry in China is made in two parts: ready-mixed concrete (RMC) and precast concrete products.

(1) Ready-mixed concrete

RMC helps to a great level improve the quality of concrete by increasing mechanical properties and reducing the variation in quality. The evolution of RMC in China can be classified into three phases [283].

Phase I Infancy Stage (1949–1978)

There was not RMC in the market at the beginning. The only use of RMC limited to iron and steel companies in Anshan and Baotou in 1950s for internal construction projects within 5 km transportation distance. Beijing 3rd and 1st Construction Company built one RMC station each in April 1970 and May 1970 respectively. Unfortunately RMC was not widely spread then due to the poor market accessibility until in 1978 Changzhou Construction Engineering Company in Jiangsu province started marketing RMC to satisfy the increasing demand of local infrastructure construction. The successful use of RMC in Changzhou was highly appreciated and soon promoted by the Ministry of Construction, which leads to the flourishing of RMC in China since then.

Phase II Slow Progressing Stage (1978–1986)

RMC progressed in a low pace during this 8-year stage mainly as a result of the lack for driving force of infrastructure construction. Yet a number of concrete standards such as the Conformity Assessment Code for Con-

Annual RMC output since 1980, Chihna Data [283, 284]

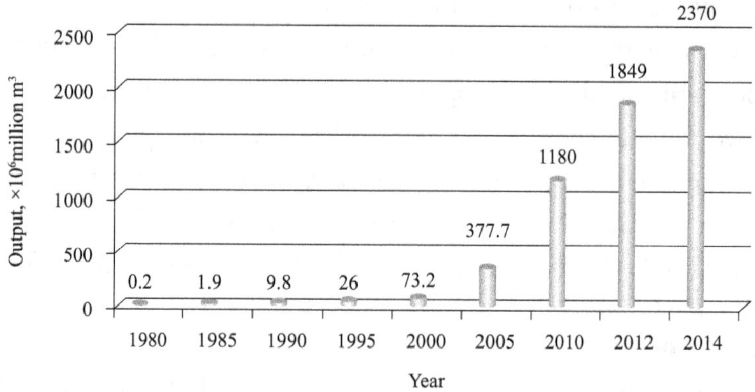

crete Strength, Unified Standards for Testing Methods of Concrete Properties, Testing Methods for Advanced Presumption of Concrete Strength, and Codes for Concrete Quality Control, which lay foundation for the rapid progress of RMC in next stage.

Phase III Booming Stage (1987–now)

RMC annual output in China during this stage tremendously increased from 4.3 million m^3 in 1987 to 2.37 billion m^3 in 2014, which satisfies the huge infrastructure such as building dams, railway, highway, airports, ports, bridges, etc. and residential housing investment. At the same time many cement giant players in China extend their production and value chain into concrete business through company merging and acquisition. The largest RMC company-China Building Materials Group produced 87 million m^3 in 2014 with a total capacity of 206.5 million m^3.

It is also important to mention that chemical and mineral admixtures with different functions have contributed greatly to progress of RMC. Another important message is that China RMC industry, just like the overcapacity cement industry at present however, is another over-heated sector to be re-structured in the coming years.

(2) Precast Concrete Products

Similar as the evolution of ready-mixed concrete, development of precast concrete products in China can also be classifies as three Phases [285, 286]:

Phase I Development of reinforced precast concrete products

It happened during the 1950s. China started production of reinforced concrete pipe and concrete tiles containing asbestos, concrete drain pipes, concrete electrical poles, concrete sleepers, concrete pit barring, concrete ships, etc.

Phase II Development of pre-stressed precast concrete products

This mostly started and furthered developed in China in 1960s during which pre-stressed concrete pipes, poles, piles, sleepers and self-stressed concrete pipes, etc. though it was already very popular in Europe using pre-stressed concrete for bridge construction in 1950s and 1960s.

Phase III Application promoting

In 1970s pressure concrete pipes entered a stage of fast progress due to the government incentive both in policy and funding.

Phase IV Fast and full progress

Thanks to the implementation of opening policy and fast economic growth which led to a dynamic progress of the precast concrete products in China after 1980. This has been manifested by the increase of products types and volume, the improvement of manufacturing technology, equipment and product quality.

A comparison data after the year 2000 can give a clear picture of the fast progress of the sector [285, 286]. The statistic data comes from an aggregation of the precast concrete products enterprises with annual gross industrial output value of more than 20 million RMB Yuan.

Selected data of Chinese precast concrete products sector

Year	Number of precast concrete products enterprises	Percentage of the number of precast concrete products enterprises among China building materials industry	Number of employees	Rank among 32 sectors under China building materials industry based on gross industrial output value
2003	1312	9.25%	198000	No.3
2007	2421	11.61%	292600	No.2
2008	3523	11.90%	414700	No.2

The following are accumulated output of some of the main precast concrete products in China by the end of 2007:

- *Self-stressed concrete pipes and pre-stressed concrete pipes*: about 112 thousand km and 51 thousand km respectively, for water transportation and Construction of farmland irrigation and water conservancy;
- *Concrete poles with a length of 6–15m*: 194 million, for transmission lines of

power, lighting, communication and broadcasting;

- *Pre-stressed concrete sleepers*: 305 million;
- *Pre-stressed concrete piles*: 1.15 million km;
- Large volume of various reinforced and pre-stressed concrete products for industrial and civil buildings.

……

A. Concrete pipes

Concrete pipes have been for long used for transmission of water, oil and gas. It normally contains plain concrete pipe, ordinary reinforced concrete pipe, self-stressed concrete pipe, and pre-stressed concrete pipe with different inner diameter (small diameter pipe less than 400mm, medium diameter pipe 400-1400mm, and large diameter pipe more than 1400mm) and with varying pressure-bearing ranging from 0.4MPa to 1.2MPa [286, 287].

Based on the statistical data of Chinese Precast Cement and Concrete Products Association, China annually produces 30,000 km (about 4.5 million tons) of reinforced concrete drainage pipes, 1000 km of self-stressed concrete pipes, 2500–3000 km of pre-stressed concrete pipes including 1200 km of Pre-stressed Concrete Cylinder Pipes (PCCP).

Taking PCCP as an example, use of PCCP in the world was initiated by Bonna Pipe Company of France in 1939 and experiences a history of more than 70 years. PCCP does not appear in China until in 1984 and 1985 PCCP as lined cylinder pipe (PCCPL) with a diameter of 600mm and 1200mm respectively was first developed. In spite of the simplified process used in these attempts it was proved to a feasible and successful solution to pipe making solution and laid good foundation for later progress of PCCP in China.

The full development of PCCP in China took place after 1989 initiated by Shandong Electric Power Pipeline Engineering Company who imported key equipment from Ameren, USA. Later in 1997 PCCP was laid using for the first time Chinese PCCP manufacturing equipment. Another type of PCCP as embedded cylinder pipe (PCCPE) was developed and used at the same time.

When the world enters new millennium a number of large water infrastructure projects and agricultural irrigation projects have been launched, which let to remarkable improvement of the PCCP technology and equipment and rocket high increase in the production capacity. By the end of 2007 the accumulated laying length of PCCP with a diameter ranging from 600mm to 4000mm reached 5300km. The annual output of PCCP in 2007 is more than 1200km [287].

B. Concrete Piles

China started developing pre-stressed concrete piles from 1960s. Post-tension pre-stressed concrete piles with centrifugal vibration roller press process were explored in 1970s. Pre-stressed high strength concrete piles with a dimension of 400–1000mm were started to production in late 1980s and promoted by Ministry of Construction of China in 1990s.

In 1987 annual output of concrete piles was only less than half a million. Now this sector has grown up so fast by producing over 300 million of concrete piles. So far there are about 500 enterprises in China in this sector located in over 10 provinces and autonomous regions. The top three concrete piles companies feature an annual capacity of 130 thousand km, 55 thousand km and 10 thousand km respectively, adding up to 50% of the total Chinese concrete piles sector.

The dimension of the concrete piles produced is mainly 300–1000mm in diameter and 5–20m in length. The maximum dimension of concrete piles is ϕ1200mm×30m produced by 3rd Harbor Engineering Co. of China Communication Construction Group, Shanghai and ϕ1000mm×50m produced in Jiangmen Yuda Pile Co., Guangdong province [288].

C. Concrete ships

A concrete boat was made in the first time by Joseph Louis Lambot Southern France in 1848. The boat, being made by steel wires covered with a sand and cement plaster-called Ferro-cement, was featured in the 1855 World's Fair in France [289]. Reinforced concrete ship and ferro-cement vessels were widely manufactured in Europe and North America during the both the First and the Second World War due to the lack of raw materials and lack of steel in particular.

China after 1958 had built quite many concrete ships as barges, crafts, tugboats, fish boats and cargo ships with a capacity of several million tons. In particular after 1970 there was a surge in China making concrete ships due to lack of iron and steel. Some of the concrete ships had been in service

Picture of the news report in 1973 on Gutian cement ship's successful maiden voyage [290]

Picture from Xinjing News of China reporting the end of Gutian cement ship's final destiny on January 16, 2013 [291].

for more than 20 years.

As said a world largest (probably it was not) load of 3000-ton concrete ship Gutian was built in 1974 with a dimension (length × width × depth) of 105.2m × 14.5m × 8.1m and a waterline of 5.7m and a displacement of 5773 tons [290]. The ship was designed for all types of packaging cargo and bulk cargo transport. The ship, after finishing its maiden voyage was abandoned for long term voyage due to its too big self-weight and high fuel consumption. It was finally demolished in early 2013 after nearly 40 years' abandonment.

D. Concrete Poles

Concrete poles can be in made in different shapes. Before 1949, concrete poles were mainly hand-made solid poles with rectangular, triangular and circular design. Rectangular and circular concrete poles were used in Beijing in 1924 and 1926 respectively [292]. In 1930s through 1950s this was replaced by circular hollow concrete poles through centrifugal process. The first use of centrifugal hollow concrete poles took place in 1938 and further improved in Beijing in 1952 [292]. With the improved quality and the benefits of saving steel and lowering cost and maintenance, concrete poles were spread out in China first for 35kV power transmission line and then for 110kV and 220 kV in 1960s [292]. Pre-stressed concrete poles appeared in Wuhan, Hubei province and were used for 330 kV in early 1970s [292, 293], unfortunately cracking were found during the service period which restricted the extensive application due to the concern of safety of power transmission line.

The concrete poles sector then almost stagnated during the period of 1964–2002 due to the lack of technical advancement and the limitation of the quality of raw materials such as cement, steel, and resultant concrete.

The development of high strength concrete with strength grade C60 and C80 together with the advantages of spinning and pre-stressing technology makes it possible to manufacture concrete poles with the higher density, lower weight and better strength-to-weight ratio in order to compete with steel structure and to meet the expanding need of a supporting structure for; street, highway and regional lighting, athletic lighting, signs, overhead power distribution and power transmission as well as power substations and traffic signals.

High load-bearing partial pre-stressed concrete poles were manufactured in 2003 in Guangdong province symbolizing a technical upgrading of concrete poles in China. The concrete poles sector has been entering into another round of dynamic development. In 2012, 8.95 million concrete poles were manufactured with 17.8% of growth rate compared with 2011 to meet the new round of power grid renovation project for rural areas.

E. Concrete sleepers

Concrete sleepers made by steel reinforced concrete for railway track were first used on the Alford and Sutton Tramway in 1884. Yet only during the World War II the major progress of concrete sleeper technology was achieved and pre-stressed concrete sleepers were developed.

China started developing pre-stressed concrete sleepers in 1953. Trail production and laying were made in 1954. The first concrete sleeper plant began to set up in 1954 and was put into batch production in 1961, and the total annual production exceeded 4 million in 1983 with accumulated concrete sleepers laid in China up to 64.45 million. Statistic data shows that since 1956 a total accumulated amount of 162.5 million concrete sleepers have been placed accounting for 76% among the various types of railway sleepers. Moreover, the quality and types of concrete sleepers have been remarkably increased with the increase in the volume, load and train speed. At present China annually produces more than 10 million of pre-stressed concrete sleepers [294].

Construction of CRTS Ⅲ ballastless concrete tracks for rail transit, a kind of low-carbon, energy-efficient, environment-friendly green high speed transportation, have been developed and applied in China as modern ballastless concrete tracks technology [295]. One of the innovative processes is to replace CA mortar with self-compacting concrete as packed adjustment layer so as to form a more stable compound slab consisting of effective combination of self-compacting concrete layer and track slab.

CRTS Ⅲ ballastless concrete tracks used for Chengdu-Dujiangyan passenger line [295]

F. Fiber cement products

Fiber cement products in

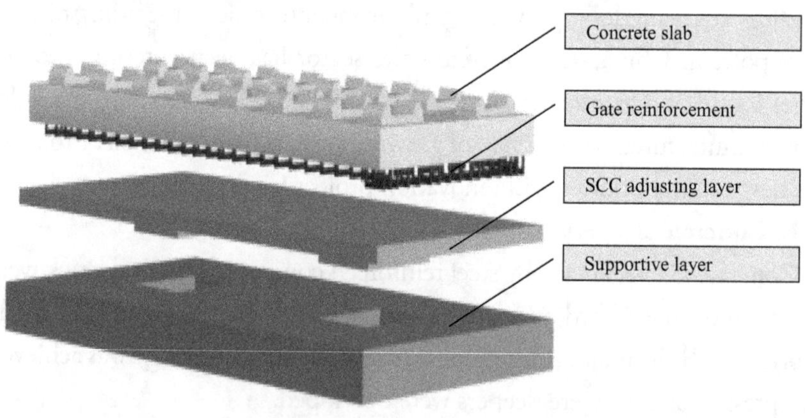

Compound slab consisting of effective combination of self-compacting concrete layer and concrete track slab for CRTS Ⅲ ballastless concrete tracks [295]

China includes autoclaved calcium silicate board and fiber cement products. Glass fiber reinforced cement (GRC) products, generally also as fiber reinforced cement products is discussed here as a specialty.

The first use of GRC in China in 1950s and 1960s proved to be unsuccessful because of the poor durability resulting from the corrosive reaction between the glass fibers and high alkaline $Ca(OH)_2$ in normal Portland cement pastes. Thanks to the contribution of China Building Materials Academy in providing an innovative "Double Guarantee" solution that push the large-scale production and use in China in 1980s to 1990s. The principle for enhancing the durability of GRC products lies in developing low alkalinity special cements (pH less than 11.5 in pore solution) and anti-alkalinity glass fibers containing zirconia, combining both new raw materials and thus producing a higher durability of GRC products. The annual output of various GRC boards is more than 100 million square meters.

In view of chemistry, both autoclaved fiber cement and calcium silicate products are produced with the same or similar composition of the calcium silicate hydrates in their final products through autoclave process. The fibers normally used include asbestos and non-asbestos inorganic fibers and organic fibers such as cellulose fibers and synthetic chemical fibers.

Asbestos cement products have existed in China with few alternatives for a very long time but that situation has changed into an increasing percentage of non-asbestos cement products due to the growing demand for green and environment-friendly building materials. The time span from 2006 to 2013 witnessed not only an absolute growth of total output of fiber cement products, but also an increase of the percentage of non-asbestos cement products from 8% to 18% [296]. More and more companies are joining the track on manufacturing and using environment-friendly fiber cement products.

Statistic number of China fiber cement production in 2006 and 2013 (million m^2) [296]

Year	2006	2013
Asbestos corrugated sheet	300	200
Asbestos flat sheet	50	360
Non-asbestos flat sheet	30	120
Total	380	680

Breakdown of China fibre cement production in 2006 and 2013 [296]

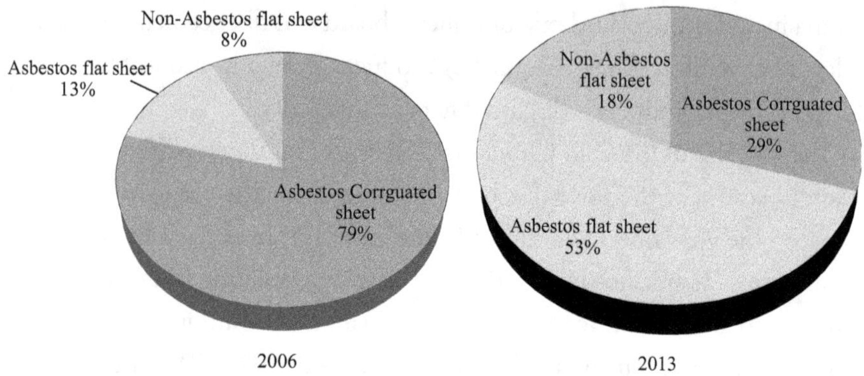

The need for fiber cement products in general has been remarkably increasing with an annual increase rate of more than 30% in recent years due to the higher strength, better moisture resistance and dimensional stability compared with gypsum based walling materials.

Considering the dynamic increase of gypsum board from 173 million square meters in 2003 to 2.7 billion square meters in 2013, we have the reason to expect that fiber cement board will have a brighter future.

F. Aerated concrete

Aerated concrete as a kind of porous light weight building materials made by autoclave process was first used in China in 1930s in Shanghai. Early 1960s the establishment of Beijing Silicate Plant and Guiyang Lime-Sand Brick Plant created the beginning of aerated concrete in China.

The statistical data for the year 2010 shows that 72 million m^3 in capacity and 40 million m^3 in output were realized for aerated concrete block, which is about 5 times larger than those in 2002.

7 What will happen in the future?—afterthoughts

We will be careful in predicting the future of concrete. We are certain that great changes would have taken place when the history of concrete is written hundred years from now.

Simplified a bit, the history will read thus;

- 10 000 years ago, the center of development was east and southeast of the Mediterranean.
- 2 000 years ago, it was in Rome.
- 1 000 years ago, very little development went on.
- 200 years ago, much of the development was in England, with some branches across the channel, in particular, to France.
- 100 years ago, the center of the development was somewhere in the middle of the Atlantic Ocean, or at least a bit out in the ocean.

Today, a volume of concrete equal to a mountain with an area of 1×1 km and a height equivalent to the height of Mount Everest has been produced across the world annually. The center of development is no longer Europe

The cement consumption in the world, in 2007, according to Cembureau [297]

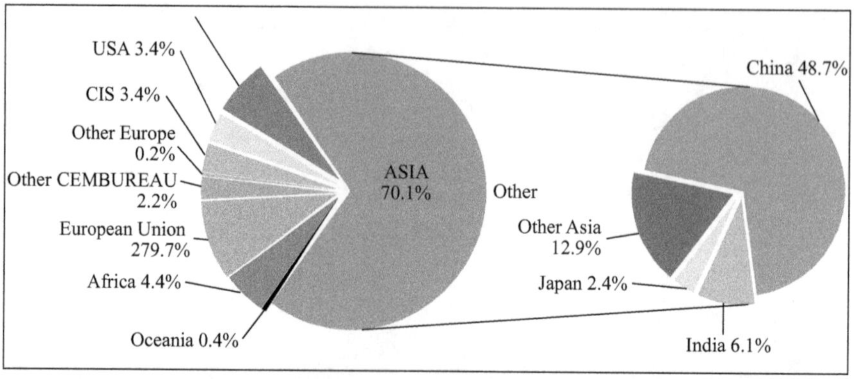

The cement consumption in the world in the future [298]

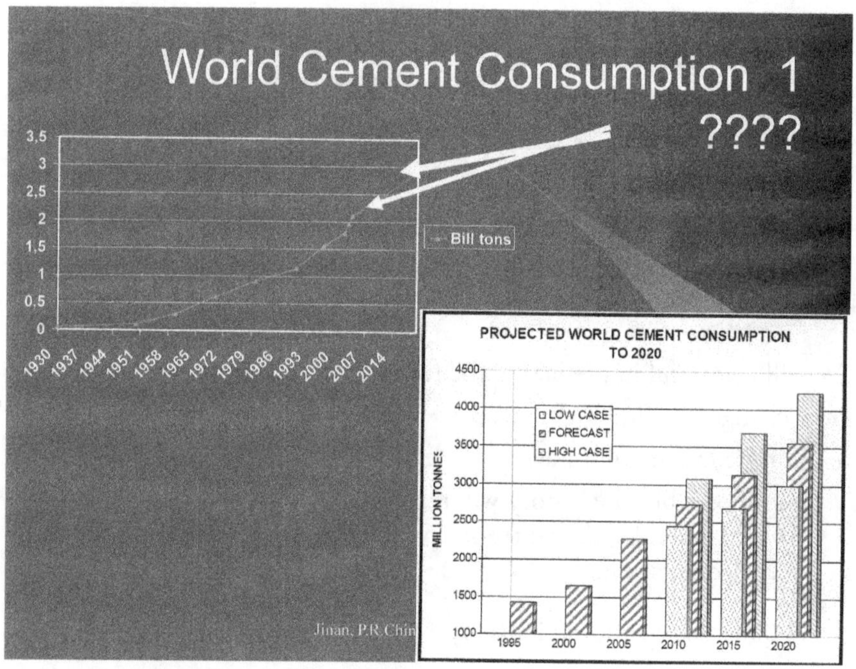

or America, but Asia. Today, China produces and consumes more cement and concrete alone than twice the amount Europe and America together do. The consumption per capita in China is three times as high as in the USA.

In 2005, India took over the second place in the cement statistics from USA and now they have a higher rate of growth in their consumption than China. Countries like Vietnam and Brazil also have very steep growth curves.

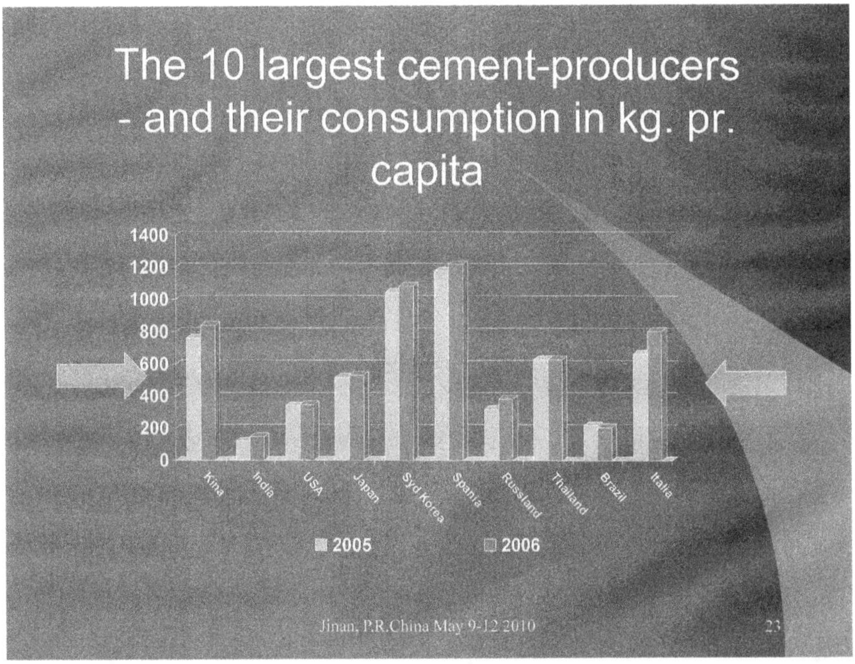

Cement consumption per capita in the 10 largest cement producing countries. The arrows show a consumption of half a ton per capita. The high consumption in South Korea, Spain, and Italy is partly due to the fact that these countries are major exporters of cement [297].

The extremely high consumption in China is due to a number of reasons. In addition to social improvement, the fact that half a billion people, in a few decades, shall move from agricultural activity to other activities is an extremely important factor. This leads to a demand for new living facilities, infrastructure, and increased energy production at a rate that the world has never seen before.

Projects such as the Three Gorges Dam, Qinghai–Tibet-jernbanen, Hangzhou Bay Cross Sea Bridge, the constructions for the Summer Olympics in 2008, the constructions for the World Expo in Shanghai, and many more have been completed. New challenges such as the gas pipe line from

The Three Gorges Dam was built in 1993–2009. The dam will produce 84.68 BkWt. For the dam project, 28 million m^3 of concrete was used, with 16 million m^3 for the dam itself [299].

31 km long offshore bridge from Shanghai to harbors on the islands outside [300].

Wanxian Bridge (1997) with a length of 846 meters and an arch span of 420 meters (the longest in the world of its kind). For comparison (See Chapter 3.1.3)

west to east, the water transfer project from north to south, new super speed railways in several directions, and more are under construction and are reasons for further high consumption of cement.

In 1978, 10% of China's total energy production was consumed by buildings. This increased to 27% in 2005, and is expected to be 35% in 2020.

China's cement requirements were met by 5 200 cement plants with about 7000 cement kilns in 2007. This is an industry with very variable efficiency; from old vertical kilns with capacities of 300 to 500 tons per day through wet process kilns, regular dry process kilns, to new highly advanced dry process kilns with capacities up to 10 000 tons per day are used. This industry is also under modernization and has improved remarkably with respect to the reduction of greenhouse gas emission at a very fast pace [301].

As indicated in Chapter 6, China cement has kept increasing till 2014. The total annual production of cement in China in 2014 reaches 2.48 billion tons with an increase rate 1.8% compared with the output of 2013, this is also the lowest growth rate in the last 30 years. It is predicted that Chinese cement is going to achieve the highest production within the next 5 years and then enters into an era of gradual decrease.

The Chinese themselves think that the cement production will be stabilized in the years 2012–2015.

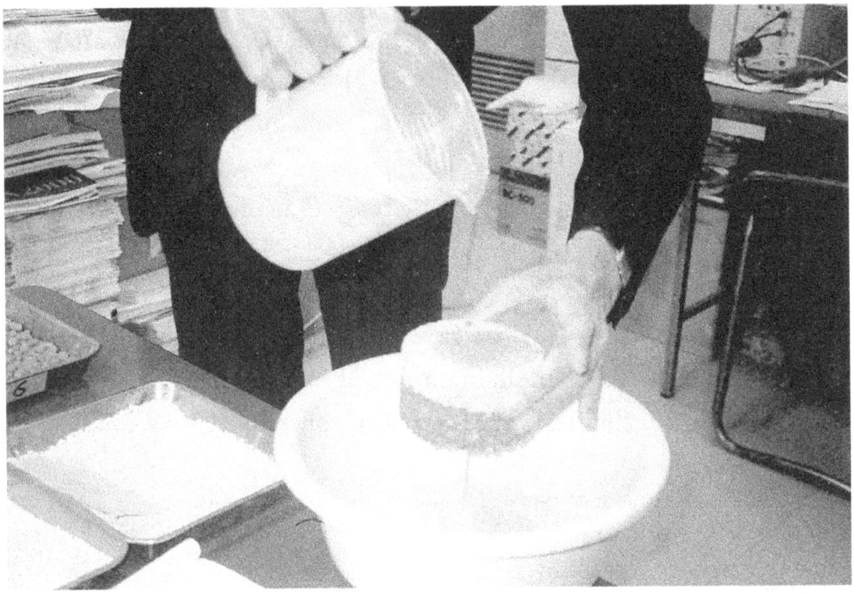

Permeable concrete

Sustainable development

Increased focus on the environment and on sustainable development will create considerable changes, both in the use and the production of binders for concrete and the way we use concrete in the future.

A number of investigations and analyses have shown that concrete is a very sustainable building material and considering the way it has been used and will be used in the future, there is scope and necessity for improvising on the production methods, keeping in tune with the future demands for emission reductions.

New energy capacity in EU in 2008, total installed: 23.852 MW

Concrete is one of the most important elements for sustainable development in the future, in

Spain is today number two in Europe in wind power, but has ambitions to be the leading country. From left to right: Wind–Gas–Solar panels; Coal–Biomass–Others-Atomic energy. Wind power in Europe according to [302]

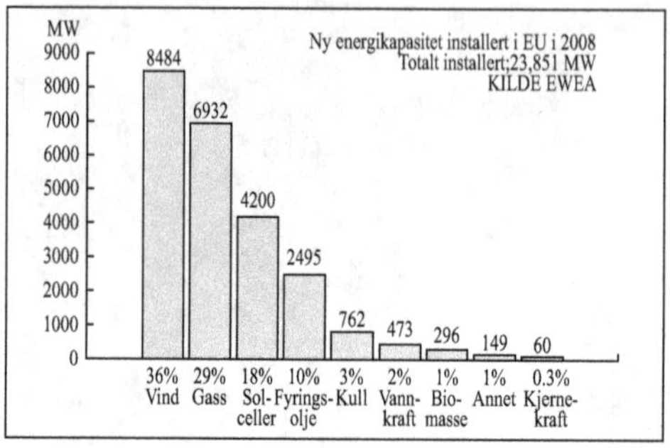

particular, as a raw material for improvements in infrastructure and the building of new renewable energy sources.

One of the many examples of the "new" use of concrete is the increased use of permeable concrete. Permeable concrete has already got a good foothold in important industrial countries.

Some of the possibilities with this type of concrete are;

- Improved drainage of pathways and traffic spaces.
- Improved erosion protection.

Another example of the new area of use of concrete in line with increased focus on a sustainable development is structures for new renewable energy.

An article in *Teknisk Ukeblad* in Norway, 43/09, [302], states that 36% of all new energy capacity installed in EU in 2008 was wind power.

Windmill with tower in precast concrete, Spain 2008.

Windmill towers in concrete are already in use and new producers have announced intensified development work in the area of both insitu cast and precast concrete. Some of the towers have been built to a height that is unheard of so far.

Concrete, because of its ability to store heat, is also a very interesting alternative for energy savings in buildings, with a corresponding reduction in the emission of CO_2, especially when approached from a long-term perspective[303]. More effective use of the "heavy" building materials is of extreme importance and will mean modifications to the way we build in the future.

The demand from the society for reduction in the CO_2 emission will also mean that the cement and concrete industries must also play their role in the reduction.

Concrete is the second biggest single commodity used after water. This huge volume also implies that production of concrete represents 6%–7% of the total man made CO_2 emission. On a global basis, the emission of CO_2 is approximately 0.85 tons per ton of cement used. The emission during cement production represents about 85% of the emissions from the production of concrete. This means that concrete production today emits 1 ton of CO_2 per ton of cement that is consumed [304]. It is important to remark that these numbers vary considerably from one country to the other and even within regions in the same country. The numbers for Norway, for example, are (about two thirds to three fourths of the global averages) some of the lowest in the world.

It seems like there will be an ecological and political demand for these

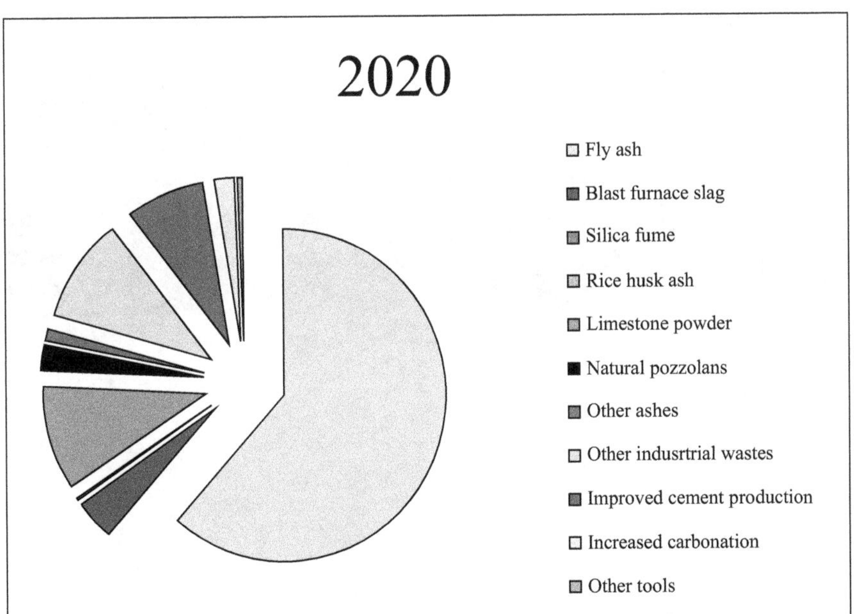

Potential emission reduction possibilities 2020

Alternative energy sources are important, but the reality is that in the many years ahead, one new coal fired power plant will be opened every week. Coal-based electricity production will in several decades produce more than half of the total electricity produced in the world. The ash from these plants is a very important source for the reduction of the emission from concrete production.

sectors to reduce the emission by a minimum of 20% before 2020 and one could envision a reduction of 50% before 2050. At the same time, we see an increased consumption of cement, in particular, in the countries in the south. A reasonable estimate is that the amount of binder we will need to produce concrete in 2020 would have increased to 4 billion tons from the 2.77 billion tons that was registered in 2007.

The consequence is that the level of emission we have today, 0.85 ton per ton, must be reduced to below 0.47 ton per ton before 2020. The technology and the raw materials to carry out this change exist, but education and willingness to act is needed. As an example of the possibilities, we mention that increased use of the so-called substituting materials such as fly ash, granulated blast furnace slag, limestone powder, silica fume etc alone, is theoretically enough to bring about this change in technology.

But increased use of substituting materials is not enough if we want to

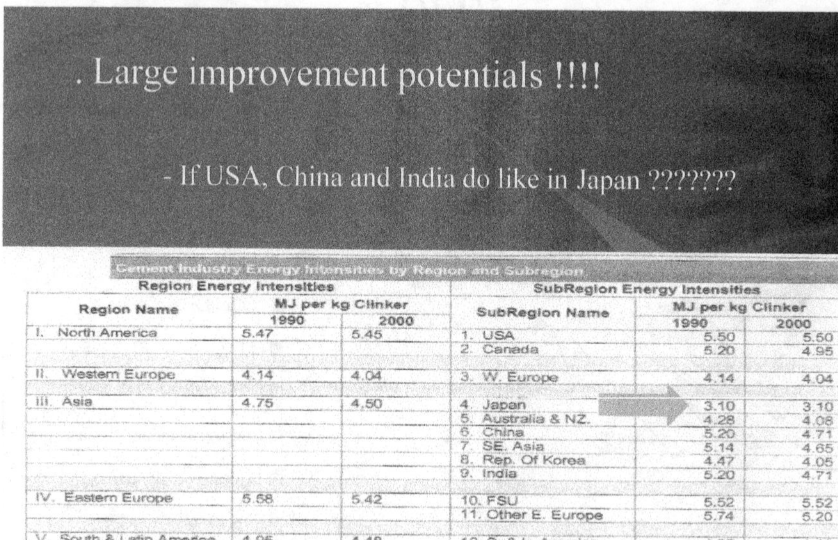

Cement Industry Energy Intensities by Region and Subregion

Region Energy Intensities			SubRegion Energy Intensities		
Region Name	MJ per kg Clinker 1990	2000	SubRegion Name	MJ per kg Clinker 1990	2000
I. North America	5.47	5.45	1. USA	5.50	5.50
			2. Canada	5.20	4.95
II. Western Europe	4.14	4.04	3. W. Europe	4.14	4.04
III. Asia	4.75	4.50	4. Japan	3.10	3.10
			5. Australia & NZ.	4.28	4.08
			6. China	5.20	4.71
			7. SE. Asia	5.14	4.65
			8. Rep. Of Korea	4.47	4.05
			9. India	5.20	4.71
IV. Eastern Europe	5.58	5.42	10. FSU	5.52	5.52
			11. Other E. Europe	5.74	5.20
V. South & Latin America	4.95	4.48	12. S. & L. America	4.95	4.48

The energy consumption per ton is highly variable; the table is from [305].

meet the goal of IPCC's anticipation of a minimum 50% reduction in greenhouse gas emission by 2050. However, we have a number of other alternatives;

There is still a lot to be achieved by modernizing the cement production in the world.

Significant reduction can be achieved by converting the energy used to emission-neutral energy. At the Dalen plant in Norway, Norcem started to use alternative fuel in 1992. From 1995 to 2004, Norcem invested 350 million NOK to be able to do this, and in 2007, 50% of the thermal energy came from emission neutral energy sources [303].

If the cement clinker production is modernized and the energy consumption reduced to 3 to 3.5 MJ/kg clinker (Damtoft claims in a lecture at Lillehammer in 2007 that a theoretical minimum is less than at 3000 kJ per ton), and one third of the energy comes from alternative fuel, it is possible that the emissions from cement production across the world can be reduced by 20% per ton of clinker produced [306].

There are other technical reduction potentials also, but not in the same order of reduction as the ones mentioned above. However, it is important to remember that the various potential methods vary considerably from country to country, and within regions of the same country, partly depending on the technologies that are already in use and partly on the availability of raw materials. Probably, we have to use all the available potential tools to be able to reach the final goal.

From a long-term perspective, CCS (Carbon Capture and Storage) is an interesting alternative. The European Union expects that 10–12 full scale projects will be in operation in 2015. It is hoped the technological breakthrough of the technology shall come in the 2030s. The cement industry is of particular interest in this connection because it in this industry we find concentrated emission [307].

The dream scenario half a century from now is that concrete, due to the use of CCS, and the absorption of CO_2 through carbonization in the finished product, has a negative emission effect.

The research activity in most industrial countries is very high in the emission sector to facilitate the development of new technology that will reduce the emission during the production of cement and concrete.

The use of concrete will increase, but the increased focus on the environment and sustainable development will change the technology we utilize in the years to come.

It has been said that "history repeats itself." Does this mean that this will happen in the case of concrete too?

Yes, without doubt. We see in many cases that ideas which have been

In Finland, oval pipes were produced on a regular basis from the beginning of the 2000s.

Artists' view of a living quarter on the moon.

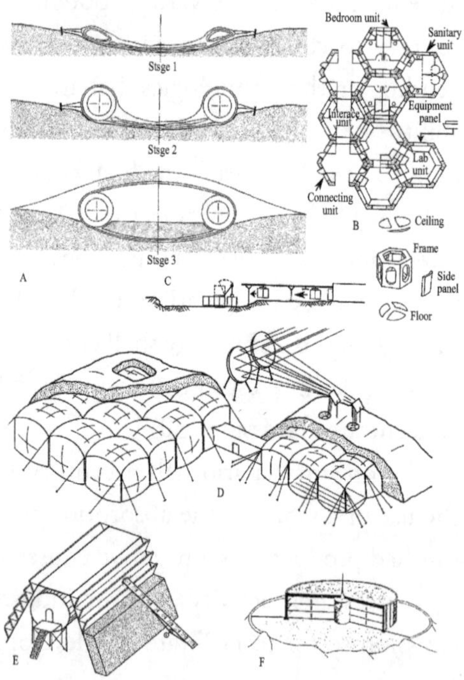

Various suggestions for building designs on the moon.

A. Erection principles for a fabric dome, stiffened by structural foam materials. (T.Y.Lin International)

B. And B & C

C. Concrete components with hexagonal modules mounted together by crane (Shimuzu Corp)

D. Fabric modules "tufted" by tensile columns (Colorado State University)

E. "Construction shack" on top of a previously deployed unmanned platform and enveloped by truss structure (NASA)

F. Prestressed concrete structure (T.D. Lin, CTL)

put on the shelf are brought forward later because as the demand changes the conditions also change. We will probably see more of this in the case of concrete. Here is an example;

Oval pipes;

The early concrete pipes that were produced had an oval cross section. This gives hydraulic advantages because the small diameter at the bottom provided good self-cleaning properties in the dry season. As a result of modern machinery, casting technology, and production economy, the pipes over time became circular in shape. Changes in climate will necessitate new ways of leading the flow of water through pipes. Probably, this will lead to the use of oval pipes again.

When someone writes about the history of concrete 50 or 100 years from now, we might even read about **the first concrete structure on the moon.** Research work has already been carried out for a number of years now to face this challenge. When Neil Armstrong and Buzz Aldrin put their feet on the surface of the moon on 20 July 1969, thoughts about establishing permanent bases there already existed. President Bush announced 20 years later, on 20 July 1989, that a permanent base on the moon should be built in the twenty first century. If human beings aim to reach other planets, it is important to use the moon as a test station for a number of functions. Among others, it is important to learn how to utilize the local raw materials and to recirculate water and air.

In the PCI Journal of July–August 1989, we find a special report about the building of concrete houses on the moon. George Nasser has given a general introduction to the topic [308] and T.D. Lin et.al. elaborate on the study that has been done for the construction of a living quarter in prestressed concrete [309].

An important point in the design of a permanent living quarter is to use local building materials. Several hundred samples of materials from the moon have been analyzed, and experts opine that the samples are not non-representative.

ACI, American Concrete Institute, established a committee to work with the task "Concrete on the Moon" 20 years ago. The committee was

An astronaut with a large block of stone. Smaller samples of this typical moon material, regolith, are seen in the foreground.

Lin's proposal for a three story building.

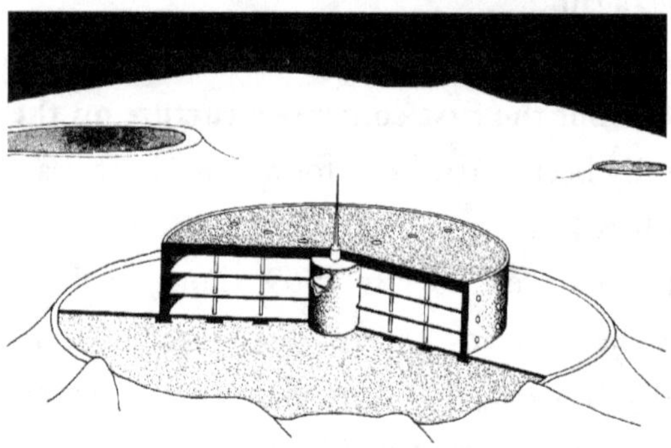

from the start headed by T.D. Lin from Construction Technology Laboratories in Skokie, Illinois, and the committee had a broad representation, also with representatives from NASA.

A structure on the moon must be designed like a pressure vessel, with breathable air inside, a minimum of pressure on the outside, and leak proof walls. Most experts are of the opinion that the structure must be protected by a thick layer of moondust to protect against the deadly sun radiation. Since there is no atmosphere, wind, snow, and ice loadings can be omitted. The main problem is the vacuum on the outside and therefore the walls must be leakage free. There are of course gravitational loads, but they are about one sixth of the loads on the earth.

"Imported" building materials will probably cost something in the order of 6 000–10 000 dollar per kg, so it is important to use local materials.

Scientific studies show that it is possible to produce cement and high performance concrete using materials from the moon.

All the materials in T. D. Lin's proposal can be produced from local raw materials, with the exemption of the hydrogen.

Lin's proposal has a diameter of 36.6 meters, is 7.3 meters high, and has a working area of 3060 m^2.

Lin has, among others, based his design of the ideal concrete station on the fuel tank, of a space ship, that has a diameter of 8.2 meters and a length of 46.9 meters. The structure is further based on the production of concrete with a compressive strength of about 75 Mpa. This is a fascinating design in precast concrete components that are stressed together afterwards. The amount of material needed, with the exemption of the foundation, should weigh 8 500 tons on the earth and 1400 tons on the moon. 36 tons of hydrogen for the production was planned to be transported from the earth.

In November 2009, NASA published that they had found water in the moon. This definitely makes concrete production an even more interesting alternative.

More than 100 years ago, Jules Verne fascinated readers by talking about a trip to the moon. Maybe some people will one day witness concrete production there too.

I have in the book tried to give some glimpses of the history of concrete. Though I have approached the topic from my perspective, I hope that the book provides an overview of the manifold uses and production methods of the most important building material in the world.

References

[1] Patterson R.F.: *English Dictionary*, The No. 1 Publishing Co. Ltd., London.

[2] Uvarov E.B.: *A Dictionary of Science*, Penguin Reference Books, Harmondsworth, England, 1961.

[3] Parker Sybil P.: *McGraw-Hill Encyclopedia of Science & Technology*, 1982.

[4] Fokus, *Illustrert familieleksikon*, H. Aschehoug & Co., Oslo, Norway, 1958.

[5] Lea F.M.: *The Chemistry of Cement and Concrete*, Third edition (first edition 1935), Edward Arnold Ltd., Glasgow, 1970.

[6] Gartmann Frithjof: *Sement i Norge i 100 år*, Norcem As, Oslo, Norway, 1990.

[7] Thaulow Sven: Cement, *Teknisk Ukeblad*, Nr. 35, 1967, Oslo, Norway.

[8] Aïtcin Pierre-Claude: *Binders for Durable and Sustainable Concrete*, Taylor & Francis, New York, 2008.

[9] Malinowski R., Garfinkel Y.: Prehistory of Concrete, *Concrete International*, March 1991.

[10] Patterson R.F.: *English Dictionary*, The No. 1 Publishing Co. Ltd, London, England.

[11] Malinowski Roman, Garfinkel Yosi: Betongens forhistorie, *Nordisk Betong* 5, 1988.

[12] Ying Bao: Looking for the Ark of the Covenant, *Tenth ACI international Conference on Recent Advances in Concrete Technology and Sustainability Issues*, Supplementary papers, pp. 783–800, Seville, Spain, 14–16 October 2009.

[13] Bibelen: *Den hellige skrift- bibelen (The Holy Bible)*, Det Norske Bibelselskap, 9.opplag, Oslo, Norway, 1995.

[14] Internet 21.12.09: Mesopothamia – http://en.wikipedia.org/wiki/mesopotamia

[15] Rivera-Villarreal Raymundo, Krayer Stafan: Ancient Structural Concrete in Mesoamerica, *Concrete International*, June 1996.

[16] Campbell James W.P., Pruce Will: *Brick – A World History*, Thames & Hudson Ltd., London, 2004.

[17] Sui Tongbo: History of Cementitious Materials for Building in Ancient China, E-mail to the author, 16.03.10.

[18] China Intercontinental Press: *China in Diagrams*, State Council Information Office, ISBN 7-5085-0842-4/K 726, Beijing, 2005.

[19] *History of Concrete*, Cement and Concrete Association, London, 1974.

[20] AS Norcem Betongdivisjonen (Concrete Division): Litt om betongens historie, og vår egen (A Bit about Concrete History, and our Own), Broshure, Norcem, 1977.

[21] Internet 31.10.09, Lepenski Vir, http://en.wikipedia.org/wiki/Lipenski Vir

[22] Whitehouse Ruth & Wilkins John, translated by Gunnar Bureid: *Sivilisasjonens røtter (The Roofs of Civilization)*, Cappelen, Oslo, 1989.

[23] Eco Umberto, Zorzoli G.B., translated by T. Dannevig Hauge: *Oppfinnelsenes historie (The History of Inventions)*, Gyldendal Norsk Forlag , 1964.

[24] Lorange Per: Cementfremstilling (Cement Production), Kåseri (Lecture) 22.12.1975, Norcem, Oslo.

[25] Canpbell Donald H., Folk Robert L.: The Ancient Egyptian Pyramids – Concrete or Rock?, *Concrete International*, August 1991.

[26] Davidovits Joseph: Ancient and Modern Concretes. What is the Real Difference?, *Concrete International*, December 1987.

[27] Davidovits J., Morris M.: *The Pyramids: An Engima Solved*, Hipposcrene Books, Inc., New York, 1988, p. 263.

[28] Morris Margie: The Cast-in-Place Theory of Pyramid Construction, *Concrete International*, August 1997.

[29] Moors Bob, debate contribution to /26 and 27/, *Concrete International*, February 1992.

[30] Christiansen Per A.: Faraos fjell (Pharao' Mountains), *Aftenpostens søndagsutgave*, ca 1970, Oslo, Norway.

[31] Hellstrøm Bo, Granholm Hjalmar, Eriksson Axel: *Betong (Concrete)*, Bokforlaget Natur och Kultur, Stockholm, Sweden, 1945.

[32] Internet 28.11.09. The History of Concrete and the Nabataeans, http://nabateaea.net/cement.html

[33] Idorn Gunnar M.: Tilbageblik og fremskridt i Betongteknologien (Reflections and Inventions in Concrete Technology), Festskrift til finsk Betongforenings 50-års jubileum (Jubilum-Script to Finish Concrete Society), Aalborg Portland, Karlslunde, June 1975.

[34] Boardman John, Griffin Jasper, Murray Oswyn: *The Roman World*, Oxford University Press, 1989.

[35] Hover Kenneth C.: The Roman Empire Fell – Its Concrete Endures, *Concrete International*, November 2004.

[36] Vitruvius: *Ten Books of Architecture*, translated by Morris Hickey Morgan, Dover publications, Inc., New York, 1960.

[37] www.Wikipedia/nero, 05.10.09.

[38] Redaksjonelt: Boligforhold u*æ*ndret gennom 2000 år, (Living Conditions Unchanged through 2000 Years), *Byggeindustrien* 24, 1967, Danmark.

[39] White K.D.: *Greek and Roman Technology*, Thames and Hudson, London, 1984.

[40] Idorn Gunnar: *Concrete Progress – From Antiquity to the Third Millennium*, Thomas Telford, London, 1997.

[41] Hill Donald: *A History of Engineering in Classical and Medieval Times*, Routledge, 1996.

[42] Internet 01.01.10. Pont du Gard, http://no.wikipedia.org/wiki/Pont_du_Gard

[43] Thaulow Sven: *Bad og Svømmeanlegg (Baths and Swimming Pools)*, Norges Svømmeforbund og Norsk Cementforening, 1965, Oslo, Norway.

[44] Pørtner R.: 200 Jahre Frieden am Rhein – Rømerzeit, Blutezeit, bauzeit, *The Role of Concrete in Conservation of Historic Buildings*, Cembureau, Paris, 1976.

[45] Pørtner R.: 200 Jahre frieden am Rhein-Rømerzeit-Bluezeit, bauzei, The Role of Concrete in Conservation of Historic Concrete, Cembureau, Paris, 1976.

[46] Harries Kent A.: Quake Resistance in Ancient Rome, *Concrete International*, January 1997.

[47] Connolly Peter and Dodge Hazel translated by Bjarte Kaldhol og Lisbeth Lyngås: *Antikkens storbyer, (Anscient Large Cities)*, Aschehoug, Oslo, 1998.

[48] Circus_Maximus https://en.wikipedia.org/wiki/Circus_Maximus, accessed on August 20, 2014.

[49] A Quote for Hadrian's Wall, *Concrete International*, January 1975.

[50] Concrete Editorial; Out of Warranty? *Concrete International*, September/October 1995.

[51] McCall Colin: Cement Ancient and Modern, *Concrete International*, September 1975.

[52] Pedersen Per Helge: *Fra Farris til Troll (From Farris to Troll)*, Jubileumsbok Norsk Betonforening, Norwegian Concrete Society, Oslo, 1995.

[53] Internet 01.11.09; History of Concrete, http://concretecontractor.com/concrete-history/

[54] Tavernor Robert, Schofield Richard: *Andrea Palladio-The Four Books on Architecture*, Massachusetts Institute of Technology, USA, 2001.

[55] Bakken Asbjørn: Farrisdammen- *Et bidrag til sementens historie (The Farris Dam – A Contribution to the History of Cement)*, Særtrykk av Vestfoldminne, Norway, 1959.

[56] Jahren Per: Historisk betong – fortsatt meget god (Historical Concrete – Still Very Good), *Betong Produkter* 1/92, Oslo.

[57] Advertisement PFA, *Concrete International*, London, October 1975.

[58] Internet 24.11.09: www.eutopeanconcrete.eu/members/cembureau

[59] Corley W.G.: Protecting the public from Fools and Rascals: Building Codes for the Millennium, *Concrete International*, September 1992.

[60] Åberg Alf: *Cement i 100 år. En krønka om Skånska Cementaktiebolaget (A History about SC)* – AB Cementa, Malmø, Sweden, 1972.

[61] Internet 01.11.09: The History of Concrete, http://matse1.mse.uiuc.edu/concrete/hist.html

[62] Harper Roger H.: Concrete's Battle for London, 1867–1886, *Concrete International*, October 1976.

[63] Markestad Asbjørn: *Materialer og materialprøving (Materials and Material Testing)*, 2 Del, Tapir forlag, Trondheim, Norway, 1961.

[64] Internet 24.11.09: FortuneMagazine, 1933. http://www.usbr.gov/lc/hooverdam/History/articles/fortune1933.html

[65] Wallis Shani: Eight Wonder of the World? Historic British Tunnel Renovated, *Concrete International*, December 1997.

[66] Barfoot R,J.: Joseph, James and William – The Aspdin Jigsaw, *Concrete*, August 1974.

[67] Idorn G.M., Thaulow N.: Examination of 136 Years Old Portland Cement Concrete, *Cement and Concrete Research* 13, 739–743, 1983, USA.

[68] Carroll Rutter, Cabrera J.G.: Reinforced Concrete in Northeast England, *Concrete International*, June 1998.

[69a] Barfoot R.J.: Tall's Testimonial, *Concrete*, September 1976.

[69b] Hurme Riitta et al.: *Betoni Suomessa (Finish Concrete History)*, 1860–1960, Finsk Betongforening, Helsinki , Finland, 1960.

[70] Fidler John: American Pioneers of Concrete, *Concrete*, May 1979.

[71] Heun Raymond C.: Ponckhockie Union Chappel – 1870, *Concrete International*, October 1993.

[72] Dabney Charles M.: Historic Concrete in California, *Concrete International*, April 1990.

[73] Kasai Yohio, Nagataki Shigeyoshi: *One Hundred People Who Supported the Concrete Technologies of Japan*, Cement Newspaper Co. Ltd.. (Japanese). Some parts translated by Sumida Takashi.

[74] Lyse, Wiig: *Betong – Uarmert og Armert (Unreinforced and Reinforced Concrete)*, F. Bruns Bokhandels Forlag, Trondheim, Norway, 1957.

[75] Eika Albrecht: Et hundreårsjubileum (A Hundred Years Celebration), *Teknisk Ukeblad*, Nr. 35,Oslo, Norway, 1967.

[76] Sutherland James: Historic Reinforced Concrete Drawings – An Appeal, *Concrete*, March/April 1992.

[77] Internet 29.09.09: www.westminstercathedral.org.uk

[78] Hunt Martin: Westminster Cathedral Used Mass Concrete, *Concree International*, July 1979.

[79] Kolderup Edvard: *Monierkonstruktionene (Monier Structures)*, Kristiania (Oslo), Norway, 1893, H. Aschehoug & Co.s Forlag.

[80] Hamilton-Baillie: Nineteenth-Century Concrete and the Royal Engineers, Part 1, *Concrete*, March 1980.

[81] Hamilton-Baillie: Nineteenth-Century Concrete and the Royal Engineers, Part 2, *Concrete*, April 1980.

[82] Association of the Concrete Industry in Finland: *Concrete in Finish Architecture 1929–1989*, Helsinki, 1989.

[83] Suonio Jukka: *Concrete – A Developing Construction Material, Fifty Years of Finish Civil Engineering. Suomen rakennusiniøørien Liitto*, Helsinki, Finland, 1984.

[84] Morley Jane: Frank Bunker Gilbreth's Concrete System, *Concrete International*, November 1990.

[85] Mather Bryant: Robert E. Lee and Concrete Innovations, ACI Committee 120, History of Concrete Session, Baltimore, November 1999.

[86] Goodheart Adam: Why Dolores Chumsky Hates Thomas Edison. Internet 30.10.08: http://flyingmoose.org/thruthfic/edison.htm.

[87] Greene Brian: Failure of a Concrete Dam in 1911 at Austin, Pennsylvania, ACI Committee 120, History of Concrete Session, Spring 1999.

[88] Internet 21.11.09: www.pancanal.com/eng/history.html

[89] Loov Robert E.: Reinforced Concrete at the Turn of the Century, *Concrete International*, December 1991.

[90] Internet 25.10.09: htt://en.wikipedia.org – Eugene Freyssinet.

[91] Anderson Athur R.: *Advances in Concrete Construction*, Norwegian Concrete Day, Oslo, Norvegian Concrete Society, Oslo.

[92] Zollmann Charles C.: The End of the "Beginnings", *PCI 25 Years 1979, Reflections on the Beginning of Prestessed Concrete in America*, PCI, 1981.

[93] Zollmann Charles C.: Magnel's Impact on the Advent of Prestressed Concrete, *PCI 25 Years 1979, Reflections on the Beginning of Prestessed Concrete in America*, PCI, 1981.

[94] Cembureau: World Cement Market in Fugures 1913–1977, Statistical Review No. 35, Paris, December 1978.

[95] Kuarni Vijai R.: Historical Development of Cement and Concrete in India, Abstract for ACI Committee 120, History of Concrete Session, Spring 2004.

[96] Korhonen Arto: Concrete Technology Development in Finland, *Nordisk Betong* 5, 1988.

[97] Tekniska Nomenklaturcentralen: *Betongteknisk ordlista*, Stockhlolm, Sweden, April 1971.

[98] FIP: *Multilingual Dictionary of Concrete*, Elsevier, Amsterdam, 1976.

[99] Norsk Betongforenings Miljøkomite, *CO_2-utslipp – sement og betong, Utfordringer og Perspektiver (CO_2-Emission – Cement and Concrete, Challenges and Perspectives)*, Norwegian Concrete Society, Oslo, April 2009.

[100] Erlemann Gustav G.: Steel Reinforcing Bar Specifications in Old Structures, *Concrete International*, April 1999.

[101] Hattori Kenichi: Experience with MIGHTY Superplasticizers in Japan, Report from KAO Corporation, Japan, 1978.

[102] KAO: Technical Bulletin P-5001, 1891, Tokyo, Japan.

[103] Jahren Per: *Eksempler på høyfast betong i utlandet (Examples of High Strengths Concrete Abroad)*, Norwegian Concrete Day, Norwegian Concrete Society, Oslo, 31 October 1986.

[104] Taisei Corporation: Brochure – D –Runway, Tokyo International Airport (Haneda Airport).

[105] Hoff George: Evolution of Concrete Technology. *Tenth ACI International Conference on Recent Advances in Concrete Technology and Sustainability Issues*, Seville, Spain, 14–16 October 2009, PPt-presentation.

[106] Snell Luke and Billie: Concrete Tree Surgery. ACI Committee 120, History of Concrete Session, Spring 2004.

[107] Fasting Kåre, Gartmann Frithjof: Dalens Egne år , 1916–1966, AS Dalen Portland-Cementfabrik, Brevik, 1968.

[108] Rye D.: Anvisninger til Udførelse af Bygningsarbeider af Konkret (Recommendations Regarding Building Structures Using Concrete), W.C. Fabritius, Kristiania, Norway, 1875.

[109] Møst Aage : Grundere gjennom et sekel (Grunders through a Secel), *Byggeindustrien*, Oslo, Norway, Nr. 7, 1985.

[110] Eeg-Henriksen: Fra de gode gamle dager(From the Good Old Days), *Byggeindustrien*, Oslo, Nr. 7 1985.

[111] Vigerust Gunnar: Sven Thaulow er død, (Sven Thaulow is Dead), Teknisk Ukeblad, Oslo, Norway, Nr. 35, 1967.

[112] Abrahamsen Helge: *Hus før og nå (Houises before and Now)*, Aschehoug, Oslo, 1961.

[113] Bernhardt Carl Johan: Forskningsinstituttet for Cement og Betong ved (The Research Institute for Cement and Concrete) NTH – FCB. *Ingeniør-Nytt*, Oslo, Norway, Nr. 40, 1968.

[114] Gartland Kirsten Joveig, Lien Synnøve, L'Orange Olav: *Fra bredd til bredd-Bruhistorie fra Namdalen (Bridge History from Namdalen)*, Statens Vegvesen, Nord – Trøndelag, Norway, 2001.

[115] Andersen Aksel: Forelesninger over Trekonstruksjoner (Lectures in Timber Structures), *Tapir Forlag*, Trondheim, Norway, 1956.

[116] Løvs Gudmund: Hatbrev mot tyskerne murt inn i bro (Hate-Letter to the German's Cast into a Bridge), *Aftenposten*, Oslo, Norway, 07.03.1999.

[117] Nagell Arne: *Streiftog gjennom vårt tekniske samfunn, (Glimps through our Technical Society)*,Ingeniørforlaget, Oslo, November 1973.

[118] Steen Øyvind: *Varige Spor (Permanent Footprints)* – Dr. Ing. A. Aas-Jakobsen AS, September 2007.

[119] Jensen Jens Jacob: Broer – historikk, miljø og høyteknologi, Cement Nå, Oslo, 1/93.

[120] Pedersen Per Helge: En guru om betong gjennom 50 år. – Den grensesprengende betongen (A Guru about Concrete), Special Addition to *Byggeindustrien*, Oslo, October 2004.

[121] Borchgrevink T.: *Veidekker i betong – utviklingen i Norge (Concrete Roads – the Development in Norway)*, Norwegian Concrete Day 1972, Norwegian Concrete Society, Oslo, 1972.

[122] Sylte O.K.: Betongveier, Norsk Betongdag 1974, Norsk Betongforening, Oslo, 1974.

[123] Hafskjold Sigvard: *Betongkonstruksjoner på kontinentalsokkelen, Grunnlag og muligheter (Concrete Structures on the Continental Shelf, Basics and Possibilities)*, Norwegian Concrete Day 1974, Norwegian Concrete Society, Oslo.

[124] Pedersen Per Helge, Steen Øyvind: *Skaperevne og Initiativ – Byggenæringen i 100 år (Crativity and Initiatives – The Building Trade 100 years) 1905–2005*, Byggeindustrien/Bygg og Anlegg Media AS, Oslo, 2005.

[125] Hafskjold P.S.: *Forsknings- og utviklingsaktivitet i bedriftens egen regi (Research and Developmemt by the Company)*, Norwegian Concrete Day 1978, Norwegian Concrete Society, Oslo.

[126] Jacobsen Bent: *Betong i Offshorekonstruksjoner – Nyutvikling nye konsepter, faste og flytende plattformer (Concrete in Offshorestructures – New Developments, New Consepts, Fixed and Floating Platforms)*, Norwgian Concrete Day 1985, Norwegian Concrete Society, Oslo.

[127] Gudmestad Ove T.: Betong i Offshorekonstruksjoner – En byggherres syn på valg av betongplattformer kontra andre løsninger (Concrete in Offshore Structures), Norwegian Concrete Day 1985, Norwegian Concrete Society, Oslo, Norway, 1985.

[128] Historic Concrete Number 25, *Concrete*, September 1976.

[129] Historic Concrete Number 13, *Concrete*, London, September 1975.

[130] www.seatonvisitorcentretrust.org

[131] Waine N.D., Long G.R.: An Early Concrete Structure Axmouth Bridge, Seaton, *Concrete*, December 1990.

[132] Barfot R.J.: A Monument of Concrete, *Concrete*, London, October 1975.

[133] Historic Concrete Number 6, *Concrete*, February 1975.

[134] Internet 24.10.09: http://bwtas.blogspot.com, British Water Tower Appreciation Society.

[135] Historic concrete Number 5, *Concretel*, London, January 1975.

[136] Historic concrete Number 4, *Concrete*, London, December 1974.

[137] Historic Concrete Number 59, *Concrete*, London,

[138] Budek Andrew: A Time to Build Up, a Time to Tear Down – Barnes Wallis vs. the Nazis, ACI Committee 120 – History of Concrete Session, Baltimore November 2004.

[139] Snell Luke M., Snell Billie G.: The Erie Canal – America's First Concrete Classroom, *Concrete*, December 1996.

[140] Amon Josep A, Amon Debby: An Oil Tycoon and Railroad Baron Build Florida Dreams of Portland Cement Concrete in 1885, ACI Committee 120 – History of Concrete Session, Baltimore, 2004.

[141] Internet 29.11.09: Alvord Lake Bridge, http://en.wikipedia.org/wiki/Alvord-Lake-Bridge

[142] Editorial: Concrete Beginnings – History of Materials Set in Ohio, *Concrete*, September 1995.

[143] Internet 25.11.09: Hoover Dam, http://en.wikipedia.org/wiki/Hoover-Dam

[144] Internet 24.11.09: Fortune Magazine, 1933, http://www.usbr.gov/lc/hooverdam/History/articles/fortune1933/html

[145] Milestones of Events and Developments in North American Prestressed Concrete Industry, *PCI 25 Years 1979, Reflections on the Beginning of Prestessed Concrete in America*, PCI, 1981.

[146] Anderson Arthur R.: *An adventure in Presteessed Concrete, PCI 25 Years 1979, Reflections on the Beginning of Prestessed Concrete in America – Part 7*, PCI, USA.

[147] Silfwerbrand Johan, Hult Patrik: Swedish Concrete Association, *Concrete*, November 2009.

[148] Finish Association of Civil Engineers: *History of Finish Bridges*, Helsinki, November 2004.

[149] Seppænen Lauri: *Valmisbetonia 50 vuotta*, Betonitieto Oy, Saarijærvi; Finland, 2008.

[150] Internet 15.01.10: http://www.stadion/fi

[151] Jetsonen Sirkka: *Alvar Alto Houses*, Rakennustieto, Helsinki, 2005.

[152] Internet 24.01.10: www.peab.no, PEAN-Journalen Nr. 1, March 2003.

[153] The Kaknas Tower, *Brochure*, Stockholms Tourist Traffic Association, Sweden, 1967.

[154] Jahren Per: *Limfjordtunnelen (the Limfjord Tunnel), Ingeniørnytt, Oslo, Norway*, Nr. 8, 1967.

[155] Editorial: Arbejdet på Limfjordtunnelen styres med sukces af EDB-anlæg (The Work on the Limfjord Tunnel is Ruled by Data), *Stads & Havneingeniøren*, Copenhagen, Denmark, 9/67.

[156] Redaksjonelt: Boligforhold uændret gennom 2000 år, (Living Conditions Unchanged through 2000 Years), *Byggeindustrien* 24, 1967, Danmark.

[157] www.centennialhall.eu, 25.09.09.

[158] http://en.wikipedia.org/wiki/Centennial_Hall, 25.09.09.

[159] Jerzy Illosz: *Max Berg's Centennial Hall – and Exhibition Hall in Wrocklaw*, Museum of Architecture in Wrocklaw, 2006.

[160] Tietz Jurgen: *Arkitekturens historie i det 20. Århundret (The History of Architecture in the 20th Century)*, Lønemann, Køln, 1998.

[161] Collins Michael P.: In Search of Elegance, *Concrete International*, July 2001.

[162] Sharp Dennis: *Illustrated Encyclopedia of Architects and Architecture*, New York, Quatro Publishing, 1991.

[163] Schlaich Jørg: The Bridges of Robert Maillart, *Concrete International*, June 1993.

[164] "Habitat 68": http://en.wikipedia.org/wiki/Habitat_67, 20.09.09.

[165] Philleo Robert E.: Back to the Concrete Future, *Concrete International*, March 1988.

[166] Jahren Per: Morgendagens Bebyggelse, – eller arkitektens lek med esker (The Building of Tomorrow, or the Architects Play with Boxes?), *Ingeniør-Nytt* Nr. 5, 1967, Oslo, Norway.

[167] *Icons of Architecture – The 20th Century*, Edited by Sabine Thiel Siling, Prestel-Verlag, Munic, 1949.

[168] Internet 30.01.10: http://deltawerken.com

[169] Internet 30.01.10: http://no.wikipedia.org/wiki/Oosterscheldekering.

[170] Internet 29.11.09: The Millau Bridge.

[171] Wright Frank Lloyd: *On Architecture – Selected Wrightings*, Edited by Fredrik Gutheim, The Universal Library, New York, 1941.

[172] Otani Shunsuke: The Role of Analysis in Earthquake-Resistance Design – A Retrospective, Opening Session and Hardy Cross Lecture, *ACI Fall Convention*, New Orleans, 8 November 2009.

[173] Knight Caroline: *Essential Frank Lloyd Wright*, Paragon Book, London, 2001.

[174] Storrer William Allin: *The Architecture of Frank Lloyd Wright – A Complete Catalogue*, MIT Press, 1974.

[175] Jahren Per: Frank Lloyd Wright's Richard Lloyd Jones House, Tulsa, Oklahoma. A Look at Some Historic Pictures from the Construction, ACI Fall Convention, Baltimore, USA, November 5. 1999.

[176] Hølaas Odd: *Jeg skal bli verdens største arkitekt! Sa Frank Lloyd Wright – og ble det. (I am Going to the Greatest Architect in the World, Frank Lloyd Wright Said, and so he Became)*, Dagbladet, Oslo, 11 April 1959.

[177] Hauser Ernest: *Kunstner i betong (Artist in Concrete)*, Byggmesteren, Oslo, 12/1964.

[178] Frampton Kenneth: *Le Corbusier*, Thames & Hudson, London, 2001.

[179] Jahren Per: *En tankevekke (A Reminder). Ingeniør-Nytt*, Oslo, Norway, Nr. 16, 1968.

[180] Gunnarsjaa Arne: *Arkitektur leksikon*, Abstrakt Forlag, Oslo, 1999.

[181] Grønvold Ulf: *Hundre års nasjonsbygging (Hundred Years of Nation-Building)*, Pax Forlag, Oslo, 2005.

[182] Nordberg-Schulz Christian: *Modern Norwegian Architecture*, Universitetsforlaget, Oslo, 1986.

[183] *Linje og Rom – 20 norske arkitekter (Line and Space-20 Norwegian Architects)*, Grøndahl Dreyer, Oslo, 1993.

[184] Grønvold Ulf: *Lund & Slaatto*, Universitetsforlaget, Oslo, 1988.

[185] Abrahamsen Helge: *Hus før ogna (Houises before and Now)*, Aschehoug, Oslo, 1961.

[186] Anonsen Carl: Notater fra en samtale med statsminister Krusjtsjov om russisk boligbygging. (Notes from Conversation with Prime Minister Krustsjov about Russian Apartment Building), *BD- orientering*, Melding fra Boligdirektoratet, Nr. 3, 1959, Oslo, 15 September 1959.

[187] https://en.wikipedia.org/wiki/Fiber-reinforced_concrete; Accessed on September 2, 2014.

[188] Standal Per Cato: Ikke metallisk armering. (Non-metallic Reinforcement), *Norwegian Concrete Day, 12 October 2009, Trondheim, PPt-presentation*, Norwegian Concrete Society.

[189] Leivestad Steinar: Betongstandardisering – hva skjer, hva kommer? (Concrete Standardization – What Happens, What is Coming?), *Norwegian Concrete Day, 12 oktober 2009, Trondheim, PPt-presentation*, Norwegian Concrete Society.

[190] Hellan Steinar: Derfor bør norsk industry delta. (That is Why Norwegian Industry should Participate), *Norwegian Concrete Day, Trondheim*, Norwegian Concrete Society, October 1995.

[191] Fougner N.K.: *Seagoing and other Concrete Ships*, Oxford Technical Publications, London, 1922.

[192] Bremner T.W., Ries J.: Sephen J. Hayde: Father of the Lightweight Concrete Industry, *Concrete International*, August 2009.

[193] Expanded Shale Clay and Slate Institute: Lightweight Concrete, Washington, DC, October 1971.

[194a] Editorial: Betongbåt før tropptransporter återvænds til hantverkslokaler. (Concrete Boat for Troop Transport Used for Craftmanship-Localities), *Byggmästaren*, Stocholm, Sweden, 11/1982.

[194b] Wyles John: The Shipshape Concrete, *Financial Times*, 04.09.1976.

[195] Internet 12.01.10: Mulberry Harbour, http://www.concretebarge.co.uk

[196] Internet 12.01.10: D-Day Harbor Masters, http://www.nce.co.uk7cafe-nce/d-day-harbour-masters/5202934.article

[197] Berger ABAM, edited by Lightbourne Alesa: Ideas into Reality, The Berger/ ABAM Story; Federal Way, Washington, USA, 2006.

[198] Mast Robert F.: My Adventures in Prestessed Concrete, *PCI Journal*, November–December 2001.

[199] Rustad F., Holst J.D., Irgens P.: Sluttrapport betonglektervurdering (Final Report Concrete Barge Evaluation), (Not published report), Norcem, Oslo, 05.03.1974.

[200] Sigfridsson K., Haug T.: 100 000 T.DWT. Concrete Barge, Det Norske Veritas, *Report no. 73-155-C*, 1 August 1973.

[201] Steensen Anders J.: Sikrer gass til Italia, (Secure Gas in Italy), *Teknisk Ukeblad*, Oslo, Norway, 37/2009.

[202] Ellen Peter E.: *"Helsal" Racing Sloop – An Australian Development*, Publikasjon Peter Ellen & Ass. Ltd, Hong Kong (undated).

[203] Cupial Jennie: A Six-Year Labour of Love, *Concrete*, May 1979.

[204] Editorial: Berkeley Takes First Place in Concrete Canoe Competition, *Concrete International*, September 2009.

[205] Sandvik Knut, Eie Rolf, Advocaat Ja-Diedrik, Godejord Arnstein, Hæreid Kåre O. Høyland Kolbjørn, Olsen Tor Ole: Offshore Structures – A New Challenge, *XIV national Conference on Structural Engineering*, Acapulco, Mexico, 2004.

[206] Magerøy Hallvard: Fabrikkbetongens opprinnelse, (The Origin of the Ready Mixed Concrete Industry), *Byggeindustrien*, Oslo, Norway, Nr. 1, 1997.

[207] FABEKO: Norsk Fabrikkbetongforening Gjennom 30 år, (Norwegian Ready Mixed Concrete Association through 30 Years), Oslo, Norway, March 1994.

[208] FABEKO: 1964–2004. Betong uten grenser. Utviklingen av norsk fabrikkbetong gjennom 40 år. (Concrete witout Limits. The Development in the Norwegian Ready Mixed Concrete Industry through 40 Years), FABEKO, Oslo, Norway, 2004.

[209] Sui Tongbo, Wen Zhaijun, Wang Jing, Fan Lei: A Study on the Performance of High Belite Cement and Concrete. *International Conference on Sustainability in the Cement and Concrete Industry*, PPt- presentation, Lillehammer, Norway, 19 September 2007.

[210] Ein altes Vorfabriziertes Betonhaus, Cementbulletin nummer 9, September 1976, Der Sweizerischen Zementindustrie, Wildegg, Sveits.

[211] Scott Norman L. Precast Prestressed Concrete Beyond the Year 2000 in the United States, Keynote Address – XII FIP Congress, *PCI Journal*, July–August 1994.

[212] Edwards Harry: The Innovators of Prestressed Concrete in Florida, *PCI 25 Years 1979, Reflections on the Beginning of Prestressed Concrete in America*, PCI, 1981.

[213] Brekke E.: Elementbygging i Oslo (Precast Concrete in Oslo), *BD-orientering*,

Melding fra Boligdirektoratet, Nr. 3, 1959, Oslo, 15 September 1959.

[214] Jahren Per: Betongelementer, (Precast Concrete Components), Beongelementforeningen, NBVF, Oslo, 03.11.1972.

[215] Jahren Per: Betongelementindustriens utviklingsmuligheter. (The Development Possibilty of the Precast Industry), *Ingeniør-Nytt*. Nr. 32, 1968, Oslo, Norway.

[216] Janhunen Matti: *Stages in Prefabricated Technology. Fifty Years of Finish Civil Engineering*, Suomen rakennusiniøørien, Liitto, Helsinki, Finland, 1984.

[217] Hytønen Yki, *Seppænen Matti: Suomalaisen betoielementtirakentasen historia*, SKB, Betoniyieto Oy, Helsinki, Finland, 2009.

[218] Jahren Per: Betongelementindustrien i Norge, (The Precast Industry in Norway), Report 5. May 1972, Elementgruppen, NBVF, Oslo, Norway.

[219] Richardson John G.: Scandinavian Precasters Set High Standard, *Precast Concrete*, London, April 1973.

[220] Sørli Ivar: Armerte konstruksjoner av betong med lett tilslag, (Reinforced Structures with Lightweight Aggregate), *Teknisk Ukeblad*, Nr. 35, 1967, Oslo, Norway.

[221] Jahren Per: Verdens høyeste hus med lett tilslag. (The Worlds highest Building with Lightweight Aggregate), *Ingeniør-nytt*, Nr. 28, Oslo, Norway, 1967.

[222] Mather Bryant: Concrete First, *Concrete International*, June 1996.

[223] McKibbin Leo: Cementing the Building Blocks of History, *Concrete*, October 1998.

[224] Internet 24.12.09: The Chinese Wall. http://no.wikipedia.org/wiki/Den_kinesiske_mur

[225] Sui Tongbo: E-mail 16.03.10, with Information from; China State Cultural Relics Administration and State Survey Administration, 2009.

[226] Jahren Per: High Quality Concrete, *11th International Congress of the Precast Industry*, BIBM, Brussels, Belgium, 1984.

[227] Adams James W.: Using the Other Stuff from Wood Pulping, Unpublished article, June 1998.

[228] McKeough Pat: *The Things were...*, Weyhaeuser Paper Co., Rothschild, USA, 1993.

[229] Mail from Kåre Reknes, 23.11.09.

[230] Prusinski R.C., Bodea C.: Polymer Concrete: A Historical Artfact? The Past Thirty Years, Polymers in Concrete. The First Thirty Years, Editors: Albert O. Kaeding & Richard C. Prusinski, ACI International, SP-214, Farming Hills, Michigan, USA, 2003.

[231] Wahby W.S.: Fifty Years History of Polymers in Concrete in Review: The Past Thirty Years, Polymers in Concrete. The first Thirty Years, Editors: Albert O.

Kaeding & Richard C. Prusinski, ACI International, SP-214, Farming Hills, Michigan, USA, 2003.

[232] Dikeo J.T.: History of ACI Committee 548: The Past thirty Years, Polymers in Concrete. The first Thirty Years, Editors: Albert O. Kaeding & Richard C. Prusinski, ACI International, SP-214, Farming Hills, Michigan, USA, 2003.

[233] Per Jahren, "Long term experience with Silica Fume-Silica Fume in the field, more than 20 years' experience in real structures, presentation at Jinan University, 2015"

[234] Dåstøl Magne, Fidjestøl Per: "50 Years of Silica Fume in Concrete, *Ninth CANMET/ACI Conference on Fly Ash, Silica Fume, Slag and Natural Pozzolans in Concrete*, Warsaw, May 2007.

[235] Jahren Per: Long-term Experience with Silica Fume in Concrete; *Ninth CANMET/ACI Conference on Fly Ash, Silica Fume, Slag and Natural Pozzolans in Concrete*, Warsaw, Poland, 20–25 May 2007.

[236] Malhotra V.M. et al: *Condensed Silica Fume in Concrete, (221 s.)*, CRC Press, Inc., Boca Raton, Florida, USA, 1987.

[237] ASTM Publ. 99, ACI 21, No. 5, January 1950.

[238] Bernhard: SiO_2-støv som sementtilsetning (SiO_2 – Fume as Cement Admixture), *Betongen Idag*, Oslo, April 1952.

[239] Markestad: Tilsetning av SiO_2-støv til betong. (Adding SiO_2 – Fume to Concrete), *Nordic Concrete Congress*, Gotherburg, Sweden, August 1968.

[240] Jahren Per: Use of Silica fume in Concrete, *First CANMET International Conference on Fly Ash, Condensed Silica Fume and Slag*, Montebello, Canada, 31.07–05.08.1983.

[241] ACI Publ. ACI 234R-96: Guide for the Use of Silica Fume in Concrete, Reported by ACI Committee 234, Farming Hills, Michigan, USA, June 1996.

[242] Internet 07.01.2010: Key Tower, http://en.wikipedia.org/wiki/Key_tower

[243] Mast Robert T.: My Adventures in Precast Concrete, *PCI Journal*, November–December 2001.

[244] PCI Journal, Editorial: Medal of Honor – Robert F. Mast, *PCI Journal*, November–December 2001.

[245] Barfoot R.J.: – Aspdin's Missing Statuette!, *Concrete*, May 1975.

[246] Neville Adam: Point of View; What Everyone Who is "in" Concrete Should Know About Concrete, *Concrete International*, April 1999.

[247] Neville Adam: How Useful is the Water–Cement Ratio? *Concrete International*, September 1999.

[248] Neville Adam: In My Judgement: There is More to Concrete than Cement,

Concrete International, January 2000.

[249] Neville Adam: In My Judgement: Water and Concrete: A Love–Hate Relationship, *Concrete International*, December 2000.

[250] Neville Adam: In My Judgement: Core Tests: Easy to Perform, Not Easy to Interpret, *Concrete International*, November 2001.

[251] Neville Adam: In My Judgement: Core Tests: Autogenous Healing – A Concrete Miracle? *Concrete International*, November 2002.

[252] Broshure Cementmuseet Slemmestad, Norway.

[253] WWF-A blueprint for a climate friendly cement industry, A report prepared for the WWF-Lafarge Conservation Partnership

[254] IPCC-2005-Special Report on Carbon Dioxide Capture and Storage

[255] Sui Tongbo, Kong Xiangzhong, Li Juan et al, A discussion on technological roadmap of low carbon cement industry[J], China Cement, No.3 2010, p26–30

[256] Yang Baijun, Annotation of Spring & Autumn Zuo Zhuan[M], China Book Bureau, Beijing, 1st Edition, 1981, p801

[257] Miao Jisheng, Li Xiuying, Cheng Rongkui and Lu Baoshan, Journal of Chinese Ceramic Society, No. 2, Vol. 9, June 1981, p234–240

[258] YANG Guo-zhong, ZHI Chang-yun, The Role and Status of the Earth in Ancient Chinese Architecture, Journal of Henan University, No.6, Vol. 50, Nov. 2010, p94–99

[259] Zhao Xiping, Zhao Fangzhou, Liu Jiaping, et al, Construction technology of residential wall in Qingling Mt. area., Journal of Xi'an University of Technology, 2005, 25(1): p114–115

[260] Li Mingzhong (or Li Jie, 1065-1110, Qing Dynasty),Ying Zao Fa Shi, "Treatise on Architectural Methods" or "State Building Standards", 1st edition published in 1103.

[261] Song Yingxing (Ming Dynasty), Tian Gong Kai Wu, Ha'erbin Publishing House, 2009, p249–251

[262] Zhang Huyuan, Zhao Tianyu and Wang Xudong, Earth construction technology in ancient China., Dunhuang Research, No.5, 2005, p81–90

[263] Accessed on March 1, 2015. http://image.haosou.com/i?ie=utf-8&src=hao_360so&q=%E7%9F%B3%E7%81%B0%E7%AA%91

[264] Shen Kuo (Song Dynasty), Meng Xi Bi Tan, Vol. 12. Times Art Publisher, Changchun, China, 2001, p110.

[265] Tang Changru, General Reference Materials for Chinese History-Ancient China Part, Vol. 3, China Book Bureau, Beijing, 1983

[266] Yang Fuwei, Zhang Bingjian, Pan Changchu and Zeng Yuyao, Traditional lime mortar modified with sticky rice starch-One of the most important invention in ancient China., Technological Science, Edition E of SCIENCE IN CHINA, Vol. 39 (1), 2009, p1–7

[267] Liu Q, Zhang B J. Syntheses of a novel nanomaterial for conservation of historic stones inspired by nature. Mater Lett, 2007, 61(28): p4976–4979

[268] Zeng Y Y, Zhang B J, Liang X L. A study on the characteristics and consolidating mechanism of Chinese traditional mortars used on historical architecture. Sci Conserv Archaeol, 2008, 20(2): p1–7

[269] Yang L, Ding W J, An Y G, et al. Control synthesis of aragonite calcium carbonate with glucan as the template. Chem J Chin Univ., 2004, 8: p1403–1406

[270] Liu Qiang, Zhang Bingjian, Biomimetic synthesis of bio-mineralized protective materials for protecting the stone-based cultural relics, Chem J Chin Univ., 2006, 64 (15): p1601–1605

[271] Accessed on January 8, 2015 from http://cq.QQ.com, Chongqing Evening News, September, 2009

[272] Joseph Needham, History of Chinese Science and Civilization. Shanghai People's Publisher, Feb. 2001

[273] Wang Yanmou, *History of the Development of China Cement*, China Building Materials Publishing House, Beijing, August, 2005

[274] Zhou Zuitian and Han Changkai, *History of China Cement* series, Cement Technology, No.1–6, 2011 and No. 1–6, 2012

[275] Zhu Xiaoqiu, The past and present of Generalissimo Sun Yat-sen's Mansion. *China Charm*, No. 3, 2015, p71–80

[276] http://news.163.com/09/1209/10/5Q3937P700011SM9.html, access on Oct. 1, 2014

[277] http://rabe.nju.edu.cn/sec/5international/aqq/07.htm, access on Oct.1, 2014

[278] http://www.chinadaily.com.cn/china/2011-02/21/content_12049645.htm, access on Oct.1, 2014

[279] T. Sui and Y. Yao. Recent Progress in Special Cements in China [C]. Proceeding of the 11th International Congress on the Chemistry of Cement, Vol. 4, Durban, South Africa, 11-16 May, (2003), p2028–2032.

[280] Tongbo Sui, Lei Fan, Zhaijun Wen et al. Study on the Properties of High Strength Concrete using High Belite Cement [J]. Journal of Advanced Concrete Technology, Japan Concrete Institute.,Vol.2, No.2, June (2004) p201–206.

[281] Sui Tongbo, Technological Innovation for Cement Sustainability-China's effort, 4-12, ALITINFORM International Analytical Review Cement, Concrete and Dry Mixtures. ISSN1998–1295, No.3 (20), 2011

[282] Sui, T., Li, J., Peng, X., Li, W., Wen, Z., Wang, J., and Fan, L. "A Comparison of HBC and MHC Massive Concrete for Three Gorges Project in China." In Proceedings of Measuring, Monitoring and Modeling Concrete Properties, Greece, 2006, p341–346.

[283] Liu Liangji, Development of RMC in China for the Last 20 Years, 2007' China RMC Worhshop, 2007, p1–24

[284] RMC output and increase ratio of China in 2014, Access on May 1, 2015. http://www.cbc.net.cn/news/hyzx/2015/0430/91575.html

[285] Zhang Shukai, Developing from zero and growing up to form prefabricated cement and concrete products sector with Chinese characteristics, p21–28

[286] Zhang Shukai, A review and prospect of 60-year history of prefabricated Chinese cement products, 21st CENTURY BUILDING MATERIALS, No. 1, Vol.2, Feb. 2010, p1–11

[287] Zhang Shukai, A review and prospect of PCCP in China, China Cement and Concrete Products Net, Oct. 29, 2009. Access on March 20, 2015 via http://www.concrete365.com/news/2009/10-29/H17400705.htm

[288] Current situation and future development of Chinese concrete piles sector. China Quality News Web., Oct. 16, 2013

[289] A Brief History of Concrete Ships, http://www.concreteships.org/history/, access on March 1, 2015

[290] Access on June 7, 2014. http://news.163.com/photoview/ 00AP0001/21657.html

[291] Access on June 7, 2014. http://epaper.bjnews.com.cn/html/2013-01/16/content_403989.htm?div=-1

[292] History, present situation and future development of reinforced concrete poles of China, http://www.qzhaiyu.com/industry/116.html, accessed on March 1, 2015

[293] Wang Gaoyi, High load-bearing partially pre-stressed concrete poles, http://www.chinabaike.com/z/dianlijishu/607437.html, access on March 1, 2015

[294] Analysis on application of concrete sleepers, Report from Infrastructure Department, Railway Transportation Administration of China, June, 2003

[295] New Ideas of the Construction of CRTSIII Ballastless Track, Internal communication, China Railway 23rd Bureau Group Co., Ltd., January 2011

[296] Zhen LIN, Yi ZHANG and Xinmeng LI, The Development of Non-Asbestos Fibre Cement in China, Proceedings of International Inorganic-bonded Fiber Composites Conference, Danang City, Vietnam 15–19 September, 2014, p29–37

[297] Jahren Per: CO_2 – Emission – Triple Focus, *Tenth ACI International Conference on Recent Advances in Concrete Technology and Sustainability Issues*, Seville,

Spain, 15–17 October 2009.

[298] Cembureau Statistics, www.cembureau.be

[299] Tongbo Sui, Wen Zhaijun, Wang Jing, Fan Lei: A Study on the Performance of High Belite Cement and Concrete, (PPt-presentation), *International Conference on Sustainability in the Cement and Concrete Industry*, Lillehammer, 16–19 September 2007, Norsk Betongforening, Norway.

[300] Tongbo Sui: High Performance Low Heat Portland Cement – High Belite Cement, Preparation & Performance, Application and Perspective, 2007, *International Workshop on Cement & Concrete Technology for Sustainable Development*, Beijing, Nanjing and Lhasa, August 2007, China Building Material Society, Beijing, China.

[301] Yan Peiyu: The Current Status of Cement and Concrete Technology in China, 2007, *International Workshop on Cement & Concrete Technology for Sustainable Development*, Beijing, August 2007, China Building Material Academy, Beijing, China.

[302] Hovland Kjetil Malkenes: Norge griper ikke sjansen. (Norway do not Take the Opportunity), Teknisk Ukeblad, Oslo, 43/2009.

[303] Norsk Betongforeing, Miljøkomiteeen: CO_2-utslipp – sement og betong. Utfordringer og perspektiver (CO_2-Emission – Cement and Concrete. Challenges and Perspectives), *Report*, Norwegian Concrete Society, Oslo, March 2009.

[304] Jahren Per: CO_2 – Emission, Triple Focus, *Tenth ACI International Conference on Recent Advances in Concrete Technology and Sustainabilty Issues*, Seville, Spain, October 2009.

[305] WBCSD: The Cement Sustainability Initiative (CSI) Climate Actions, Sveits, November 2008, www.wbcdcement.org

[306] Damtoft Jesper Sand: The Sustainable Cement Production of the Future. *International Conference on Sustainability in the Cement and Concrete Industry*, Lillehammer, 16–19 September 2007, Norsk Betongforening.

[307] Bellona: Carbon Dioxide Storage: Geological Security and Environmental Issues – Case Study on the Sleipner Gas field in Norway, *Bellona Report*, May 2007, www.bellona.no/filearhieve/fil-co2-storage-rep.final.pdf

[308] Nasser George D.: The Next Frontier – Building Permanent Bases on the Moon, *PCI Journal*, July–August 1989.

[309] Lin T.D., Senseney Jonathan A., Arp Larry Dean, Lindbergh Charles: A Feasibility Study of Prestressed Concrete Construction on the Moon, *PCI Journal*, July–August 1989.

Names

The years

20 600 BC	– Mortar Core, Turkey (1)
20 000 – 10 000 BC	– Gypsum and lime glues are known
7000 BC	– Concrete in Southen Gallilea (1.1)
5600 BC	– Floor in Lepenski-Vir (1.3)
5 000	– 3 000 BC. – Baihuimian- mortar in China (1.2)
3 000 BC	– Floor in Dadinsan in China (1)
2 800–2 300 BC	– Plastering of walls and floors in China (1.2)
2500 BC	–Concrete in South and Middle America (1.4)
2500 BC	– The Pyramids in Egypt (1.5)
	– Concrete in Mesopothamia (1.2)
1950 BC	_ Egyptian illustrations (1.5)
1 600 BC	_"Yellow mud" – mortar in China (1.2)
1 100–800 BC	– Lighweight concrete in Middle America (1.4)
1000 BC	– King Salome's water channels (1.2)

---

700 BC	– Lime burning from mussel shells in China (1.2)
500 BC	– The Phoenicians use concrete.
500–300 BC	– The Greeks utilizes Santorin (1.6)
200 BC	– Mortar in walls in Bali in Indonesia (1.1)
150 BC	– Etruscans produces concrete (1.6)
100 BC	– The first Roman concrete. (1.6)
55 BC	– The theater in Pompeii, first known building with pozzolan cement concrete (1.6)

25 BC	– Vitruvius (1.6)
18 BC	– Pont du Gard (1.6)

70–80	– Eifel-Köln – culvert (1.6)
80	– Coliseum in Rome is opened (1.6)
122	– Hadrian's Wall (1.6)
127	– The Pantheon temple (1.6)
200	– Concrete floor in Gloucestershire (1.6)
1100	– The Holy Catherine statue in Salzburg, Austria (1.6)
1127	– Reading Cathedral (1.6.)
1413	– Concrete recipe in Durham, England (1.6)
1414 (1500?)	– Vitruvius manuscript discovered.
1499	– Notre Dame, Paris (1.6)
1534	– Charles 1 of Spain order investigation of the possibility for a channel a channel in Middle America.
1563	– Philbert de L'Orme writes about concrete (1.6)
1570	– Andrea Palladio's "The four book of architecture"
	– End of the 1600's - Use of trass to hydrate lime mortar in Holland. (1.6)
1676	– Foundations for the fort on Montrael Island, Canada (1.6)
1678	– Moxon about hydraulic lim (1.6)
1718	– Trollhätte sluises / locks utilizes Alun cement (2.1)
	– About 1750 – Parker Cement (2)
1756	– John Smeaton build the Eddystone Lighthouse (2.1)
1764	– Work at the Farris Dam starts, The first Norwegian concrete (1.7)
1779	– Bry Higgin's cement patent (2.1)
	– Start production of alun cement in Garphyttan, Sweden (2.1)
1780	– Edgar Dobbs cement patent (2.1)
1796	– Parkers cement patent (2.1)

1816	– Concrete used in bridges in France (2.1)
1817	– The German professor John develops his special cement (2.1)
1818	– The French engineer Vicat develops his type of cement (2.1)(2.3)
	– The first cement production in USA (White-cement) (3.1.3)

1822	– James Frost "British Cement" /(1)
1824	– Joseph Aspdin takes out the patent on Portland Cement
1825–1943	– The tunnel under the Thames (2.2)
1830	– Lafarge establishes a factory for hydraulic lime in
	– Le Teil in South of France (2.1)
	– Hydraulic cement production in Canada
1840	– Cement production starts at Nordre Langøy and
	– Bakaas Brug near Oslo in Norway (2.2)
	– France, first cement factory (2.2)
1844	– Johnson makes the modern Portland cement (2.2)
	– First Portland concrete house in USA, Milton House, Wisconsin (3.1.3)
1846	– Nikolai Church in Hamburg built with med trass cement (2.2)
1848	– Lambot's concrete boat (4.1)
1849–1852	– Robert E. Lee developed the "tremie" for under water concreting (2.3)

1854	– Williamson's reinforced concrete floor patent (2.2)
1855	– Cement factory in Stettin (2.2)
	– Six "Roman Cement"- factories at the island of Bornholm in Denmark (2.2)
	– French François Coignet gets patent on concrete floors (2.3)
1860	– Lafarge factory in Le Teil develops slow hardening cement.
1863	–The first Swedish Concrete structure, - a harbor pier in Halsingborg.
	– Denmark's first concrete structure with Portland cement, Fort Prøvensten (2.2)
1864	– Tall's patent on concrete houses in standardized moulds (2.2)
1865	– Le Teil – factory, France, gets an order of 120 000 tons of cement to the Suez channel.
1866	– Tall's house in Bexleyheath (3.1.2)
1867	– Monier's flower pot patent (2.2)
	– Discussion meeting about cement in the Polytechnic Society in Norway (2.4)

1868 – Denmark's first cement factory in Ringsted
– Reed's "water - law" (2.3)
1869 – Campbell's Farm (3.1.2)
– Fahnehjelm investigates lime at Gotland (5.1)
– Finland's first Portland cement factory (2.2)
1870 – Monier produces water cisterns in France (2.2)
– First reinforced concrete structure in USA (3.1.3)
1871 – USA's first concrete house (3.1.3)
– USA's first Portland cement patent
– Cement factory in Visby, Sweden (5.1)
1872 – Brannons patent on fire safe houses in concrete, England.
– Deakon's guide for ready mixed concrete. (4.2)
1873 – Swedens firs cement factory in Lomma (2)
1875 – Axemouth Bridge (3.1.2)
– Start of Portland concrete production in USA (3.1.3)
– Lt. D. Rye's Anvisninger (Advices about) om "Konkret" (2.4)
– Precast concrete (4.3)
– Cement production starts in Japan (2.2)
1877 – Thaddeus Hyatt's reinforcement report (2.3)
– German cement industry establish the first trade organization (VDZ) (2.2)
1882 – Rye becomes Director for the Lighthouse Administration in Norway (2.4)
1883 – Le Chatelier introduces petrography testing (2.3)
1884 – Gaudi starts the work on Sagrada Familia, Barcelona, Spain. (3.1.5)
– 8 000 barrels of cement from Germany for the foundations of the Statue of Liberty in New York (2.2)
1885 – The English Fr. Ransone takes out patent on the rotary kiln. (2.2)
– Flagler builds Hotel Ponce de Leon in Florida (3.1.3)
1886 – Building regulations for concrete in London (2.2)
– US Patent on prestressed concrete by Jackson (2.3)
1887 – Swedish cement industry finds it necessary to start their own contractor company – SKANSKA (2.1)

	– The Prussian concrete norm (2.2)
1888	– Christiania Cement Aktiebolag starts cements production at Slemmestad, Norway.
	– Doering – prestressing patent (2.3)
1889	– The first reinforced concrete bridge in USA (3.1.3)
1891	– The first concrete road in USA (3.1.3)
1892	– Passow, Germany, - produces iron Portland cement and slag cement. (2)
1895	– The first Rotary kiln built by Atlas Cement in USA. (2.2)
	– Captain Kolderup's book on reinforced concrete (2.4)
1897	– C.W. Stevens takes out patent on artificial stone made of concrete.
	– Hennebique makes reinforced concrete frames (2.2)
1898	– Glenfinnan Viaduct, England (3.1.2)
1899	– Ehemass, Germany, - produces a white cement (2)
	– First concrete bridge in Finland (3.1.4)

1900	– French cement industry establish a trade organization (SPC)
1901	– The first blended cement
1902	– The first skyscraper in reinforced concrete, Ohio, USA (3.1.3)
	– Cement trade organization in USA (ACA) (2.2)
1903	– Westminster Cathedral, London (2.3)
	– The first concrete standard, Switzerland (4)
	– Jurgen-Heinrich Magen's first patent on ready mixed concrete (4.2)
	– The first reinforced concrete bridge in Japan (2.3)
1904	– Newton-le-Willows Water tower (3.1.2)
1905	– Tuckton Bridge opened (3.1.2)
	– The first reinforced concrete building in Japan (2.3)
1906	– Stanford Bridge opened (3.1.2)
	– The world's first runway in concrete in Denmark (3.1.4)
	– Thomas Edison presents his concrete house (2.3)
	– 6.400 tons of cement in one shipment from Limhamn, Sweden to San Francisco (5.1)
1909	– Edison's patent on long rotary kilns (2.3)

1910	– Use of steel fibers (4.5.1)
	– Cerrigdrudion Dam (3.1.2)
	– The first concrete standard in USA
	– NTH (Norwegian Technical University) opened by
	– King Haakon VII (2.4)
1911	– Gilbreth's concrete system (2.3)
	– Failure of the dam in Austin, USA (2.3)
1912	– Jahrhunderthalle in Wroclaw, Polen (3.1.5)
	– The Swedish Concrete Society established (first in the Nordic countries)
	– NK in Stockholm opened (3.1.4)
1913	– French Bied makes High Alumina Cement (2)
1915	– Austrian Spindel makes Rapid Hardening Cement (2)
	– Le Corbusier's Do-mino – system (3.5.3)
1916	– The First design code for earthquakes (3.3.1)
	– Frost & Bjørnstad teaching book about concrete (2.4)
	– Dalen Portland Cement, Norway factory established (5.2)
	– Bridge with precast concrete components in Finland (3.1.5)
	– First textbook on concrete in Japan (2.3)
1917	– The cement factory Ce-No, Norway starts up (5.2)
	– VNamsenfjord, The first seagoing concrete ship (4.1)
1918;	– Hayde's lightweight concrete patent (4.4)
	– The first international cement chemistry conference (2.3)
	– The first mile with concrete road USA (3.1.3)
	– The railway station in Helsinki, Finland (3.1.5)
1920	– First full scale production of lightweight aggregate (4.4)
	– Oslo Havnelager, Norway (3.1.1)
1921–1923	– Freyssinet's hangar at Orly, France (2.3)
	– Wettstein's prestressing patent (2.3)
1922	– The Namsen Bridge opened, Norway (3.1.1)
	– The first building with lightweight aggregate concrete (4.4)
1924	– Japan Urban Building Law (3.3.1)
1926	– Norwegian code for concrete works (4)
1928	– The first Norwegian production of reinforcing steel (2.4)
	– Freisinnet's prestressing patent (2.3)

1929	– The first skyscraper with lightweight aggregate concrete (4.4)
1930	– Engene Freyssinet designs the first prestressed concrete structure.
	– Maillart's Salginatobel Bridge (3.1.5)
	– First Norwegian professor in Concrete (2.3)
1931	– House in Tulsa, Oklahoma designed by Frank Lloyd Wright (3.3.1)
	– Norway's first ready mixed concrete factory, - Ludvig Lorentzen Ferdigbetong,Lysaker near Oslo. (4.2.)
1932	– Swedish Forsen makes silicate cement (2)
1933	– The competition about the Olympic stadium in Helsinki 1952 (3.1.5)
1936	– Hoover Dam (3.1.3)
1937	– F.L.Wrights's Falling water (3.3.1)
	– The Lignin patent (4.6)
1940	– Yoshida produces concrete with compressive strength above 100 MPa (2.3)
1942	– Stavå bridge, Norway (3.1.1)
	– Grand Coulee Dam, USA (3.1.3)
	– Concrete barge in Japan (4.1)
	– Mulberry harbor (4.1)

1950	– Peace Museum in Hiroshima, Japan (3.5.4)
	– Notre Dame-du-Haut, France (3.5.3)
	– Walnut Lane Bridge, USA (3.1.3)
1953	– First Norwegian production of deformed reinforcing bars
1957	– First international conference on prestressed concrete
1958	– Phillips Pavillion, Brussel, Belgium (3.5.3)
	– First Hollow core machine (4.3)
1960	– Olympic stadiums in Rome, Italy (3.5.2)
	– Brasilia (3.1.5)
	– Tromsø Bridge, Norway (3.1.1)
1961	– Palazzo del Lavoro, Torino (3.5.2)
1962	– Hattori develops superplasticizers (4.6)
1964	– Tokyo's Olympic stadiums (3.3.4)

	– Norwegian Ready Mixed Concrete Association established (4.2)
	– Alvar Aalto's auditorium in Otaniemi, Finland
1967	– Kaknäs tower, Stockholm, Sweden opens (3.1.4)
	– Habitat 67, Canada (3.1.5)
1969	– Limfjord tunnel, Denmark (3.1.4)
1974	– Akkaganse Bashi, Japan, a railway bridge with compressive concrete strength 96 MPa.
1975	– First international polymer concrete conference (4.6)
1991	– Town Hall in Tokyo (3.3.4)
1992	– Skarnsundet Bridge, Norway (3.1.1)
	– Bergsøysundet bridge, Norway (3.1.1)
2004	– Millau Bridge, France, World highest (3.1.5)

www.ingramcontent.com/pod-product-compliance
Lightning Source LLC
LaVergne TN
LVHW081934200926
848484LV00009B/2430